DATE DUE			
Dec3 81			

EVOLVING STRATEGIES AND TACTICS IN MEMBRANE RESEARCH

EVOLVING STRATEGIES AND TACTICS IN MEMBRANE RESEARCH

Donald F. H. Wallach

and

Richard J. Winzler

Springer-Verlag New York Heidelberg Berlin

1974

Donald F. H. Wallach
Director, Radiobiology Division
Tufts University School of Medicine
Boston, Massachusetts

Richard J. Winzler
1914–1972

Library of Congress Cataloging in Publication Data

Wallach, Donald Francis Hoelzl, 1926–
Evolving strategies and tactics in membrane research.

Includes bibliographical references.
1. Membranes (Biology) I. Winzler, Richard J., 1914–1972 joint author. II. Title. [DNLM: 1. Cell membrane. QH601 W195e 1974]
QH601.W27 574.8'75 73-21715
ISBN 0-387-06576-8

Printed in the United States of America.

ISBN: 0-387-06576-8 Springer-Verlag New York Heidelberg Berlin

ISBN: 3-540-06576-8 Springer-Verlag Berlin Heidelberg New York

PREFACE

Richard J. Winzler died of a heart attack on September 28, 1972. He was a versatile, internationally renowned biochemist, gifted with boundless energy and a great breadth of view. After early work on respiratory enzymes, he developed a lifelong interest in complex carbohydrates and, particularly, glycoproteins. He early recognized the important role of the latter substances in membrane function and developed the now widely accepted model for the structure of major erythrocyte membrane glycoproteins. He expanded this model to include the possible general roles of membrane glycoproteins in specific membrane functions, particularly cell recognition phenomena.

Richard Winzler possessed a broad and critical perception of the flux of biomedical progress. He thus appreciated very early that general recognition of the complexity of membrane-associated phenomena would foster a new, rapidly expanding, interdisciplinary membrane technology. He also recognized the need for means to bridge the communication barriers inevitably generated by such an evolution, and this conviction largely stimulated us to write this book.

Membrane technology is a vast field and we cannot treat all its aspects. Instead, we concentrate on those areas that, in our view, have evolved most rapidly. In this, we pay primary but not exclusive attention to the plasma membrane, but where relevant, consider other biomembranes and diverse model systems. Moreover, we focus on animal cells, but include information derived from microbial and artificial membranes where germane.

We deal first with the principles and techniques used in the isolation, purification, and characterization of diverse membrane entities and their macromolecular components. We then turn to the modern spectroscopic techniques used increasingly by membrane biologists to explore the organization of biomembranes and to extract molecularly specific structural information from intact membranes in diverse physiological states. In discussing the field of "membrane spectroscopy," we deal in turn with spectroscopic techniques utilizing primarily signals from intrinsic membrane components (i.e., fluorescence spectroscopy, infrared spectroscopy, laser Raman spectroscopy, nuclear magnetic resonance, optical activity) and methods using signals from "reporter groups," primarily extrinsic components of "probes," which might be optically absorbing, fluorescent, paramagnetic, optically active, or Raman active.

We do not treat all domains of membrane technology in the depth that separate specialists might provide. Instead, we present the material critically from the perspective of membrane biologists and provide an extensive list of references for those who wish to read further.

We do not deal extensively with membrane models. However, in the final chapter, we address certain cardinal working hypotheses, which will prove useful in clarifying some burning issues in membrane biology.

DONALD F. H. WALLACH, M.D.
Boston, Massachusetts
November 1973

CONTENTS

EVOLVING STRATEGIES AND TACTICS IN MEMBRANE RESEARCH

GENERAL INTRODUCTION

The structure and function of biomembranes are under intense and widespread study. Continuing interest in this field derives largely from an increasing awareness that membranes serve many functions in addition to the partitioning of cellular compartments and the mediation of transport between them. Examples include the spatial organization of biosynthetic systems and oxidative phosphorylation, the communication between cells by fusion, by contact, or by molecular mediators, and the recognition of cellular individuality. Moreover, cellular membranes are fundamentally involved in many pathologic processes, prominently neoplasia and viral disease.

Recognition of the complex diversity of membrane-associated phenomena is forcing the reassessment of what had become traditional views of membrane structure, in terms of modern principles of genetics, molecular biology, and macromolecular organization. Membrane structure has received much attention, but there is still little consensus concerning membrane architecture in the literature. The lipid bilayer hypothesis of membrane structure proposed in 1935 by Danielli and Davson (1) was one of the first models of "molecular biology." The hypothesis was elegant in its day, appealing in its direct simplicity, and was consistent with many of the properties of biological membranes. Nevertheless, it has become increasingly evident that many of the membrane phenomena recognized today cannot be explained by the lipid bilayer model or its simple variations. However, the search for satisfactory alternatives represents a Herculean task: Membranes exhibit great diversity (in time, between and within cells, and among different domains of the same surface); techniques of membrane manipulation require much additional development, and there is ambiguity as to which are the "real" components of a membrane and which are adventitiously lost or gained during experimentation.

In this book, we shall address some burning issues in membrane technology. We shall pay primary, but not exclusive, attention to the *plasma membrane*, i.e., the boundary between the extra- and intracellular milieux; however, where relevant, we shall consider other biomembranes and diverse model systems. Similarly, we shall focus on animal cells, but shall include information derived from microbial and artificial membranes where germane. Finally, we aim for a critical review, rather than an all-inclusive summary of pertinent literature.

We aim to understand the architectural and functional principles that characterize membranes in their diverse physiological states. This is an enormous task and will require at least a tripartite interdisciplinary effort involving very diverse technologic domains.

The *first* embraces the methodologies of cell biology and physiology, developmental biology, embryology, immunology, neurosciences, etc.; we cannot treat these here.

The *second* comprises approaches that lie closer to the disciplines of biochemistry and biophysics. These concern the principles and techniques used in the isolation, purification, and characterization of diverse membrane entities and their macromolecular components. We shall summarize this field in the first part of this book, although we cannot cover every worthy publication in the area. However, we shall critically appraise the strengths and weaknesses, facts and fancies, realities and ambiguities, as well as the failures and hopes of this field of endeavor.

This brings us to the *third*, perhaps most complex domain of membrane biology, which concerns what is perhaps the greatest obstacle facing membrane biologists—*the fact that biomembranes comprise essentially solid and/or liquid-crystalline arrays of probably highly cooperative components* (2–4); thus, the techniques and principles that are appropriate for solution systems cannot easily be transferred to membrane situations. For instance, the proteins and lipids of membranes exist at very high local concentrations, and their mutual, reciprocal interactions are likely to overshadow solvent contacts. Concordantly, membrane components commonly dissolve poorly in aqueous systems and change character drastically when artificially solubilized.

The latter predicaments, in particular, have led membrane biologists to employ modern spectroscopic techniques to explore the organization of biomembranes and to extract molecularly specific structural information from intact membranes in diverse physiological states. This difficult strategy holds considerable promise and has already yielded a large and important literature. We shall consider this in some detail in the second part of the book.

Understandably, this field of membrane biology easily merges into considerations of membrane models. We shall not deal with these more than absolutely necessary; too many membrane models—all seductive, but none very satisfactory—have been proposed. One of us has critically appraised this situation recently (5), calling for less speculation and a major effort to integrate available facts into meaningful working hypotheses, which might solve the many remaining problems of membrane biology.

In concentrating considerable effort into the field of "membrane spectroscopy," we shall proceed as logically as possible, dealing in turn with (a) spectroscopic techniques utilizing primarily signals from intrinsic membrane components (i.e., fluorescence spectroscopy, infrared spectroscopy, laser Raman spectroscopy, nuclear magnetic resonance, optical activity) and (b) methods using signals from "reporter groups," primarily extrinsic components or "probes"—these might be optically absorbing, fluorescent, paramagnetic, optically active, or Raman active. We must omit some topics that might be germane to membrane biology, either because they have not been applied to the field sufficiently, e.g., UV-difference spectroscopy and microwave spectroscopy, or because they are in a major state of transition, e.g., X-ray techniques (6).

Certain basic, general principles apply to all the spectroscopic techniques and will be introduced in Chapter 3.

References

1. Danielli, J. F., and Davson, H. *J. Cell Comp. Physiol.* **5:** 495, 1935.
2. Changeux, J. P., Tung, Y., and Kittel, C. *Proc. Natl. Acad. Sci* (U.S.) **57:** 335, 1967.
3. Blumenthal, R., Changeux, J. P., and LeFever, R. *J. Memb. Biol.* **2:** 351, 1971.
4. Changeux, J. P., Blumenthal, R., Kasai, M., and Podleski, T. In *Molecular Properties of Drug Receptors*, R. Porter and M. O'Connor, eds. London: J. & A. Churchill, 1970, p. 197.
5. Wallach, D. F. H. *The Plasma Membrane: Dynamic Perspectives, Genetics and Pathology*. New York: Springer, 1972.
6. Shipley, G. G. In *Biomembranes*, D. Chapman and D. F. H. Wallach, eds. London: Academic Press, 1973, p. 1.

CHAPTER 1

ISOLATION OF MEMBRANES

Introduction

Many approaches to the study of the composition and molecular architecture of membranes require their isolation free of other cellular components. Several uncertainties inherent in such isolations must be considered in the interpretation of any experimental data. Some sources of ambiguity are as follows.

(a) Isolated membranes may have lost large or small molecular components during the process of preparation. Graham and Wallach (1) illustrate the importance of this matter, as well as its relevance to (c) below in their infrared spectroscopic studies of erythrocyte ghosts. Normally isolated ghosts give no infrared evidence of β-conformation in the membrane protein, but ATP + Mg^{++} cause rapid spectral alterations, indicating a conformational change toward the antiparallel β-structure. This is accentuated by the simultaneous addition of $Na^{+}+K^{+}$ (cf. infrared spectroscopy). ATP + Mg^{++} also causes endocytotic vesiculation of erythrocyte ghosts (2). Of course, ATP, Mg^{++}, Na^{+}, and K^{+} are normal constituents of the membrane microenvironment, and a membrane can hardly be considered "native" in their absence.

(b) Membrane isolates are likely to be deficient in normal biological controls.

(c) One must suspect that membranes may undergo structural rearrangements during their isolation from the cell. In erythrocyte membranes, such alterations may be reflected in increased susceptibility to proteolytic and lipolytic enzymes (3,4) iodination in the presence of lactoperoxidase (5) and labeling with poorly permeant, covalent reagents (6–8).

(d) The possibility must always be considered that isolated membranes may contain molecules not normally membrane associated and also that the membranes under study may have interchanged molecular components with other membranes of the cell during the process of isolation.

(e) Membrane components associate with each other, as well as with environmental molecules, through multiple interacting equilibria. These will depend upon a host of physicochemical parameters including temperature and are likely to be perturbed during isolation procedures, which may consequently select against *labile membrane* states.

A major difficulty in this connection may be the failure to regulate temperature at the *levels required for normal membrane topology and function.* Thus, it is standard practice in membrane isolations to disrupt cells at 0–4°C. However, recent data (9–12) show quite clearly that the surface immunoglobulins, as well as θ and some H2 antigens of mouse lymphoid cells, can show random distribution near 0°C, compared with a high degree of polarization ("capping") at 37°C. Possible micromorphologic correlates of this phenomenon are shown in Fig. 1.1. At 20–37°C much of the surface of peripheral lymphocytes is elaborated into long, branching, undulating microvilli, but these drastically diminish in number and increase in diameter near 0–4°C (12). Other recent work indicates that surface membrane thermotropism is not restricted to lymphocytes (13).

(f) *Cell heterogeneity* introduces a generally overlooked complication into membrane purifications and, indeed, all efforts to fractionate subcellular components. Restricting our comments to plasma membranes, the problem is obvious in multicellular tissues such as liver. In this organ, macrophages constitute an appreciable portion of the tissue mass. Moreover, their cell surfaces are differently specialized from those of hepatocytes. Still, although such cells can be separated by adsorption on glass and more elaborate methods, these cautions are rarely applied. But the problem cannot be neglected even in simpler situations. Thus, ascites tumors always contain an appreciable proportion of macrophages; erythrocytes cannot be obtained as a homogeneous age population, and lymphocytes are well known to represent numerous subcategories of cells with different membrane functions. The recognition of this last fact has generated a major impetus for the development of methods that purify lymphoid cells into biologically meaningful classes and subclasses. Separations based on cell size, density, charge, and surface properties have been employed as have highly sophisticated immunoadsorption methods.*

(g) The *physiologic state* of a cell can introduce complications. For example, important plasma membrane alterations, reflected by changed lectin-induced agglutinability, occur during *mitosis* of cultured fibroblasts. (Intracellular membranes are obviously also involved.) This matter may be unimportant in membrane studies on tissues with low cell turnover,

* Cell separation techniques and their relevance to membrane fractionation have been recently reviewed by one of us (D. F. H. Wallach and P. S. Lin, *Biochim. Biophys. Acta*, in press).

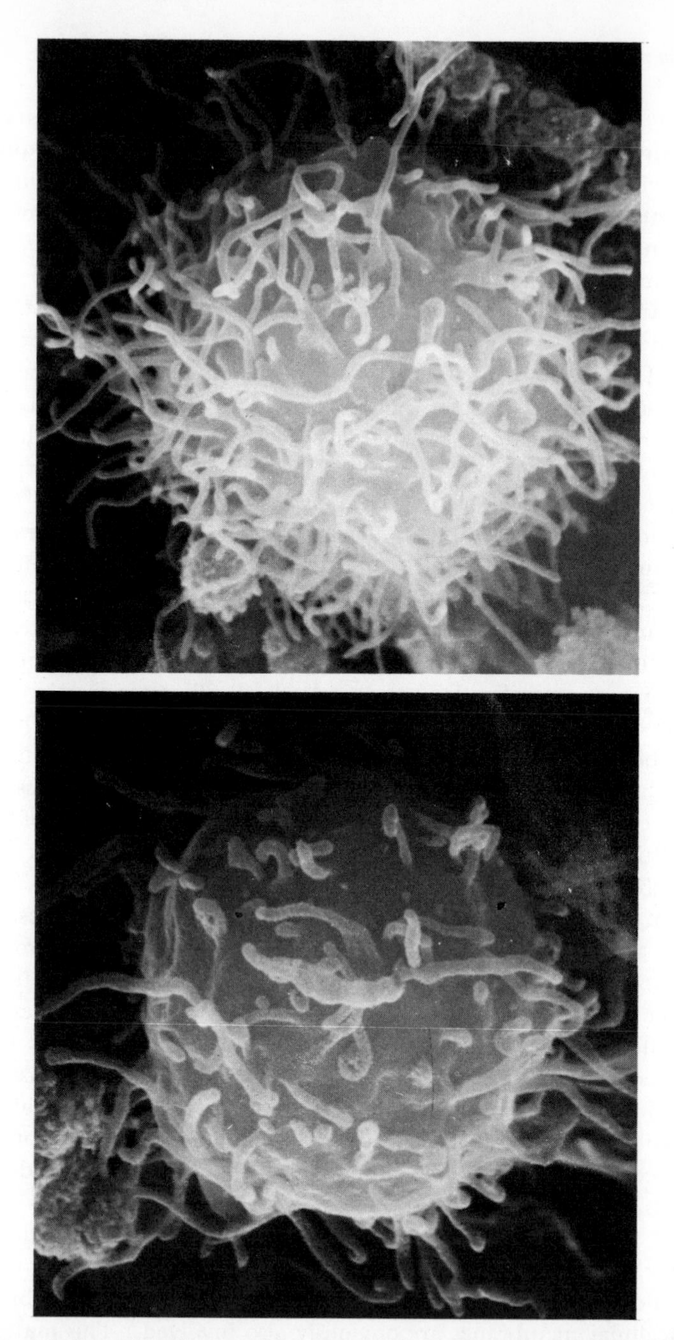

Fig. 1.1 Scanning electron micrograph of lymphocytes at two different temperatures. At 37°C (above), much of the cell's surface is elaborated into long, undulating branching microvilli. At 0–4°C (below), the number of microvilli declines drastically, the more so the longer they remain at low temperature. The process is reversible in intact cells, but clearly influences plasma membrane isolation procedures. Magnification 10,000. (P. S. Lin and D. F. H. Wallach, to be published.)

e.g., normal liver, but could be critical in experiments on regenerating liver, cultured cells, lymphocytes, thymocytes, tumor cells, etc. It is crucial for the progress of membrane biology that cell membranes be isolated in as pure and physiologically representative a state as possible, but this goal requires far greater attention to the above-listed complications, as well as other possible problems, than is generally applied.

Most fractionation schemes involve sequentially disruption of cells, centrifugal (or other) separation of membrane fragments, and identification of membrane fractions using membrane markers. We shall now critically assess these steps.

Disruption of Cells

Since cells contain widely diverse types of membranous structures, it is usually necessary to design a fractionation scheme for only one type of membrane. However, unlike other subcellular organelles, the cell surface membrane must always be disrupted to isolate it; at the same time nuclei, mitochondria, and lysosomes must be kept intact, if only to prevent their contents from adsorbing to, aggregating, or degrading other membrane components. Release of basic proteins from broken nuclei is undesirable, because these aggregate negatively charged membrane fragments. Lysosomal breakdown can cause membrane damage by the released hydrolases.

We have discussed the general uncertainties involved in membrane isolations and now address the physical fate of the plasma membrane upon disruption. Depending on the cell surface, its interaction with other cells and extracellular components, as well as the conditions of cell rupture, the disrupted surface membrane may form large, closed or "sealed" ghosts, large and small "open" pieces, and small sealed or "leaky" vesicles. In most cases, probably all species are generated, but the larger ones tend to be unstable. These facts are inadequately appreciated, but seriously complicate plasma membrane fractionations.

Three major stratagems have been developed to deal with this matter.

(a) Fractionations are tailored to isolate some stable, readily recognized, specialized subdomain of the cell surface, e.g., liver bile fronts, intestinal or kidney brush borders. It must be appreciated, however, that such are small nonrepresentative surface regions, whose characteristics should not be equated with those of other portions of the surface membrane. Nevertheless, it is of major interest to compare morphologically specialized membrane areas with other surface domains.

(b) The plasma membrane is assumed to be a homogeneous entity (14), which requires chemical stabilization to prevent disruption. The surface membranes of some cells (e.g., HeLa) appear "tough" enough to resist disruption without special maneuvers (15).

(c) Oppositely, the plasma membrane is taken to be a plastic, heterogeneous array of "organelles" arrayed at the cell periphery, which tends to fragment into vesicular subunits. We favor this view, believing it amply supported by biologic evidence. In any event, most membranes tend to vesiculate, and one must always come to grips with the fractionation of small membrane sacs, unless one chooses to stabilize the membranes chemically.

Mechanical Homogenization

Cell disruption usually depends on shear forces, which distort the plasma membrane to the point of fragmentation (16–19). Mechanical shearing effects of the widely used Dounce type homogenizers depend on the clearance and relative motion between the pestle and mortar, as well as medium viscosity. The first is difficult to control even with precision-bore homogenizers. Nevertheless, this technique remains favored for cell disruption.*

Rapid decompression of an inert gas from 50–75 atm down to ambient pressure selectively ruptures the surface membranes of many cells (19) by formation of gas bubbles at cell surfaces. This procedure can be used to homogenize cells under controlled isoosmotic conditions in an inert atmosphere without heating and without rupture of nuclei, mitochondria, or lysosomes.

Sonication and high liquid shear devices, such as the French press, produce heat and free radicals locally and often damage the structure of plasma membranes as well as those of nuclei, mitochondria, and lysosomes (20).†

Unless membranes are toughened or "fixed," physical shear tends to cause them to vesiculate by pinching off, entrapping soluble cytoplasmic components and, in some cases, such as the pinching off of nerve endings into synaptosomes, mitochondria, and synaptic vesicles. Mem-

*A useful analysis of shearing effects on erythrocytes and leucocytes has been presented recently (R. M. Hochmuth, N. Mohandas, E. E. Spaeth, J. R. Williamson, P. L. Blackshear and D. W. Johnson, *Trans. Amer. Soc. Artif. Int. Organs* **18**: 325, 1972). This and other studies indicate that cell rupture occurs primarily through interactions between the cell periphery and diverse surfaces.

† Prolonged sonication of egg yolk lecithin produces appreciable chemical degradation of the lipid even under hypoxic conditions (H. O. Hauser, *Biochem. Biophys. Res. Comm.* **45:** 1049, 1971). The conditions required to avoid this problem have been reported recently (H. Hauser and L. Irons, Hoppe Seyler *Ztschr. Physiol. Chem.* **353:** 1579, 1972). When the duration and intensity of sonic irradiation exceed certain limits, the vesicles become uniform in size, yet exhibit some chemical deterioration.

branes may also "endocytose" and shed vesicles into the compartment they enclosed, yielding particles with an inverted orientation. This has been observed in inner mitochondrial membranes (21) and erythrocyte ghosts (22).

When suspensions of large cells, such as adipocytes, are forced through perforated screens, having apertures with diameters only slightly larger than those of the cells, selective disruption of the cell membrane can be achieved (23).

Chemical Methods

Cell disruption can be accomplished by manipulation of ionic strength and pH, as well as surface active agents, lipases, and proteases.

The use of hypoosmotic lysis of erythrocytes has been employed by most investigators for the isolation of membranes from nonnucleated erythrocytes. The procedure most widely used at present is that of Dodge et al. (24), or variations thereof in which lysis is produced by 20 millimolar (mM) phosphate at pH 7.4–8.0.

However, simple hypoosmotic lysis of cells other than erythrocytes has not been generally satisfactory for membrane isolation. Moreover, hypotonic media, pH-extremes, lipases, proteases, and surface-active agents are too indiscriminate in their action, perturbing diverse intracellular membranes as well as the plasma membrane.

Centrifugal Separation of Membrane Fragments from Disrupted Cells

General

Isolation of membranes from lysed nonnucleated erythrocytes is relatively simple, since these cells contain no other cellular organelles. It is accomplished by repeated centrifugation at low ionic strength until the membranes are free of hemoglobin.

Cells other than mammalian erythrocytes contain a variety of membrane organelles, and the isolation of plasma membranes from such cells continues to present numerous problems. The most satisfactory procedures thus far have involved centrifugal techniques and have been carried out in a number of different ways with varying cell types.

Differential Rate Centrifugation

This approach separates membrane fragments according to density and/or size.

The behavior of spherical particles in a centrifugal field is described by the equation

$$s = 2r^2 (\rho_p - \rho_s) 9\eta \qquad (1.1)$$

where s is the sedimentation coefficient of the particle (its sedimentation velocity per unit centrifugal field); r is its radius; and ρ_p its density. ρ_s and η are the density and viscosity of the medium. All units are cgs. The sedimentation coefficient is usually given in Svedberg units ($S = s \times 10^{13}$ sec).

Equation (1.1) cannot be simply applied to centrifugal membrane fractionations, and it is more useful in differential centrifugation, to classify subcellular particles in terms of their sedimentation behavior upon application of a given time-integrated centrifugal force (g-min). Usually large particles (e.g., nuclei) are pelleted at $1–2 \times 10^4$ g-min, the supernatant recentrifuged at $1.5–3 \times 10^5$ g-min, to sediment organelles such as mitochondria and the second supernatant spun at $5 \times 10^6–10^7$ g-min, to sediment small particles. Unfortunately, in simple differential centrifugation, slowly sedimenting particles at the bottom of the centrifuge tube can pellet in the time required for more rapidly migrating particles to sediment from the top of the tube. This type of contamination increases the closer the sedimentation coefficients and can be reduced by recycling the pellets.

Differential Density Gradient Centrifugation

Much better results are obtained when buffered density gradients are used to prevent convective artifacts [cf. reviews in references (25–27) and for membranes (28)]. Excellent separations of membranes can be achieved by *rate-zonal* techniques. These procedures involve layering the sample atop a density gradient and subjecting it to high-speed centrifugation to sediment membrane particles into or through the gradients. At any convenient point in time—depending on the centrifugal force, medium viscosity, density, and osmotic activity, as well as particle density, shapes and size—the samples can be collected by puncturing the centrifuge tube and collecting fractions. Sucrose, glycerol, cesium chloride, rubidium chloride, and potassium fluoride are commonly used for gradients, but macromolecules such as polyglucose or polysucrose have been highly useful in membrane applications (28–30).* Density gradient separation can be adapted to large-scale isolation using zonal rotors (31–36).

* Centrifugal transfer of subcellular organelles through silicone fluid layers finds increasing application in the measurement of solute accumulation in such particles (R. E. Gaensslen and R. E. McCarty, *Anal. Biochem.* **48**: 504, 1972).

Isopycnic Density Gradient Centrifugation

Membrane particles can also be separated according to their densities in density gradient systems, by centrifuging until equilibrium is achieved. In this *isopycnic technique*, separation of membranes depends principally on the overall density of the membrane fragments. Moreover, the selection of a density gradient system can markedly influence the behavior of the many membrane fragments, which migrate as closed, charged vesicles, whose behavior varies with pH as well as ionic strength and composition and which may show selective permeability to different gradient solutes (28).

Carbohydrates constitute the favored gradient solutes for separation of membranes, but their osmotic properties are insufficiently considered. Thus, although sucrose has been most widely used for this purpose, polymeric sucrose (or glucose) offers the major advantage in that it minimizes osmotic effects (28, 29). The more viscous polymers, however, require higher speeds and longer centrifugation times than does sucrose.

Glycerol generally permeates readily into membrane vesicles, and glycerol gradient thus yields higher densities with most membranes than sucrose or polymer systems.

Gradients of CsCl, RuCl, and KBr can also be used for equilibrium (or rate-zonal) centrifugation without inhibition of marker enzymes or extraction of membrane proteins (19). Ionic strength in excess of 0.15M, low concentration of multivalent cations, and pH between 7 and 8.5 are required to prevent aggregation. Table 1.1 lists the densities of commonly employed gradient media.

General Centrifugation Procedures

The detailed application of gradient centrifugation to a given membrane fractionation will, of course, depend on a large number of variables in any given membrane isolation. The following summarizes some points, which should always be considered.

Density gradients

The length and slope of the gradient can be varied according to the requirements of a given separation. An extended shallow gradient must be employed if the mixture to be fractionated contains a large number of components with closely similar densities and/or sizes. Short gradients are satisfactory where few well-separated components are expected. In equilibrium density gradient fractionations, separations are most rapid using short gradient columns.

Table 1.1 Densities at 25°C of Density Gradient Solutions

Solute	Concentration (wt%)[a]					
	10	20	30	40	50	60
LiCl	1.054	1.113	1.178	1.250		
LiBr	1.073	1.160	1.261	1.281	1.529	1.716
KBr	1.072	1.158	1.257	1.371		
NaBr	1.078	1.172	1.281	1.410		
RbBr	1.079	1.174	1.285	1.419	1.582	
CsCl	1.079	1.174	1.286	1.420	1.582	1.785
CsBr	1.081	1.180	1.297	1.440	1.616	
Potassium acetate	1.048	1.100	1.155	1.213	1.242	1.333
Potassium citrate	1.066	1.140	1.221			
Potassium tartrate[b]	1.066	1.139	1.218	1.305	1.400	
Glycerol	1.021	1.045	1.071	1.097	1.124	1.151
Sucrose	1.038	1.081	1.127	1.176	1.230	1.289
Polysucrose[c]	1.034	1.068	1.102	1.136		
Polyglucose[c]	1.038	1.076	1.114	1.152		

[a] From International Critical Tables.
[b] At 20°C.
[c] From Mach and Lacko (37).

There are many methods for the formation of density gradients. A very satisfactory one employs a simple three or more channel peristaltic pump used in conjunction with a mixing chamber (38). When the mixing chamber contains one density extreme and the medium at the other density extreme is pumped into it at rate k_1, a linear density gradient is achieved when the material in the mixing chamber is simultaneously pumped *out* into the centrifuge tubes at rate $2k_1$. If pumping is continued until the mixing chamber is empty, the density limits of the formed gradients are those of the two solution densities used.

A problem inherent in the isolation of plasma membranes is the fact that membranes often break into fragments of varying size and shape during and after cell rupture. Most of these fragments reseal into small semipermeable sacs or vesicles (19, 28, 29). However, some surface membrane areas, such as bile fronts and brush borders, tend to be stabilized by terminal bars, desmosomes, as well as various junctional complexes. Such membrane fragments usually sediment in the "nuclear" fraction, whereas small sealed vesicles tend to sediment with the microsomes, and intermediate particles tend to distribute between these two fractions. Membrane fragments that are permeant to the gradient solutes show isopycnic densities of 1.15 to lower than 1.02, depending on the conditions of centrifugation (19, 28, 29), in particular, the osmotic gradient in the centrifuge tube, and possible ionic effects.

The sedimentation of sealed vesicles will depend upon the density of the membrane per se, the hydration of its components, and the concentration of permeant and nonpermeant solute in the medium and in the vesicle (19, 29, 30). Gradients of larger polymers such as polysucrose (Ficoll) have a limited osmotic effect and do not penetrate the vesicles. This results in a low buoyant density (1.04–1.08). On the other hand, sucrose generally permeates more slowly than water, and thus exerts a large osmotic effect. The density of sealed vesicles in sucrose is, therefore, higher (1.12–1.17), and sucrose gradients do not readily distinguish small differences in membrane properties. In gradients of permeant solutes such as glycerol and some salts, the solvent space in the vesicle assumes the density of the surrounding medium, and the particle reaches equilibrium at a density equal to that of the membrane itself (1.15–1.19).

Permeant electrolytes will tend to accumulate within the vesicles due to the charges (net negative at neutral pH) fixed within the vesicles, on their walls and/or their nonpermeant contents. At low ionic strength, these tend to draw water into the sacs and reduce their density. This effect can be counteracted by "titrating" the fixed charges by changing pH or the concentrations of permeant polyvalent cations such as Mg^{2+} (19, 29, 30).

Sample addition

In rate-zonal centrifugation, samples are usually added by floating them on the surface of the gradient with a micropipette. The zone width of the membrane fraction is primarily determined by the amount of membrane per centrifuge tube with best results obtained at low membrane concentration.

In isopycnic gradient centrifugation, the tendency of membrane fragments to aggregate can be reduced by mixing the material into the gradient during its formation.

Centrifugation

Rate-zonal centrifugation must be initiated immediately on the addition of the sample in order to avoid convective mixing. The time required for desired membrane zones to move to the center of the gradient varies inversely with the sedimentation coefficient of the particles and is, of course, more rapid at higher gravitational fields. In equilibrium gradient centrifugation, a longer time at higher speeds is required to reach equilibrium. True equilibrium is assured if the same pattern is obtained in tubes in which the sample mixed throughout the gradient is the same as that in which the sample is applied to the surface of the gradient.

Collection of samples

Following separation by rate-zonal centrifugation or by equilibrium density centrifugation, the samples can be collected by piercing the bottom of the centrifuge tube and pumping in a dense solution slowly at the bottom of the tube to force the gradient upward, or by pumping in a low-density solution at the top of the tube to push the materials out through the bottom of the tube. The samples can then be passed through an appropriate flow detector and collected in a fraction collector.

If desired, the shape of the density gradient can be determined with calibrated glass beads, by pycnometry, by refractometry, or by calibrated immiscible density gradients.

Syringes with their needles bent into a U-shape can be inserted, so that their orifices at the bottom of a desired zone can be used to collect membrane fractions when the zones are easily visible. Commonly, the bottom of the centrifuge tube is punctured with a 21 gauge needle, and the drops collected. However, the flow rate is not constant, and the drop size changes as the density and surface activity of the gradient solution change.

Density perturbation methods

In a number of cases, closed membrane vesicles can be selectively isolated and/or fractionated according to their contents. This approach has been quite widely used in the purification of *lysosomes*: Rats or mice are given about 80 mg Triton WR 1339 per 100 g body weight, either intravenously or intraperitoneally, 2 to 4 days before sacrifice. During this interval, the low-density detergent accumulates in the lysosomes. Cells are then disrupted conventionally, nuclei and "debris" removed by low-speed centrifugation, [mitochondria + lysosomes + peroxisomes] pelleted at about 340,000 g-min, and these three organelles separated in buffered sucrose gradients. In the case of rat liver, the Triton procedure lowers lysosomal density from 1.210 to 1.117 and in Ehrlich ascites carcinoma from 1.155 to 1.113 (36).

Wetzel and Korn (39) employ a similar principle to isolate phagosomal membranes from *Acanthamoeba*. The organisms are fed low-density polystyrene particles, which they rapidly engulf. After cell rupture, these particles, plus their surrounding membranes, are isolated by virtue of the polystyrene's low density.

We have cited cases where membrane particles were made less dense than normal. The opposite direction has also been employed. Thus, the density of rat liver mitochondria *increases* in proportion to their sequestration of calcium phosphate during respiration-driven calcium transport (40). Similarly, accumulation of insoluble lead phosphates within microsomal vesicles during hydrolysis of glucose-6-phosphate raises the density of the vesicles containing this enzyme (41). This approach would also be of use for other membrane vesicles endowed with specific phosphatases.*

Tsukagoshi and Fox (42) used *bromostearic acid*, whose density is much greater than the natural lipid, to separate newly synthesized membranes from *E. coli*. However, they found that this label exchanges easily between membranes. This may be because the massive bromine atom prevents proper incorporation of the analogue into the membranes (Fig. 1.2). In

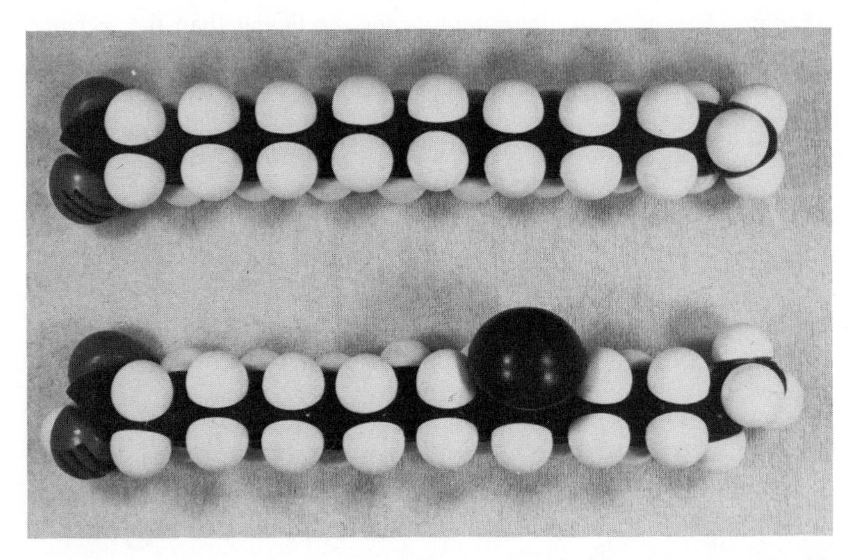

Fig. 1.2 A space filling model of stearic acid (above) compared with bromostearic (below). Note that the volume of the bromine atom (and presumably its polarity) would not be expected to promote close packing with normal membrane lipids.

principle, this is a promising approach to the study of membrane biogenesis, but it is clear that whatever dense isotopes are utilized, they must closely approach the natural in molecular dimension and configuration and need to be presented in high abundance.

A more general approach of great potential and versatility has been developed recently (43); it is termed *affinity density perturbation*. Here, membrane fragments bearing a specific receptor are separated centrifu-

* A new method, involving density perturbation, much improves the separation of synaptic membrane vesicles from mitochondria (G. T. Davis and F. E. Bloom, *Anal. Biochem.* **51:** 429, 1973). Mitochondrial density is increased by forming insoluble, dense, iodonitrotetrazolium derivatives during the oxidation of succinic acid.

gally by the density increase, caused by the combination of the receptor with its specific ligand, *coupled to a particle of very high density*. This approach is illustrated in Fig. 1.3.

Membrane fragments bearing the specific receptor are reacted with a specific density perturbant and the complex centrifuged to its isopycnic density; this lies higher than that of the membranes, but lower than that of the perturbant. For localization and quantification, the membranes and perturbant are labeled with different radioisotopes. The formation of membrane-ligand complexes is blocked or reversed, when desired, by addition of reagents of higher affinity for the ligand than the receptor, or an excess of receptor analogues with similar affinity.

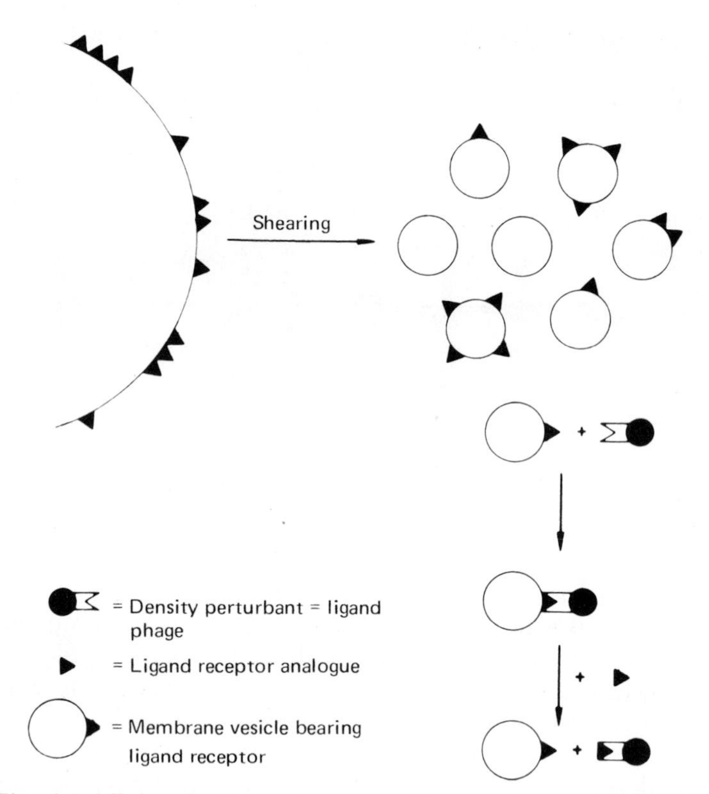

Fig. 1.3 Affinity density perturbation. The membrane, bearing specific receptors, is sheared into small vesicles. These are reacted with a ligand specific for the receptor, but coupled covalently to a dense particle, such as a small phage. Reaction of this complex with a receptor-bearing membrane particle will lower the density of the phage-ligand complex and raise that of the reacted membrane. The latter can be freed of ligand by a receptor analogue of low molecular weight (43). (Courtesy of the North Holland Publishing Co.)

Phase Separation in Aqueous Solutions of Mixed Polymers

This promising technique, first applied to the membrane field by Brunette and Till (44), represents a logical extension of the general system developed by Albertsson (45).

When aqueous solutions of two different polymers are mixed, phase separation can occur when the attractive forces between like molecules substantially exceed those between unlike species. Each phase then contains predominantly one polymer type. In contrast, when attractions between unlike species are greater than between like molecules, "complex coacervation" takes place. Then both polymers concentrate in one phase, leaving the other depleted of the macromolecules. This commonly occurs when solutions of oppositely charged polymers are mixed.

Populations of particles of diverse surface properties included in a mixed polymer system can often be resolved by one or repeated phase separations. Even particles of equivalent surface charge may be separated, if they differ in other surface properties.

Theoretically, the partition coefficient K of a particle relates to its surface properties as follows:

$$K = C_1/C_2 = e^{-A(t_1-t_2)\,kT}$$

when $$(t_1 - t_2)/t_{12} \geqq 1 \tag{1.2}$$

Here C_1 and C_2 are the concentrations in phases 1 and 2, respectively; t_1 and t_2 are the interfacial tensions between the particles and phases 1 and 2; t_{12} represents the interfacial tension between the bulk phases; A is the particle surface area; k is the Boltzmann constant, and T is the absolute temperature. When $(t_1 - t_2)/t_{12} < 1$, particles accumulate at the *interface*.

Generally small molecules partition evenly between the two phases, but proteins may have partition coefficients from 0.1 to 10, and supermolecules and other large particles tend to collect in one phase only.

When the partition coefficients of the individual components of a particle mixture differ by <10, effective separation of these components will require several sequential partitions, i.e., countercurrent extraction. In Albertsson's original approach, this was complicated by the long time required for separation of bulk phases differing little in density and/or viscosity. Accordingly, Albertsson devised a thin-layer system, where the phases are contained in layers about 2 mm thick, thereby reducing separation time to <5 min. This apparatus is now available commercially in automated form.

Brunette and Till (44) did not use the sophisticated thin-layer apparatus for their isolation of L-cell plasma membranes. Instead, they disrupt the cells by "Dounce homogenization" in a hypotonic medium, stabilized with 10^{-3} M $ZnCl_2$. After removing debris by low-speed centrifugation, the membranes are purified by repeated centrifugal partition into the *interphase* formed after mixing into and centrifuging a two-phase polymer mixture of dextran and polyethyleneglycol. The two phases were prepared by combining 200 g 20% (w/w) dextran 500 in d H_2O with 103 g 30% (w/w) polyethylene glycol (Carbowax 6000) in d H_2O, 99 ml d H_2O, 333 ml 0.22 M phosphate buffer (pH 6.5), 10^{-2} M $ZnCl_2$. After phase separation, the initial membrane isolate is dispersed in upper phase, and this mixed with lower phase, followed by centrifugal phase separation. After a suitable number of partitions, the two-phase system is disrupted by dilution with water, and the membranes isolated by low-speed centrifugation. A purified membrane isolate, consisting of large sheets, is obtained in 2 hrs. Contamination with nuclear or mitochondrial material is minimal, with considerable enrichment of several plasma membrane markers. Contamination with fragments of endoplasmic reticulum appears slight, but only a single marker was employed.*

General Quantitation of Membrane Components

Problems of quantitating the amount of protein in a gradient separated membrane fraction have been listed by Resch et al. (46). These authors have described a convenient, nondestructive fluorometric procedure involving the tryptophan fluorescence of membrane proteins.

Measurement of absorbance at 260 or 280 nm lacks sensitivity and suffers from scattering interferences by the membrane particulates. Measure of peptide absorption at 200 nm, though very sensitive and a true measure of peptide content, is unsatisfactory because of the absorbance from other solutes at this wavelength, and to the light scattering of particles and gradient solutes.

The determination of protein by the procedure of Lowry et al. (47) is usually not very satisfactory for carbohydrate-containing gradients and does not permit direct scanning of the effluents. The biuret method lacks sensitivity and generally is not suitable for membrane proteins. The

* The variables determining partitions between two aqueous polymer phases receive continued attention. Thus Walter *et al.* (H. Walter, R. Tung, L. J. Jackson, and G. V. F. Seaman, *Biochem. Biophys. Res. Comm.* **48:** 565, 1972) document that surface charge influences partition, probably because of unequal distribution of inorganic ions between phases. Moreover, their data suggest that "deep" membrane charges influence partition more than they do electrophoresis.

An aqueous two-phase polymer system has also been employed to isolate a plasma membrane fraction from rat liver (L. Lesko, M. Donlong, G. V. Marinetti and J. D. Hare, *Biochim. Biophys. Acta* **311,** 173, 1973).

ninhydrin assay can be employed if pure gradient solutes are used (30). Phospholipid and/or cholesterol estimates require removal of gradient solutes and are cumbersome as well as insensitive. At the present time, the most satisfactory general procedure for general monitoring of the effluents of gradient columns seems to be the fluorometric approach.

Membrane Markers

It is highly desirable to use multiple criteria of identity and purity of membrane preparations. However, although nearly always easy to determine *whether* plasma membrane is present, it is more difficult to tell *how much*, and very complicated to assay *purity*.

Morphology

Some membrane preparations can be identified using electron microscopy if they retain characteristic structural features such as nuclear pores, mitochondrial structure, ribosomes, "bile fronts," etc. However, many membranes tend to form smooth vesicles upon cell disruption, and morphological studies are then of little assistance. Moreover, quantitation is quite impractical. *First* of all, not all particles can be identified. *Secondly*, a sample of 1 μg dry weight might contain 10^{7-8} fragments 1 μm in size and, even with a fully random distribution of fragments, the time required to count and identify a significant proportion of the particles in a fraction is out of range. An additional, little-appreciated complication derives from the varied size of subcellular organelles and the fact that electron microscopy can only inform as to how *many* mitochondria, for example, contaminate a microsomal membrane preparation. But one mitochondrion weighs about $\sim 10^{-13}$ g compared with about $\sim 10^{-15}$ g for a typical smooth microsomal vesicle. One mitochondrion per 100 vesicles thus represents a *large* mass contamination.

Endogenous Chemical Markers

At one time *sialic acid* appeared to be a specific plasma membrane component. However, more recent evidence indicates that this or any other carbohydrate is not exclusively associated with the plasma membrane (19, 28). Consequently, determination of sialic acid may be useful but not definitive.

All the membranes we have studied so far contain *lipids*, particularly phospholipids, sphingolipids, and cholesterol. In erythrocytes, lymphocytes, and liver bile fronts, cholesterol appears high. However, there is no evidence that the composition of plasma membranes is sufficiently characteristic to permit the lipid components to be used to assess their identity or purity.

The antigenic and enzymatic diversity of plasma membranes in large part reflects the characteristics of their proteins. However, protein analyses per se have not proved generally useful in monitoring membrane fractionations. We anticipate considerable advance in this field with the advent of membrane peptide analysis by molecular sieving in SDS-laden polyacrylamide. Thus, this technique readily permits distinction between plasma membranes and endoplasmic reticulum of lymphocytes (48).

Enzyme Markers

Many enzymes are compartmentalized within the cell and frequently are associated with specific membranes and related to the membrane function. If an enzyme has a highly specialized function, it may have a unique cellular localization. This localization may be expected to be specific even in diverse tissues and species. The sensitivity and specificity of enzyme assay methods and the possibility of relating *in vitro* quantitative assays with in situ histochemical localization give these a special value for identifying an isolated membrane fraction and estimating its purity.

However, a number of major complications plague the use of enzymes as markers, e.g., (a) the possible adventitious association of enzymes with particulates with which they are not ordinarily associated in the cell; (b) the difficulty in excluding the possibility that enzymes detectable in the surface of the intact cell may also be associated with intracellular membranes; (c) the hazard that enzymes may be eluted, inactivated, or activated during fractionation; (d) the fact that a specific enzyme may not represent the entire membrane class under study. Thus, diverse mitochondrial enzymes are not uniformly distributed within these organelles, and glucose-6-phosphatase, the most widely used marker for endoplasmic reticulum, is not homogeneously distributed throughout the microsomal population; (e) the large differences in the relative intracellular abundance of various organelles, which can easily produce misleading conclusions if enzyme activities are expressed as specific activities; (f) the possibility that a membrane fraction may be contaminated by material which, lacking enzymatic specificity (or totally lacking enzymatic activity), remains undetected.

In spite of the general empiricism of the approach and other problems, the use of enzymes as markers of cell membranes has proved highly useful, and we here briefly discuss the enzymes that have been most widely used in plasma membrane fractionations.

5′-nucleotidases

These enzymes catalyze the hydrolysis of 5′-ribonucleotides to 5′-ribonucleosides and orthophosphate. Adenosine-5′-monophosphate is most rapidly hydrolyzed by membrane preparations, but most 5′-ribonucleotides and 5′-deoxyribonucleotides are also cleaved. These enzymes appear particularly concentrated in liver bile fronts (28). Coleman and Finean (49) observed the enrichment of 5′-nucleotidase in the surface membranes of several guinea pig organs, but not in brain or erythrocytes. 5′-nucleotidases are also concentrated in plasma membrane faction of HeLa cells (50), fat cells (23), and lymphocytes (51) and at present seem to be the most specific and ubiquitous enzymatic plasma membrane markers available. A surface function can be envisaged for such enzymes since nucleotides penetrate the plasma membrane poorly, and a surface enzyme, catalyzing their dephosphorylation to permeant nucleosides, would greatly facilitate their uptake.

This enzyme activity is generally found in other subcellular fractions as well, perhaps because 5′-nucleotidase is plasma membrane specific; however, these membranes fracture into particles cosedimenting with many subcellular fractions (52). Thus, Widnell and Unkeless (53) show that, while liver bile fronts have the highest specific activity, the bulk of this enzyme sediments with microsomal fractions, perhaps in plasma membrane fragments. The enzyme from both fractions appears to be the same lipoprotein, containing only sphingomyelin as phospholipid.

Alkaline phosphatases

These enzymes catalyze the hydrolysis of orthophosphoric monoesters to alcohols and orthophosphate. A variety of organic phosphates are hydrolyzed by these enzymes at slightly alkaline pH. These enzymes are associated with the brush borders of kidney and intestine, and hepatocyte bile fronts from some rodents. The specificities of the different alkaline phosphatases must be considered in using these enzymes as membrane markers. Thus, Emmelot et al. and Graham et al. (54, 55) find an alkaline phosphatase hydrolyzing *p*-nitrophenyl phosphate, but not *p*-glycerophosphate located in the bile fronts of rat liver. Lansing et al. (56) also found *p*-nitrophenyl phosphatase as well as NADase and a phosphodiesterase in plasma membrane isolated from rat liver. Coleman and Finean (49), however, found alkaline phosphatase in the brush border membranes of guinea pig intestinal mucosa and kidney, but not in liver bile fronts. Bosman et al. (50) demonstrated alkaline phosphatase (using *p*-nitrophenyl phosphate as substrate) in several membrane fractions from HeLa cells, especially in their plasma membrane fraction.

Thus, alkaline phosphatase may serve as a plasma membrane marker in some, but not all, cells. Since several subclasses of alkaline phosphatase with different Michaelis constants and specificities exist, choice of substrate and pH may be critical.

ATPases

The ATPases catalyze the hydrolysis of adenosine triphosphase to adenosine diphosphate and orthophosphate. Mg^{2+} or Ca^{2+} is usually required. Other nucleoside triphosphates may be hydrolyzed by plasma membranes, but usually at a lower rate. Some Mg^{2+} requiring ATPases depend on Na^+ and K^+ and are inhibited by cardiac glycosides such as oubain. These are considered to be involved in sodium and potassium transport and are usually present in plasma membranes at high specific activity (e.g., 54). However, they have also been demonstrated in the membranes of the endoplasmic reticulum (57).

Other hydrolytic enzymes that appear to be concentrated in plasma membranes include the following:

leucine amidases (L-leucyl-β-napththol amidase) (55)
NADase (54, 56)
NAD-pyrophosphatase (56)
nucleoside triphosphatepyrophosphohydrolase (58)
phosphodiesterase (54)
adenyl cyclase (59)

In addition to the enzymes listed above, some biosynthetic enzymes appear associated with plasma membrane isolates in some instances. Examples are the enzyme catalyzing the reaction pantotheine + acyl Co-A-S-acyl-pantotheine (60) and the enzyme transferring fucose to collagen (61).

Meaningful plasma membrane fractionations require one to establish not only the enrichment of truly specific marker enzymes, but also the impoverishment of markers characteristic of other subcellular organelles. Here we again stress that this should be on the basis of *total*, as well as specific activity, since many potential contaminants, e.g., DNA, RNA, mitochondrial, and glycolytic enzymes, and hemoglobin in erythrocytes, are often present in high abundance. We also caution once more that enzymes may be activated or inactivated in the process of cell disruption.

Immunological Markers

Membrane antigens induce formation of specific antibodies when administered to distant or related species. Such antibodies can offer a powerful means for identifying membranes and estimating their purity (15, 62–64). This approach has not yet been extensively used in studies of isolated membrane preparations, but it has had a very successful application in the study of histocompatability antigens in the mouse (65). One can prepare antibodies specific to given membrane antigens, tag them radioactively and/or with fluorescent dyes, ferritin, or small phages, thereby permitting determination of the antigen distribution on the surfaces of the intact cells. Such techniques give evidence of considerable temperature-dependent "lateral mobility" of some surface membrane antigens and illustrate the important role of *temperature* in membrane fractionation.

An important potential of the immunologic approach is that one can prepare labeled markers *specific for intracellular* membranes (63, 64). In this, antibodies are produced against whole cell homogenates, and then absorbed with intact cells, purified nuclei, mitochondria, etc., to remove antibodies specific for the absorbing membranes and leave antibodies directed against those components that have not been used for absorption. This powerful approach can also detect possible antigens common to more than one membrane type (64). Unfortunately, the methods are cumbersome and have not been widely utilized.

A closely related approach, with seemingly considerable potential, involves the use of plant "*lectins*." These have the capacity to bind to cell surfaces, reacting with terminal, nonreducing sugars in glycoproteins and/or glycolipids. Kidney bean phytagglutinin, concanavalin A, soybean phytagglutinin, and wheat germ phytagglutinin have been extensively studied, particularly with regard to the different agglutinability of normal and cancerous cells (cf. 66). The only application of lectins in membrane fractionations at this writing has been in the "*affinity density perturbation*" technique (43), and there is no assurance that lectin binding sites are restricted to the plasma membrane.

Virus Receptors

Many plasma membranes contain surface components that specifically bind certain viruses such as influenza. *Virus binding sites* can be used to identify the membranes and to study the distribution of the receptor sites on membranes (67–69).

Covalent Labels

The possibility of labeling the plasma membrane with covalently linked compounds prior to its isolation may become a powerful method for identifying the plasma membrane and localizing its components.

Useful labels for tagging *externally* exposed membrane components should be "vectorial," i.e., they should not pass through the membrane prior to reaction. They should react with membrane components under physiological conditions of pH, tonicity, and temperature; should not cause cell disruption or serious perturbation of membrane structure or function; also they should be detectable in trace amounts.

Maddy (6) used the highly fluorescent diazonium dye 4-acetamido, 4′-isothiocyano-stilbene-2, 2′-disulfonic acid for labeling ox erythrocytes.

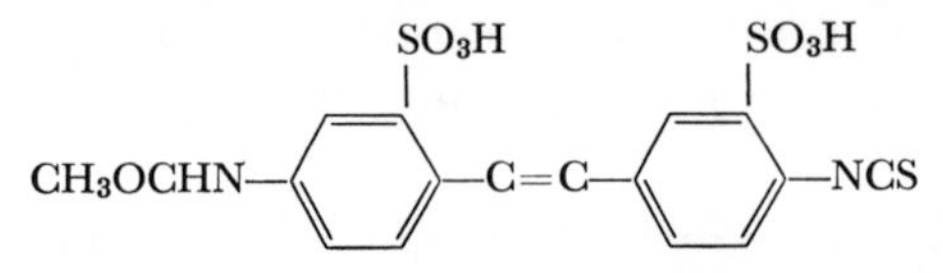

This compound is specific for amino, histidyl, and guanido groups of proteins. It was found that 4.5×10^5 molecules bind per ox erythrocyte. Membrane lipids do not react when intact cells are utilized. Maddy states that one molecule of SITS binds per 400 molecules of hemoglobin. But a typical mean corpuscular hemoglobin of 3×10^{-13} g/cell corresponds to $\sim 2.4 \times 10^7$ molecules of hemoglobin (tetramer) per cell and, if one out of 400 molecules binds SITS, one gets 6.0×10^5 molecules per cell in hemoglobin! The basis for this discrepancy is not explained, but casts some doubt on the usefulness of SITS as an unambiguously "vectorial" marker.*

Pardee and Watanabe (70) have used 7-amino-1, 3-naphthalene disulfonate in an analogous fashion, to localize the sulfate binding protein to the outside of the permeability barrier of E. coli membranes. We are not aware of its application in other membrane studies.

Berg (7) and Bender et al. (71) have employed the diazonium salt of ^{35}-Sulfanilate to label the membranes of intact erythrocytes.

They claim that this compound is essentially impermeant and that internal hemoglobin is only slightly labeled when intact cells are treated. However, the exact degree or *rate* of penetration is unclear from published data. The labeled proteins could be separated by SDS gel electrophoresis. This agent inactivated acetyl choline esterase, consistent with the probable external localization of this enzyme, but we suspect that diazonium sulfanilate does not constitute a truly "vectorial" reagent even in the erythrocyte. In any event, it rapidly permeates into lymphocytes (72) with concomitant cell lysis.

* A recent study (Z. I. Cabantchik and A. Rothstein, *J. Memb. Biol.* **10:** 311, 1972) shows that SITS binding to erythrocyte membranes is largely non-covalent. Certain other disulfonic acid stilbene derivatives react primarily covalently.

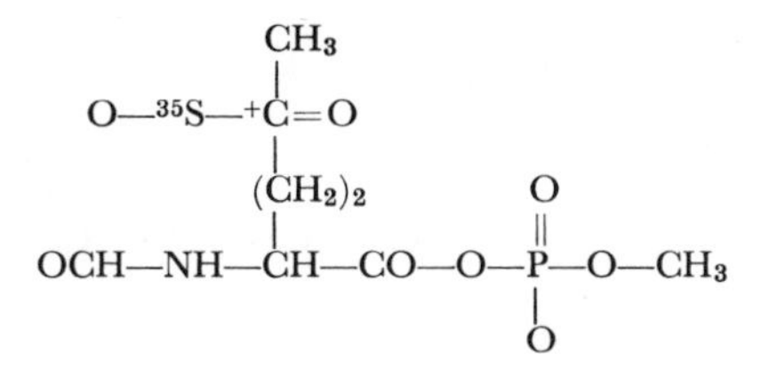

^{35}S-labeled formyl-methionyl sulfone methyl phosphate, an active acylating agent, has been used by Bretscher (8, 73) to label reactive proteins of intact human erythrocytes. Unfortunately, the reaction is carried out for 10 min at pH 10 and under such conditions the membranes of many cells deteriorate. Moreover, Bretscher's data indicate that even in the case of erythrocytes, hemoglobin labeling accounts for about 15% of the whole after 10 min. Bretscher's compound is thus also not truly "vectorial."

This reagent reacts preferentially with two *membrane* proteins with apparent molecular weights of 105,000 and 90,000, as determined by SDS gel electrophoresis. The smaller of the two contains most of the sialic acid of the erythrocytes and presumably is the glycoprotein discussed later. The larger protein has little or no carbohydrate. Strikingly, phosphatidyl ethanolamine, and phosphatidyl serine are not labeled in the intact cells, but are heavily labeled when erythrocyte ghosts are employed. Treatment of the erythrocytes with pronase prior to reaction with the reagent prevents the labeling of the glycoprotein and reduces the size of the larger protein.

Bonsall and Hunt (74), as well as Steck (75), argue that 2, 4, 6-trinitrobenzene sulfonate

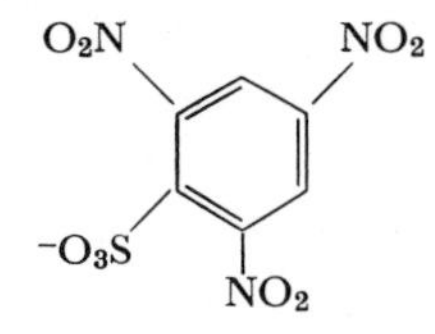

reacts preferentially with externally exposed components of erythrocyte membranes. However, Arroti and Garvin (76) clearly show that this compound reacts within the erythrocyte membrane, while diffusing from the external milieu through the membrane to the hemoglobin "sink." *We suspect that all low-molecular-weight labels utilized to date follow this pattern and that the membrane proteins labeled in intact erythrocytes are not necessarily only more "exposed," but perhaps more reactive than other membrane components.*

The labeling of the *sulfhydryl* groups of cell membranes can be accomplished with such compounds as *p*-chloromercuribenzoate (PCMB)

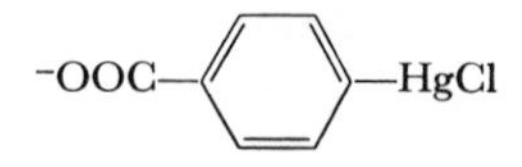

1-bromo-2-hydroxypropane, $BrCH_2$-CHOH-CH_3, or *N*-ethylmaleimide

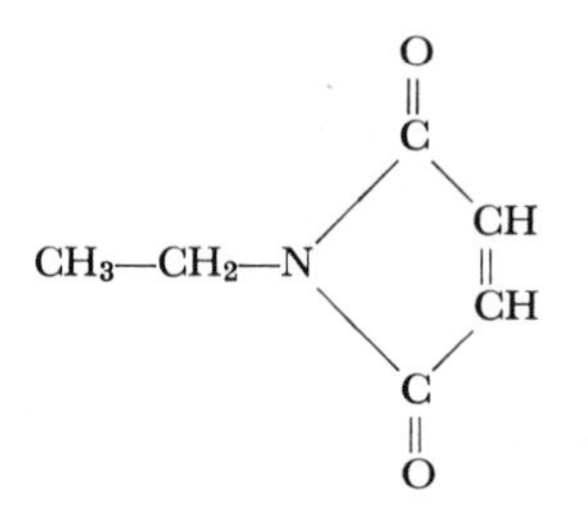

PCMB reacts with at least three types of sulfhydryl groups in the erythrocyte membranes. Some information as to the localization of these groups can be obtained from the fact that the PCMB complexes on the external membrane surfaces can be dissociated by the nonpenetrating sulfhydryl compounds such as glutathione. While the small SH-reagents described above are all permeant to a greater or lesser degree, Yariv et al. (77) have synthesized two "vectorial" membrane SH-reagents: (a) *N*-(3-mercuri-5-methoxy propyl) poly-D, L-alanyl amide-^{203}Hg, as SH-blocker, and (b) poly-D, L-alanyl cysteine, as a vectorial agent to protect or regenerate SH-groups located outside the membrane permeability barrier. These compounds have not yet been applied to studies on animal cells.*

The possibility of using affinity-labeling procedures to tag specific proteins in cell membranes is a promising avenue for future research. Indeed, aryl azides (78) have already been used to tag the acetyl choline binding sites of erythrocyte membranes (79).

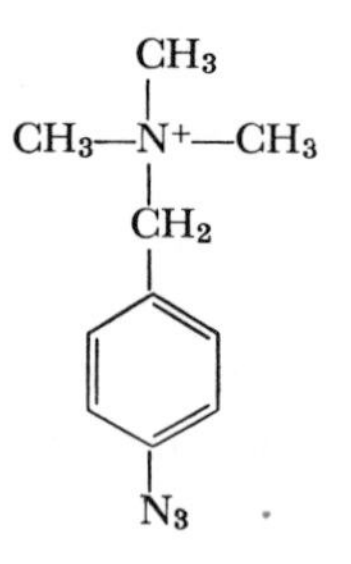

* One can couple PCMB to aminoethyl dextran to yield a polyfunctional, polydisperse SH-blocker. This inhibits the ouabain-sensitive $Na^+ - K^+$ ATPase of intact erythrocytes (H. Ohta, J. Matsumoto, K. Kagano, M. Fujita and M. Nakao, *Biochem. Biophys. Res. Commun.* **42:** 1127, 1971).

This agent associates with acetyl choline receptors and, when irradiated at 324 nm, is converted to a highly reactive nitrene, which then combines covalently with the binding sites. This reagent may be made into a highly sensitive probe by the utilization of radioactive aryl azides. Application of standard procedures of affinity labeling, e.g., protection by natural substrates, may be expected to facilitate the use of these kinds of markers.

Other membrane enzymes may also be tagged, using labeled substances or inhibitors. Thus, membrane ATPases form ^{32}P-labeled phosphoproteins from ATP$-\gamma$-^{32}P. Erythrocyte membranes hydrolyze ATP only on their inner surfaces, giving rise to labeled phosphoproteins of about 200,000 daltons and also some of about 90,000 daltons (80). The Na^+-K^+-sensitive ATPases of human erythrocyte membranes can also be labeled with oubain-^3H, and the complex fractionated in sucrose density gradients (81).

Phillips and Morrison (5, 82, 83) have developed an approach to labeling the external surfaces of diverse cells in a way that may be more reliably vectorial in some cases. They use lactoperoxidase in the presence of $Na^{125}I$ and H_2O_2 to iodinate accessible, reactive proteins on the outer surface of erythrocyte membranes. This procedure tags tyrosine residues of membrane peptides and serves as a very sensitive marker. Predominantly one protein (mol wt ~90,000 by SDS-gel electrophoresis) and another near 70,000 daltons iodinate when intact cells are employed, but most membrane peptides react when erythrocyte ghosts are used.

Hubbard and Cohn (84) have refined the procedure by using glucose oxidase to generate continuing low levels of H_2O_2. Their iodination of intact human erythrocytes causes no cell lysis. A total of 97% of the incorporated ^{125}I is associated with ghosts proteins and 3% with hemoglobin. The reactivity with hemoglobin is not explained. They find the label restricted to monoiodotyrosine. About 15% of the trichloroacetic acid-precipitable label can be extracted with lipid solvent, but represents either labeled protein or traces of ^{125}I. Electrophoresis in sodium dodecyl sulfate-laden polyacrylamide shows only two labeled proteins: one, carbohydrate poor, with an apparent molecular weight of 110,000 and containing 40% of the label; the other, with an apparent molecular weight of 74,000, containing all of the ghost sialic acid and 60% of the label.

One factor possibly not yet sufficiently evaluated in applications of the peroxidase iodination procedure is the role of hemoglobin as a *peroxidase*. This is important, since both H_2O_2 and iodine are permeant and could lead to the iodination of suitably reactive, internally exposed proteins. The peroxidase approach may also be limited in cells that (a) exhibit active pinocytosis and/or (b) contain internal peroxidases, e.g., in peroxisomes.

The enzymatic iodination technique can also be used with cells more fragile than erythrocytes. Thus, Marchalonis et al. (85) have used lactoperoxidase and H_2O_2 to iodinate intact cells from mouse spleen and several murine lymphoid tumors with ^{125}I. They argue that a number of these cells' membrane proteins are accessible to iodination by this technique, as judged by disk electrophoresis in the presence of dissociating agents or by gel filtration in the presence of urea. However, their data cannot be taken to show that *only* plasma membrane proteins react. Still, it appears that the iodination procedure may prove useful in determining the turnover of accessible plasma membrane components (86).

Rifkin *et al.* (87) have developed a novel method to label the external surfaces of influenza virus membranes, which may also offer some advantages for animal plasma membranes.

The rationale of the procedure depends on the established fact that biomembranes generally exhibit poor permeability to phosphate esters. Moreover, one such ester, pyridoxal phosphate

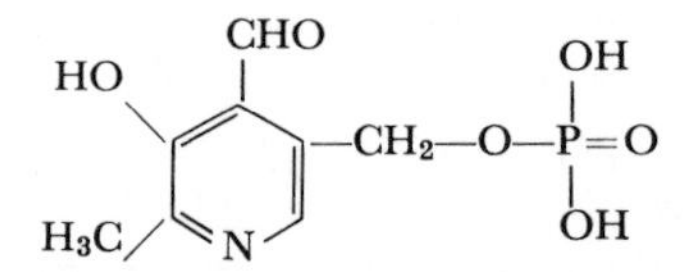

forms Schiff bases with free amino groups on proteins and can be simply tritiated to high specific activity by reduction with NaB^3H_4 (sodium borotritide) under physiologic conditions of pH, ionic strength, and temperature. Moreover, tritiated borohydride comprises an inexpensive, commercially available, rather innocuous reagent.

To label virus envelopes, Rifkin et al. reacted purified virus with 0.01 M pyridoxal phosphate in 0.05 M *N*-2-hydroxyethylpiperazine-*N*′-2-ethane sulfonate, pH 7.5, at 37°C for 30 min, cooled to 4°C and reacted with a 10-fold molar excess NaB^3H_4 (with respect to pyridoxal phosphate) at 4° for 15 min. Residual borotritide was oxidized with excess pyridoxal phosphate, and the virus purified by dialysis and isopycnic ultracentrifugation.

This treatment does not affect the infectivity, sialidase activity, or the hemagglutinating function of the virus. Further selectivity can be achieved by pretreatment with nonradioactive borohydride, to eliminate labeling that is not pyridoxal dependent.

The data presented are impressive, but indicate that this method also is no panacea, since some proteins, known to lie internally, contain approximately 20% of the label.

It remains to be established how effectively this technique applies to animal cells. Pinocytosis represents an obvious problem, and enzymatic cleavage of pyridoxal phosphate, by nonspecific phosphate-monoesterases, such as occur in some plasma membranes, might introduce artifact.

There are other promising methods by which specific labeling of some of the proteins on the external face of cell membranes may be labeled, provided pinocytosis is excluded. One of these is the procedure described by Himmelpach (88), in which polysaccharides of desired size are coupled with the cell surface by first converting the polysaccharide to a flavozole. This can then be diazotized and coupled to reactive tyrosyl groups in protein components of the cell membrane.

The procedure described by Suttajit and Winzler (89) can also be expected to be useful in labeling membrane proteins containing sialic acid. However, since permeant reagents are utilized, this does not permit ready localization of the reactive proteins. The method involves treatment of intact cells with very low concentrations of periodate to oxidize the polyhydroxyl side chain of sialic acid to aldehyde. This is followed by reduction of the aldehyde to an alcohol with tritium-labeled borohydride. Because of the ease with which the polyhydroxyl side chain is oxidized in comparison with the vicinal hydroxyl groups in the ring form of sugars, conditions can be chosen in which the oxidation of sialic acid is virtually the only demonstrable reaction. Application of this approach gives membranes specifically labeled only in those glycoproteins or glycolipids containing sialic acid.

The nature of membrane proteins lying at the inner face of erythrocyte membranes has been approached by Marfey (90), who reacted the bifunctional reagent 1, 5-difluoro-2, 4-dinitrobenzene

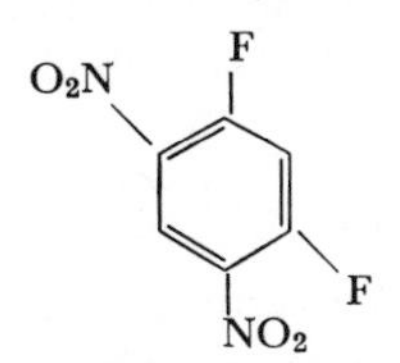

with erythrocytes and then isolated the membranes. He found that the beta chains of hemoglobin were cross-linked from cysteine-β 93 to a membrane protein. These results suggest that there may be a layer of hemoglobin oriented in a defined manner within a few angstroms of the internal surface of the erythrocyte membrane.

Recovery

The matter of recovery in plasma membrane fractionations continues basically unresolved. No one marker or even set of markers can be assumed to be representative of the entire cell surface, and a full recovery of one type of marker does not necessarily imply full recovery of the whole membrane. For this reason, one does not find much in the way of recovery data, but rather specifications as to what proportion of total cellular protein, lipid, etc., is found in a given membrane isolate.

This formidable problem could conceivably be approached morphometrically. Here, one would estimate the surface area of a cell by thin section electron microscopy, using computerized morphometric methods (91) and repeat this on an isolate of known purity.

An analogous approach has been repeatedly applied to the erythrocyte membrane and also to Ehrlich ascites carcinoma plasma membranes (92). The membrane surface area A, volume V, mass M, thickness t, and density ρ are related as follows:

$$V = A \cdot t$$
$$\rho = \frac{M}{V} \tag{1.3}$$
$$A = \frac{M}{\rho t}$$

The dry mass M of an erythrocyte ghost is 1.5×10^{-12} g, taking a protein value of 9.1×10^{-13} g/ghost (93). Various estimates suggest that the mass of a hydrated ghost would be higher, i.e., about 1.7×10^{-12} g.

Membrane thickness estimated by thin section electron microscopy lies between 75–80 A. Again allowing for dehydration, a wet membrane would be about 95 A thick. The density of erythrocyte ghosts in glycerol lies between 1.08–1.16 g/cm³ (29). In glycerol gradients, the membranes have very likely lost some of their hydration water, and the hydrated density probably approaches 1. 1. Accordingly,

$$A = \frac{1.7 \times 10^{-12}\ \text{g}}{1 \cdot 1\ \text{g/cm}^3 \times 9.5 \times 10^{-7}\ \text{cm}^3} = 1.63 \times 10^{-6}\ \text{cm}^2$$

This value is larger than usual estimates of whole cell surface areas, $1.45 \pm 0.08 \times 10^{-6}$ cm²/cell, but the uncertainties in ρ and t could account for much of the difference. Nevertheless, within these limitations, such calculations, or preferably refined morphometry, at present appear the only reasonable approach to the recovery problem. However, the resolution of such measurements lies near 100 A, i.e., not at a molecular level.

Specific Approaches to Plasma Membrane Fractionation

General

Plasma membrane purification methods have progressed less rapidly and more divergently than those for cytoplasmic membranes, because (a) unlike other organelles, surface membranes *must be disrupted* to be purified; (b) their fate after homogenization is usually difficult to monitor and to control and as yet impossible to quantify; (c) the plasma membrane usually fragments into particles of diverse size and shape; (d) the physical properties of surface fragments often overlap those of other organelles; (e) few plasma membrane markers are *assured* to be absent from all other membrane; and (f) plasma membrane properties vary among differing cell types, as well as in an individual cell. This heterogeneity often leads to unrecognized experimental selection of specialized surface elements.

Many workers envisage the plasma membrane as a homogeneous entity, but we suspect that it constitutes an assembly of organelles distributed at the cell periphery. This is a fact in the case of bacteria, but equally impelling evidence exists for animal cells (94, 95). The shearing distortions and low salt concentrations usually employed in cell fractionations tend to fragment nonspecialized membrane domains into small, sealed, semipermeable sacs, analogous to the vesiculation of the membrane sheets of the endoplasmic reticulum (19, 28, 29). However, surface structures, such as terminal bars, desmosomes, adherent "coats" found in surface specializations such as bile canaliculi and brush borders, tend to stabilize the membranes to yield large pieces.

Large membrane fragments usually sediment at $\sim 10^4$ g-min; small vesicles tend to pellet at $3–6 \times 10^6$ g-min, and intermediate particles distribute in between. "Open" membrane fragments exhibit equilibrium densities between 1.15 and 1.19 sealed membrane sacs, but may exhibit densities lower than 1.02, depending on conditions (19, 28, 29).

Large Fragments

Erythrocytes

Adult mammalian erythrocytes, being readily obtained and easily freed of their putative, nonmembrane components by osmotic lysis, provide favorite subjects for membrane studies. Under suitable conditions, the membranes reseal and recover many permeability characteristics of the intact cell (24), but really "functional" membranes usually contain some hemoglobin. Hemoglobin-free "ghosts" with native morphology are obtained by multiple washes with 20 milliosmolar phosphate (pH 7–8) conditions which engender cell lysis, but also minimize presumably nonspecific protein adsorption. Ghosts can also be converted into vesicles of "normal" and inverted orientation (22).* Some of the biochemical characteristics of isolated erythrocyte membranes are summarized in Tables 1.2 and 1.3.

* The conditions required for the production of cation-impermeable, inverted erythrocyte membrane vesicles have been more extensively specified (J. A. Kant and T. L. Steck, *Nature* **240:** 26, 1972). It appears that the centrifugal behavior of such

Continued on p. 32

Table 1.2 Overall Amino Acid Composition of the Proteins in Certain Animal Cell Plasma Membranes[a]

Amino Acid	Erythrocytes[b]	Ehrlich[c] Ascites Carcinoma	Liver (Bile Fronts)[d]	Myelin[e]
Lys	5.0	6.3	7.2	5.8
His	2.7	2.6	2.6	2.3
Arg	5.1	4.7	5.2	4.0
NH_3	—	14.7	12.4	—
Asp	8.3	8.8	9.3	6.1
Glu	13.9	10.1	12.0	7.1
Thr	5.2	5.5	5.3	6.4
Ser	6.3	6.6	6.0	10.4
Pro	5.6	5.2	4.9	1.1
Cys	0.5	trace	0.9	3.8
Met	2.1	2.7	2.3	1.0
Gly	6.8	8.5	7.8	10.9
Ala	8.1	7.8	8.0	9.6
Val	6.6	6.6	6.6	5.9
Ile	4.8	6.1	5.1	4.5
Leu	11.6	10.1	9.6	8.8
Tyr	2.4	3.1	2.7	3.1
Phe	5.0	4.8	4.5	4.0
Trp	—	1.5	—	5.1

[a] Values are in moles/100 residues.
[b] From Zwaal and Van Deenen (96).
[c] From Wallach and Zahler (97).
[d] From Takeyuchi and Terayama (98).
[e] From Hulcher (99).

membrane vesicles depends upon an asymmetrical Donnan ion distribution plus the osmotic activity of cations and other solutes trapped within them.

Inverted and normally oriented erythrocyte membrane vesicles allow one to gain some insight into the disposition of the ouabain-sensitive, (Na^+-K^+) ATPase of these membranes (J. R. Perrone and R. Blostein, *Biochim. Biophys. Acta* **291:** 680, 1973). Specific [^{3}H]-ouabain binding by intact erythrocyte ghosts and normally oriented vesicles approximates 1.1 and 0.7 picomoles/mg membrane protein, respectively, and correlates well with the degree to which the (Na^+-K^+) ATPase is inhibited by the drug. In contrast, inverted vesicles bind no ouabain and their ATPase activity is ouabain-insensitive. The data support other evidence that ouabain binding sites lie at the external membrane surface.

Inverted plasma membrane vesicles can be isolated from rat adipocytes and insight gained into the functional asymmetry of these cells' surface membranes (V. Bennett and P. Cuatrecasas, *Biochim Biophys. Acta* **311:** 362, 1973). Treatment of isolated adipocytes with buffers of low ionic strength induces endocytosis, and gentle cell lysis allows isolation of inverted plasma membrane vesicles. These sequester insulin and wheat-germ phytagglutinin added before endocytosis. However, normal insulin binding can be restored by sonication, Triton X-100, and phospholipase C.

Table 1.3 Overall Composition of Human Erythrocyte Membranes[a]

Component	% of Dry Mass
Protein	49.2
Lipid (total)	43.6
Phospholipid	32.5
Cholesterol	11.1
Carbohydrate (total)	7.2
Sialic acids	1.2
Hexosamines	2.0
Neutral sugars	4.0

[a] From Rosenberg and Guidotti (100).

Bile fronts

The purification of hepatocyte plasma membranes is complicated by the intricate structure of liver and its cells (101). Firstly, liver parenchyma comprises two major cell types: hepatocytes (60%) and macrophages. Secondly, while the membrane facing the hepatic blood channels is convoluted into prominent microvillous protrusions of enormous surface area, the surfaces of apposed hepatocytes are specialized into bile fronts. There the membranes of the adjacent cells lie about 200 A apart except in the regions of the bile canaliculi, which constitute channels indenting adjacent cells' surfaces. Their walls are the plasma membranes of the surrounding hepatocytes, specialized into small microvilli, which project into the canalicular lumen. Apposed hepatocyte surfaces are linked by desmosomes, terminal bars, junctional complexes, and interdigitating projections, all of which make these surface regions more stable than the rest of the hepatocyte membrane and allow their facile isolation.

To isolate hepatocyte bile fronts, minced rat liver is homogenized in 0.001 M $NaHCO_3$ (pH$\sim$7), filtered to remove debris, gelled nucleoprotein and large particles and the bile fronts pelleted by several cycles of low-speed differential centrifugation [see Neville (102)]. The resuspended sediment is homogenized into sucrose of density = 1.19, and the mixture overlaid with sucrose of d = 1.16 and spun (75 min) to bring the membranes to the d = 1.16/d = 1.29 interface. To eliminate residual, slowly sedimenting membrane fragments, washed bile fronts are purified in linear sucrose gradients. To improve the method, Takeuchi and Terayama (98) substitute 0.25 M sucrose. 0.005 M $CaCl_2$ for 0.001 M $NaHCO_3$ to stabilize nuclei, mitochondria, and lysosomes, while Coleman et al. (103) perfuse in situ to remove erythrocytes and also shear the canalicular membranes into small particles, which are readily separated centrifugally from large contaminants. A typical bile-front enzyme profile is given in Table 1.4. Recently, bile-front *gap junctions* have been purified by Goodenough and Stoeckenius (104).

Table 1.4 Comparison of Some Enzyme Activities in Isolated Rat Liver Bile Fronts and Lysosomal Membranes[a]

Enzyme	Bile Front	Lysosomal Membrane
5′-nucleotidase (nmoles 5′-ATP-min^{-1}-mg-prot^{-1})	1.30 (1.30)[b]	0.35 (0.08)[b]
ATPase, nonspecific (nmoles Pi-min^{-1}-mg-prot^{-1})	0.70	0.05
ATPase, Na^+, K^+-specific (nmoles Pi-min^{-1}-mg-prot^{-1})	0.38	0.0
Glucose-6-phosphatase[c] (nmoles Pi-min^{-1}-mg-prot^{-1})	0.04	0.22 (0.02)[b]
Leucine aminopeptidase (*n*-moles min^{-1}-mg-prot^{-1})	4.80	24.5
+0.1% Triton × 100	4.70	2.0
Lysolecithin-0-acyltransferase[d] (nmoles fatty acid incorporated min^{-7}-mg-prot^{-1})	9.0	0
Acyl-CoA-synthetase (nmoles lecithin formed. min^{-1}-mg-prot^{-1})	8.2	4.2

[a] From Kaulen *et al.* (112).
[b] Figures in parentheses = after addition of 20 m ML(+) tartrate.
[c] Very likely a "microsomal" contaminant.
[d] Used arachidonic acid and/or CoA derivative, and 1—stearoyl—3-glycerophosphorylcholine.

The "Neville method" has been empirically applied to nonhepatic tissues (104–107) with dubious results, since the method depends on the particular properties of bile fronts. Also, Emmelot and associates use it for hepatomas (108–111) and report differences between neoplastic and normal cell membranes. However, these may result from uncontrolled variables such as the nonuniform bile-front morphologies of the various cell types and the different media employed.

The *membrane yield*, as indicated by the marker 5′-nucleotidase, is generally below 14%, partly because bile fronts tend to vesiculate during isolation, and thus become lost as small fragments, and also because bile fronts comprise only a small fraction of the hepatocyte surface. The rest of these cells' plasma membranes and those of the macrophages have been little considered. Moreover, since bile fronts are destroyed when liver cells are dispersed, and noncanalicular surfaces do not contribute to this type of membrane isolate (19, 28, 29), it probably systematically excludes a major portion of the cell surfaces.

Plasma membranes from other cells with surface specializations

Methods for the isolation of uniquely modified membrane surfaces (e.g., brush borders, sarcolemma), as well as those of specialized cells (leukocytes, thyroid, fat cells, microorganisms), are reviewed in Steck and Wallach (28).

The membrane-stabilization approach

Hoping to maintain the plasma membrane a "complete" structural entity, Warren and associates (14) fix disperse cells with agents such as 0.1 M acetic acid, fluorescein mercuric acetate, and Zn^{++}. Cells thus treated swell, the cytoplasm adheres to the nucleus, and this mass can then be extruded by mechanical homogenization. Purification is by differential and/or isopycnic ultracentrifugation in sucrose or glycerol gradients.

About 30% of the cells (mostly small ones) stay intact. Counts of "intact ghosts" indicate a maximal recovery of 80%, but since membranes tend to shed submicroscopic vesicles, much of their mass could be lost.

"Warren isolates" are relatively free of nuclei and mitochondria, but contain a polysome-containing "rind" on the internal membrane surfaces. Because of denaturation by the fixatives, functional markers have been difficult to apply. The preparations are unlikely to be satisfactory for studies on membrane protein synthesis, but may be suitable for investigations on overall membrane lipid composition and synthesis.

In a modification of the "Warren method," Perdue et al. (113) reverse the effects of Zn^{++} fixation by chelating agents, causing the membranes to rupture into vesicles; these are centrifugally purified and contain active plasma membrane marker enzymes. This approach parallels that of Boone et al. (15), where cells are mechanically disrupted in hypotonic 0.01 M tris, pH 7.0, 0.001 M $MgCl_2$, the homogenate brought to 0.25 M sucrose to prevent membrane fragmentation, most of the membranes pelleted as large envelopes and these purified centrifugally in sucrose gradients. These sheets are then vesiculated by quick sonication and further purified by gradient ultracentrifugation.

Fractionation of membrane vesicles

As previously noted, large, unstabilized membrane surfaces (e.g., plasma membrane, endoplasmic reticulum) tend to fragment into small, semipermeable *vesicles* upon exposure to shear and/or low ionic strength.

In density gradients, such vesicles equilibrate according to the *density of the membrane per se*, the *density of the intravesicular solution*, and *the volume proportions of these two*. Membrane density is influenced by the hydration of its components, which depends on the medium. Intravesicular fluid density is a function of the permeant molecules in the medium and of nonpermeant intravesicular molecules. Accordingly, the isopycnic vesicle density will depend on the gradient material (19, 28, 29).

Gradients of *large polymers*, such as polysucrose and polyglucose, produce small osmotic loads and do not penetrate within the vesicle, which, being filled with permeant molecules (water and ions), have a low equilibrium density (1.04–1.08).

Sucrose is usually also nonpermeant, but its osmotic activity constricts the intravesicular space progressively as the particles move through the density gradient. "Equilibrium" occurs when the forces tending to expand the vesicle balance the external osmotic forces. The density of vesicles in sucrose is, thus, higher than in polymer gradients (e.g., 1.12–1.17).

In gradients of permeant solutes such as glycerol and some salts, the vesicle solvent space assumes the density of the surrounding medium, and the particle equilibrates at a density equal to that of the membrane itself (1.15–1.19).

Since many membranes have similar sizes, densities, and permeabilities to gradient media, these are fractionated according to their ionic properties, which reflect the composition of the membrane per se and its impermeant contents (19, 28, 29).

Membrane vesicles act as sacs bounded by semipermeable layers studded internally with fixed charges. With permeant electrolytes, the intravesicular ion concentration exceeds that in the medium because of the fixed charges, an effect greatest at low ionic strength. The ion excess within the vesicle tends to expand its water compartment. However, if the bulk medium contains a nonpermeant solute, its osmotic activity will shrink the vesicle. An equilibrium is reached when the effective osmotic concentrations of the two sides of the membrane are identical. At equilibrium, the density of a membranous vesicle can be described by the following equation (28):

$$\rho_T = \frac{Z + Q \cdot \rho_w}{(Z/\rho_M + Q)} \tag{1.4}$$

Here ρ_T, ρ_w, and ρ_M are the densities of the vesicle, its aqueous space, and the medium, respectively; Q is the effective charge density (meq/g membrane) on the inner membrane surface; and $Z = \sqrt{S^2 + 4mS}$, where S is the molal concentration of impermeant solute, and m the molal concentration of permeant univalent electrolyte.

Accordingly, a heterogeneous mixture of vesicles should separate most readily at low osmotic activity, ionic strength, and surface charge density, as shown in the fractionation of microsomal membrane vesicles; these vesicles derive from both plasma membrane and endoplasmic reticulum and have nearly the same densities, even under low ionic-osmotic conditions. However, when the fixed anion excess within the vesicles is titrated by lowering pH or, preferably, by adding permeant polyvalent cations (e.g., Mg^{2++}), the vesicle density increases accordingly. Because endoplasmic reticulum vesicles tend to increase in density more rapidly than those of plasma membranes as their charges are titrated, these two membrane types can be separated in polymer gradients (19, 28, 29).*

Other Cytoplasmic Membranes

Introduction

In complex cells, plasma membrane functions are so closely linked to those of cytoplasmic membranes that their study and purification inevitably involve their separation from other cellular membranes. Accordingly, we shall briefly comment on this matter and also provide representative reference data on important intracellular membranes, as well as myelin, a plasma membrane derivative.

Endoplasmic Reticulum

The endoplasmic reticulum (ER) comprises a tubular or cisternal labyrinth, often with associated vesicles. It abounds in mature cells, active in biosynthetic processes, and is impoverished in eggs, many tumor, or otherwise undifferentiated cells. The system consists of interconnected, membrane-bounded spaces, whose membranes have limited permeabilities and distinctive internal and external surfaces. The membranes exhibit the typical "unit membrane" appearance on thin section electron microscopy. There are considerable differences in ER morphology from cell to cell, and many cells have two forms of ER membranes in continuity: The *first, without associated polysomes*, is the *smooth ER*; the *second*, the *rough ER*, a system of flattened cisternae, has polysomes attached to the cytoplasmic surface of the membrane. The two ER types are likely interchangeable, and ribosomes can reversibly attach to the ER membranes. Many cells also show a clear continuity of ER with the

* A complete treatment for the isopycnic density distribution of sealed membrane vesicles in osmotically inactive gradients should include terms not only for charges fixed to the membranes but also for possible, trapped but soluble, charged molecules.

outer lamella of the nuclear envelope, which also often bears polysomes.

It appears that "export" proteins are first synthesized by the polysomes on the cytoplasmic surface of the ER membranes, thereafter penetrating to the interior of the cisternae to form "intracisternal granules." In some cells these are eventually released as membrane-enclosed "zymogen granules." The export of certain proteins, particularly glycoproteins, may proceed from ER via the Golgi apparatus, which finally fuses with the plasma membrane to release its contents.

Many enzymes are closely associated with the ER, and those of use as "markers" are listed below. In addition to its roles in protein synthesis, the ER contains many of the enzymes essential in phospholipid and cholesterol synthesis, as well as the machineries for drug detoxification.

Upon cell rupture, the ER fragments into small particles, sedimenting at $3–6 \times 10^6$g-min *and* constituting part of the "microsomal" fraction. Many of the enzyme activities in this are thought to arise exclusively from the endoplasmic reticulum, but these assignments generally assume that the "microsomal" fraction derives *solely* from the endoplasmic reticulum. In reality, however, "microsomes" are a composite population, derived from several organelles, including plasma membrane. Thus, the activity of a particular "microsomal" enzyme may not accurately reflect the properties either of the endoplasmic reticulum or of all "microsomal"

Table 1.5 Typical Overall Composition of "Smooth" Membranes Isolated from Rat Liver Microsomes[a,b]

Component	% of Dry Mass[c,d]
Protein	71.4
Lipid (total)	28.6
Cholesterol	0.7
Phospholipid	27.9
Phosphatidic acid	0.5
Phosphatidyl choline	13.6
Phosphatidyl ethanolamine	4.5
Phosphatidyl serine	2.4
Phosphatidyl inositol	3.7
Sphingomyelin	3.0

[a] Calculated from Korn (115).
[b] These membranes probably contain fragments of plasma membrane and Golgi apparatus.
[c] These values assume negligible mass contribution from carbohydrate.
[d] Calculated taking a phospholipid:cholesterol ratio of 0.025; however, this value may be as high as 0.077.

membranes. An approximate composition of ER membranes is found in Table 1.5.

The endoplasmic reticulum contains numerous enzymes, but only a few have been used as "markers," specifically:

Glucose-6-phosphatase catalyzes the reaction:

$$\text{D-glucose-6-phosphate} + H_2O \rightarrow \text{D-glucose} + \text{orthophosphate}$$

This enzyme group, thought to be located specifically in the membranes of endoplasmic reticulum, catalyzes the transfer of the phosphoryl group from nucleoside diphosphates or triphosphates to glucose or other sugars, but such reactions occur *not only* in endoplasmic reticulum.

NADH oxidases, catalyzing the reduction of reduced NAD by various acceptors (e.g., ferricytochrome C) and often called *NADH "diaphorases,"* are used as a marker for endoplasmic reticulum in cells where glucose-6-phosphatase is low.

These enzymes cannot be assumed to reside solely in ER; moreover, they vary widely between tissues.

The endoplasmic reticulum of muscle, i.e., the *sarcoplasmic reticulum*, has attracted particular interest because of its role in regulating intracellular Ca^{2+} in muscle and thereby with the contraction-relaxation cycle. For this reason also, the sarcoplasmic reticulum has become a favored object for study by diverse spectroscopic methods.

However, nearly all studies in this area assume that "microsomal" membranes from muscle comprise only sarcoplasmic reticulum. Wheeldon and Gan (114) have reviewed this matter and conclude that heart muscle microsomes are contaminated by plasma membrane fragments, but that this is not the case in skeletal muscle. The difference is attributed to the close association of the plasma membrane and extracellular layers in the case of skeletal muscle. The authors have also devised a procedure for separating sarcoplasmic reticulum from other membranes, using flotation in polyanion-containing sucrose gradients.

Golgi Membranes

The typical structure of the Golgi apparatus comprises three major constituents (Fig. 1.4):

(a) flattened, membrane-enclosed cisternae, about 115 A apart, with about 60 A intracisternal space;
(b) dense vesicles, about 600 A in diameter; and
(c) large, apparently empty vacuoles.

The Golgi apparatus generally lies near the nuclear envelope, and its membranes lack attached ribosomes. Continuities between the Golgi membranes and the endoplasmic reticulum are frequent, and many investigators consider the Golgi complex as a specialization of the endoplasmic reticulum, participating in biosynthetic and export activities, both of proteins and other substances. It is also involved in attaching

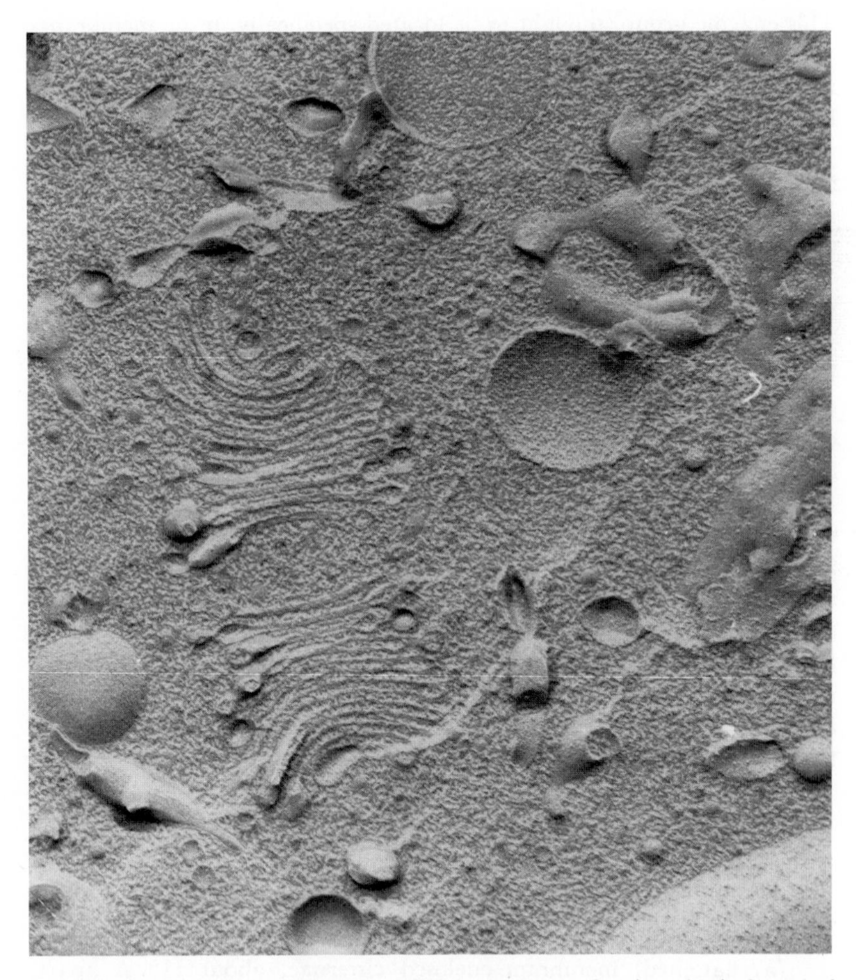

Fig. 1.4 Freeze-cleave image of Golgi apparatus, showing classical stacked lamellae with associated vesicles. Magnification 65,000. This is from Allium root tip; animal cells show less distinct Golgi apparatus. Portion of a nucleus is seen at the left border and ER cisternae at the right. (Courtesy of V. Speth.)

carbohydrates to certain export proteins and in the synthesis of large carbohydrates.

Golgi membranes vesiculate readily, and early isolation attempts therefore used gluturaldehyde-fixed cells as starting material, separating the morphologically intact apparatus in sucrose gradients. However, Fleischer et al. (116) and Morré et al. (117) now show that unfixed Golgi membrane fragments, characterized by their high UDP-galactose: *N*-acetyl glucosamine-galactosyl transferase, separate satisfactorily on sucrose density gradients because of their low densities (~ 1.13 in sucrose).

Table 1.6 Phospholipid Composition of Rat Liver Golgi Membranes[a]

Lipid	% of Total Phospholipids
Sm	12.3
PC	45.3
PE	17.0
PS	4.2
PI	8.7
LPC	5.9
LPE	6.3

[a] Quoted by Kleinig (118).

Note: Sm, sphingomyelin; PC, lecithin; PE, phosphatidylethanolamine; PS, phosphatidylserine; PI, phosphatidylinositol; LPC, lysolecithin; LPE, lysophosphatidylethanolamine; ND, not detected.

The lipid composition of Golgi membranes is given in Table 1.6. The following enzymatic markers are commonly used to identify Golgi membranes:

UDP-galactose.

N-acetyl glucosamine galactosyl transferase.

Nucleoside diphosphatase, catalyzing reactions such as:

$$\text{nucleoside diphosphate} + H_2O \rightarrow \text{nucleotide} + \text{orthophosphate}$$

A UDP-preferring enzyme may be concentrated in the Golgi apparatus, and CDP may also be split therein.

Lysosomal Membranes

Lysosomes occur in most cells and abound in hepatocytes and various phagocytic cells. Their membranes present no unusual micromorphologic features, but lysosomes, just as other closed membrane systems, exhibit physical properties related to their content of osmotically active substances and to the fixed charges on the membranes *per se*. They release their degradative enzymes upon osmotic rupture, but this can be avoided by use of inert nonpermeant solutes. Moreover, the limited degree of permeability of the membrane to metabolized solutes, e.g., glycerophosphate, confers osmotic protection as long as the rate of penetration does not exceed its rate of hydrolysis within the lysosomes. These permeability phenomena apply to the passage of substances into as well as out of closed membrane sacs.

Mechanical breakage, disruption, freezing and thawing, surface-active agents such as saponin, deoxycholate, or Triton X-100, autolysis at 37°C, phospholipase C, trypsin, chymotrypsin, and low pH all injure the lysosomal membrane. Most of the lysosomal contents diffuse out, but enzymes may remain bound to the lysosomal membranes. Thus, although Romeo and De Bernard (119) find lysosomal enzymes retained very simply within liposomes, the "free" acid phosphatase activity in lysosomal isolates varies with particle damage, and only part of the enzyme remains soluble after high-speed centrifugation.

From the accessibility of lysosomal enzymes as a function of pH and ionic strength, Sawant et al. (120) ascribe an asymmetric charge to the lysosomal membrane, allowing different enzymes to bind to its inner and outer surfaces preferentillay; however, these studies could reflect release of soluble enzyme, increased substrate penetration to bound enzyme or both.

Lysosomes vary in stability, but are always difficult to separate from mitochondria. Their hydrolytic enzymes obscure the enzymatic character of mitochondrial isolates and can degrade other cellular membranes. Of the many purification methods devised, the most successful, already described (36), uses pretreatment of animals with the detergent Triton WR 1339, which selectively lowers the equilibrium density of lysosomes to below that of mitochondria, without obvious deleterious effects. Triton WR 1339 probably concentrates in the lysosomes as a protein complex and may influence the lysosomal membrane, although no undue permeability to the marker enzymes occurs (Table 1.7). Since pretreatment of animals with dextran and sucrose can also selectively change lysosomal density, these may be preferable for the study of lysosomal *membranes*. These are isolated from the intact particles mechanically, osmotically, by freezing and thawing or by various nonionic detergents.

Table 1.7 Isopycnic Gradient Purification of Various Rat Liver Organelles in the Triton WR 1339 Procedure[a]

Enzyme	Relative Specific Activity (with Respect to Whole Liver)[b] Lysosomal Fraction	Mitochondrial Fraction	Peroxisomal Fraction
Protein	1.00	1.00	1.00
Acid phosphatase	18.9 ±5.7	0.49 ±0.2	0.27 ± 0.4
Glucose-6-phosphatase	0.31 ±0.3	0.22 ±0.1	0.09 ± 0.08
Cytochrome oxidase	0.87 ±1.3	4.25 ±0.9	0.11 ± 0.1
Glutamate dehydrogenase	1.19	4.48	0.25
Isocitrate dehydrogenase	0.76 ±0.41	0.61 ±0.11	2.57 ± 0.57
Catalase	5.8 ±2.5	0.60 ±0.4	36.3 ± 6.4
L-α-hydroxy acid oxidase	5.7 ±2.3	0.71 ±0.3	35.8 ± 7.2
D-amino acid oxidase	5.8 ±1.9	0.81 ±0.3	30.0 ± 7.7
Urate oxidase	0.43 ±0.9	0.81 ±0.2	50.0 ±14.4
Total yield (mg proteins/100g liver)	170	1100	120

[a] From Leighton *et al.* (36).
[b] Values listed = mean ± SD.

The following markers are considered lysosome specific:

Acid phosphatase, catalyzing the reaction:

orthophosphoric monoester + H_2O → an alcohol + orthophosphate at *acid pH*

Arylsulphatase, catalyzing the reaction:

a phenol sulphate + H_2O → an aromatic alcohol + sulphate

β-glucuronidase, catalyzing the reaction:

β-D-glucuronide + H_2O → an alcohol + D-glucuronate

Cathespin, hydrolyzing peptides, especially at bonds involving an aromatic amino acid adjacent to a free α-amino group.

Deoxyribonuclease, catalyzing the reaction:

$$\text{DNA} + (n-1)\,H_2O \rightarrow n \text{ oligodeoxyribonucleotides}$$

Ribonuclease, transferring the 3′-phosphates of polynucleotides from the 5′ position of pyrimidine nucleotides to the 2′ position of adjoining nucleotides. The cyclic phosphates are ultimately split to the corresponding 3′-phosphates.

Urate oxidase, located partly in lysosomes *and* peroxisomes, catalyzing the reaction:

$$\text{urate} + O_2 \rightarrow \text{unidentified products}$$

Mitochondria and Their Membranes

Animal cells contain 300–800 mitochondria. They weigh $\sim 10^{-13}$ g and are oblong to spherical, 0.2–1.0 μm in diameter, and perhaps 3–10 μm in length (Fig. 1.5). In rat liver, these organelles account for about 35% of the total protein and 22% of the cytoplasmic volume, but some algal cells contain only a single mitochondrion. Mitochondria often lie near energy sinks, such as myofibers, nuclei, and the plasma membranes of neuronal renal tubular or other cells. Such associations also depend upon the life cycle and metabolic state of the cells. Mitochondria change shape frequently in vivo; rods may fragment into several spheres, and reverse again by fusing into rods. They also alter morphology dramatically during metabolic transitions and may move about either randomly or in an orderly fashion, according to cell type, metabolism, and environment; individual mitochondria often exhibit sudden, autonomous changes in velocity or direction.

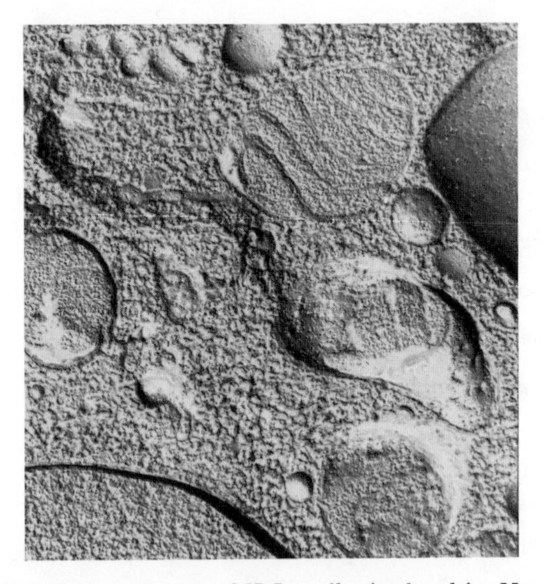

Fig. 1.5 Freeze-cleave image of HeLa-cell mitochondria. Note outer membrane and cristae. A portion of the nuclear envelope is seen lowermost. Magnification 41,000. (Courtesy V. Speth.)

Considerable evidence indicates that mitochondria, even from one-cell type, represent more than one type of organelle. Thus, it is well established that "heavy" and "light" mitochondria (i.e., particles sedimented at $\sim 5 \times 10^4$ g-min and 1×10^5 g-min, respectively) differ in enzyme content. Moreover, two peaks of cytochrome oxidase activity appear upon rate-zonal density gradient centrifugation of rat liver mitochondria (121), possibly representing two metabolic states. Finally, mouse hepatocyte mitochondria exhibit a nonrandom, functionally significant heterogeneity in size, mirrored in their sedimentation rates, but not their equilibrium densities (122). Current data suggest that ornithine amino transferase concentrates preferentially in small mitochondria and malate dehydrogenase in larger particles.

The composition of whole mitochondria varies with source and possibly metabolic state. The lipid content of liver mitochondria lies between 15 and 25%, but that in brain may approach 50%. Mitochondria normally have low-molar cholesterol:phospholipid ratios (0.015–0.03), but can be enriched in the steroid by exposure to cholesterol-containing liposomes or lipoproteins. The phospholipid composition of mitochondria is exceptional in the lack of sphingomyelin and phosphatidylserine and the high proportion of diphosphatidyl glycerol.

The prevailing view of mitochondrial micromorphology is that these organelles consist of an *outer membrane*, enclosing a separate *inner membrane*, convoluted into cristae, the latter containing most of the machinery of oxidative phosphorylation, as well as a host of other membrane functions. The outer membrane is about 45% phospholipid and contains more cholesterol than the inner membranes (cholesterol:phospholipid ratios of 0.03–0.10), but the cholesterol exchanges readily. The inner membrane contains on the average 21% phospholipid and its cholesterol:phospholipid ratio ranges from 0.015–0.045 (Table 1.8).

The protein composition of mitochondria is highly complex, and some major components are listed in Table 1.9. Although mitochondria

Table 1.8 Typical Phospholipid Composition of Whole Bovine Liver Mitochondria[a]

Component	Content (Moles %)
Diphosphatidyl glycerol	17
Choline phospholipids	35
Ethanolamine phospholipids	43
Inositol phospholipids	~3

[a] From Korn (115).

Table 1.9 Molecular Weights and Estimated Dimensions in Mitochondrial Membrane Protein[a]

Compound	Molecular Weight	Diameter[b] (Å)
Cytochrome c	12,400	12 × 25 × 37
Cytochrome b	30,000	~50
Succinate dehydrogenase	200,000	4 × 55
	49,000	~55
Coupling factor F_1	280,000	~90
Cytochrome a	360,000	5 × 60
	72,000	60
Cytochrome c_1	51,000	~55
Choline dehydrogenase	850,000	>100
NADH dehydrogenase	1,000,000	>100
α-glycerophosphate dehydrogenase	2,000,000	>100

[a] From Sjöstrand (127).
[b] The dimensions of molecules other than cytochrome c are derived from the molecular weights and an assumed spherical shape.

contain a specific DNA as well as associated protein-synthetic machinery, most mitochondrial proteins are synthesized in the endoplasmic reticulum, to be subsequently transferred to the mitochondria.

In general, whole mitochondria are sedimented from postnuclear supernatants by differential or rate-zonal gradient centrifugation at 2–5 × 10^5g-min in various media (e.g., 0.25 M sucrose, 0.1 M tris-HCl, pH 7.8, 0.001 M $MgCl_2$). The proper allocation of function of mitochondrial membranes and the correct understanding of mitochondrial organization depend upon the unambiguous separation of mitochondrial components, but this is a controversial area. In opposition to the conventional view stated above, Green and associates (123) consider what others call "outer mitochondrial membrane," as endoplasmic reticulum and argue that mitochondria are surrounded by a single membrane, with an outer and inner layer, the cristae being attached to the latter.

In conventional mitochondrial membrane isolations, the mitochondria are disrupted osmotically, separating the "outer" membrane from the "inner" membrane (cristae + matrix) and fractionating these by rate-zonal centrifugation in sucrose gradients. Alternatively, the two-membrane components and soluble material are separated centrifugally after treatment with digitonin or other nonionic detergents. These separate the matrix and membrane portions ("inner" membrane + cristae + matrix; density 1.21 in sucrose) of the membrane, retaining its typical negative staining morphology (124).

"Outer" membranes, once isolated, lack morphologic specificity and are generally identified by their low density (1.13 in sucrose) and using monoamine oxidase and rotenone-insensitive NADH-cytochrome c reductase as markers. However, the Green group argues that properly prepared mitochondria lack the latter enzyme and that the former enzyme is a "microsomal" contaminant, Beattie (125) and Shnaitman and associates (126) to the contrary. They, instead, digest with phospholipase A to obtain fractions K (outer and inner layers of the boundary membrane), R_2 (cristae + matrix), and S (loosely bound components between the inner and outer layers of the boundary membrane).

Mitochondria are rich in characteristic marker enzymes. Only the most important are listed here:

Choline dehydrogenase, catalyzing the reaction:

$$\text{choline} + \text{acceptor} \rightarrow \text{betaine aldehyde} + \text{reduced acceptor}$$

Cytochrome c, a mitochondrial membrane protein par excellence, being readily eluted, is not a satisfactory marker.

Cytochrome oxidase, composed of at least two separable components, cytochromes a and a_3, which are distinct chemically as well as spectrophotometrically, and catalyzing the reaction:

$$4 \text{ ferrocytochrome c} + O_2 \rightarrow 4 \text{ ferricytochrome c} + H_2O$$

Monoamine oxidases, exhibiting different substrate specificities. They are generally, if not universally, thought to be markers for the *outer* mitochondrial membrane and catalyze reactions such as:

$$\text{a monoamine} + H_2O + O_2 \rightarrow \text{an aldehyde} + NH_3 + H_2O_2$$

Rotenone-insensitive-NADH-cytochrome-c-reductase is a variant of cytochrome c reductase, in which oxidation of NADH by ferricytochrome c *is not* inhibited by rotenone, as are other mitochondrial enzymes of this type. It may be a marker for the "outer" mitochondrial membrane.

Succinate dehydrogenase, catalyzing the reversible reaction:

$$\text{succinate} + \text{acceptor} \rightarrow \text{fumarate} + \text{reduced acceptor (inner membrane)}$$

Glutamate dehydrogenase, catalyzing the reaction:

$$\text{L-glutamate} + H_2O + \text{NAD (P)} \rightarrow \text{2-oxoglutamate} + NH_3 + \text{reduced NAD (P) (inner membrane)}$$

Glycerol-3-phosphate dehydrogenase, catalyzing the reaction:

$$\text{L-glycerol-3-phosphate} + \text{NAD} \rightarrow \text{dihydroxyacetone phosphate} + \text{reduced NAD(inner compartments)}$$

Isocitrate dehydrogenase, catalyzing the reaction:

threo-Ds-Isocitrate + NAD → 2 oxoglutarate + CO_2
+ reduced NAD (inner compartments)

Myelin

Myelin, the multilamellar *plasma membrane derivative* ensheathing axons, is probably the most enduringly studied membrane system. Its isolation and general properties have been recently reviewed by Mokrash (128).

Myelins from various species are generally purified by centrifugation in sucrose gradients according to the approach of Autilio et al. (129). They show rather similar analyses and contain the lowest proportion of protein in biomembranes, about 20%, all of it apparently without enzymatic activities. Several distinct proteins have been identified electrophoretically, of which at least one basic component appears to be responsible for the encephalomyelinogenic properties of myelin dispersions administered in vivo.

Myelin also contains "proteolipids" (92% lipid), which are soluble in organic (but not aqueous) solvents, rich in apolar amino acids and 60–70% helical in organic solvents, according to optical activity measuremenst. About 5% of the lipid is very tightly associated with the peptide moiety and required for helicity.

The lipid compositions of various myelins are similar. That of human myelin is given in Table 1.10, which demonstrates the unusually high content of cerebrosides, cholesterol, and plasmalogens, the latter accounting for about 50% of the phosphatidyl ethanolamine. The fatty acid composition, atypical of biomembranes, shows only 10% C_{20}–C_{24}

Table 1.10 The Molar Proportions of Lipids in Human Myelin[a]

Lipid	% of Composition
Cholesterol	39
Cerebroside	14
Cerebroside sulfate	5
Sphingomyelin	5
Phosphatidyl ethanolamine	15
Phosphatidyl serine	5
Phosphatidyl choline	13
Phosphatidyl inositides	2

[a] From O'Brien and Sampson (130).

polyunsaturated acids and a high proportion of C_{18} acids. The lipid moieties of the cerebrosides are also unique, consisting primarily of α-hydroxy, C_{22}–C_{26}, saturated, or monounsaturated fatty acids (130).

Nuclear Membranes

As recently reviewed (131), the nuclear membrane participates importantly in nucleocytoplasmic interchanges. Animal and other eukaryotic cells are unique in that most of their genetic material lies within a discrete *nucleus*, bounded by a nuclear *envelope*, consisting of two membranes, 70–80 A thick, 150–300 A apart, and perforated by circular "*pores*" (Fig. 1.6). Depending upon cell type and metabolic state, the pores can occupy 3 to 25% of the nuclear surface area. The "pores" are not open gaps, but contain complex, as yet poorly defined nonmembranous structures, involved in nucleocytoplasmic exchange.

The *outer* nuclear membrane, fluid in outline and commonly continuous with the membranes of the *endoplasmic reticulum*, bears *polysomes*

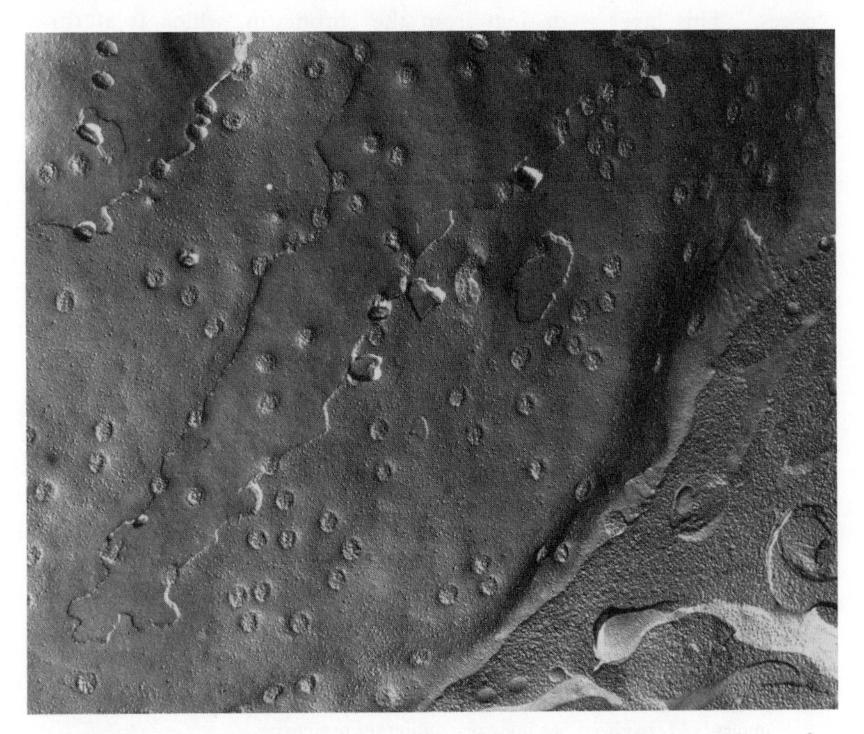

Fig. 1.6 Freeze-cleave image of the bilamellate nuclear envelope of a fibroblast. Note prominent, intricately structured "nuclear pores" penetrating simpler regions of the envelope. Magnification 36,000. (Courtesy V. Speth.)

Table 1.11 Gross Composition of Nuclear Membranes from Rat and Pig Liver[a]

	% of Dry Mass in Nuclear Membranes	
	Rat	Pig
Protein	75.5	74.8
Phospholipids	16.1	18.2
Nonpolar lipids	2.8	3.0
RNA	3.6	2.8
DNA	2.0	1.2

[a] From Franke *et al.* (132).

on its cytoplasmic surface; the perinuclear space is thus continuous with the *cisternae* of the endoplasmic reticulum. In contrast, the *inner* nuclear membrane is smooth, perhaps because of the tightly apposed, *internal dense lamella*, a compact layer, about 2500 A thick, discontinuous at the "pores." This layer is distinct from the chromatin, which is at times membrane associated.

About 75% of the membrane mass of liver nuclei is protein (132) (Table 1.11), and more than 20 peptide components can be separated by electrophoresis in polyacrylamide gels. Many of these proteins are enzymes (Table 1.12).

The nuclear envelope usually dissociates into small, double membrane units early in metaphase, begins to reform during anaphase, and is

Table 1.12 Some Enzyme Activities of Nuclear Membranes Isolated from Pig and Rat Liver[a]

Enzyme	Pig	Rat
Mg-ATPase[b]	6.9 ±1.0	5.6 ±0.5
Na^+-K^+-ATPase[b]	0.2	0.3
Glucose-6-phosphatase[b]	0.1	0.1
Alkaline phosphatase[b]	2.1 ±0.6	1.7 ±0.3
Acid phosphatase[b]	10.0 ±3.6	3.2 ±1.5
Glutamate dehydrogenase[c]	1.9 ±0.7	—
NADH-cytochrome c reductase[d]	0.045 ±0.023	0.10 ±0.05

[a] From Franke *et al.* (132).
[b] umoles inorganic phosphate released per hour per milligram protein.
[c] umoles NAD oxidized per hour per milligram protein.
[d] umoles cytochrome c reduced per minute per milligram protein. About 16–18% of the nuclear membrane mass is phospholipid and about 3% neutral lipids, mostly cholesterol.

again complete by the end of the telophase, most of the new material coming from membranes associated with the mitotic apparatus.

Isolated nuclei are permeable to small molecules and macromolecules such as trypsin, chymotrypsin, RNAse, and DNAse, but this is probably not so in vivo. Thus, fluorescein-labeled serum albumin, injected intracytoplasmically, does not invariably enter the nucleoplasm. Moreover, certain cell types exhibit potential differences of up to 15 mv across the nuclear envelope, attesting to the limited permeability of the structure and its "pores."

Most of the cellular RNA of eukaryotic cells is formed in the nuclei, and the 18S and 28S components are probably exported as protein complexes; most likely, the passage of such large particles occurs through and is regulated at the nuclear pores, while ions and other small molecules may permeate at other sites also.

Most attempts at the isolation of nuclei and nuclear membranes have been listed by Franke et al. (132).* Aware of the special difficulties involved, they have developed a superior procedure for this purpose: Crude nuclei from connective tissue-free liver homogenates are isolated by differential centrifugation in 0.4 M sucrose, 0.07 M KCl, 0.01 M tris-acetate, pH 7.2, 0.004 M octanol, and 2% gum arabic. They are purified by rehomogenization and differential centrifugation first in 1 M sucrose,

Table 1.13 Phospholipid Composition of Nuclear Membranes, and Microsomes in Percent of Total Lipid Phosphorus with Standard Deviations of Single Determinations[a]

Lipid	Pig Liver Nuclear Membranes	Pig Liver Microsomes
Sphingomyelin	2.4 ±0.4	2.9 ±0.5
Lecithin	58.2 ±2.4	59.9 ±1.4
Phosphatidylethanolamine	25.9 ±1.8	27.4 ±0.9
Phosphatidylserine	4.4 ±0.6	3.2 ±0.8
Phosphatidylinositol	8.9 ±0.9	6.5 ±1.0
Phosphatidic acid	1.0	1.0
Lysolecithin	1.0	1.0
Other lysoderivatives	ND[2]	ND[2]
Cardiolipin	ND[2]	ND[2]
Molar ratio cholesterol to phospholipid	0.104	0.092

[a] From Kleinig (118).

* Monneron *et al.* (A. Monneron, G. Blobel and G. E. Palade, *J. Cell Biol.* **55:** 104, 1972) have introduced a novel approach to nuclear membrane isolation, by dissociating nuclei in 0.3–0.5M $MgCl_2$.

1% gum arabic, 0.07 M KCl, 0.01 M tris-acetate, pH 7.2, and then in a 2.2 M sucrose solution of identical ionic composition. Nonmembrane components are then extracted first with 0.3 M sucrose in 0.135 KCl, 0.1 M tris-phosphate, pH 7.4, and then with buffered 0.3 M sucrose in 1.5 M KCl. The membranes are finally isolated as a pure band by isopycnic ultracentrifugation in a linear density gradient of buffered sucrose. The compositions of nuclear membranes are given in Tables 1.11 to 1.13.

Peroxisomes

These membrane sacs are troublesome contaminants of mitochondrial isolates, since their function is essentially oxidative. They abound in liver and kidney and typically contain a number of characteristic enzymes such as urate oxidase and D-amino acid oxidase and catalase.

They can be separated from mitochondria in sucrose density gradients, in which they exhibit a higher equilibrium density. Catalase is soluble within the particles and can be removed by osmotic shock. On the other hand, the urate oxidase may be membrane associated (Table 2.7). The following markers are characteristic (36).

Catalase (Hydrogen peroxide:hydrogen peroxide oxidoreductase), catalyzing the reaction:

$$H_2O_2 + H_2O_2 \rightarrow O_2 + 2H_2O$$

D-*amino acid oxidase*, catalyzing reactions such as

$$\text{D-amino acid} + H_2O + O_2 \rightarrow \text{2-oxoacid} + NH_3 + H_2O_2$$

L-α-*hydroxyacid oxidase*, catalyzing the reaction:

$$\text{l-}\alpha\text{-hydroxyacid} + O_2 \rightarrow \alpha\text{-oxoacid} + H_2O_2$$

References

1. Graham, J. M., and Wallach, D. F. H. *Biochim. Biophys. Acta* **241:** 180, 1971.
2. Penniston, T., and Green, D. E. *Arch. Biochem. Biophys.* **128:** 339, 1968.
3. Steck, T. L., Fairbanks, G., and Wallach, D. F. H. *Biochemistry* **10:** 2617, 1971.
4. Zwaal, R. F. A., Roelofsen, B., Comfurius, P., and van Deenen, L. L. M. *Biochim. Biophys. Acta* **233:** 474, 1971.
5. Phillips, D. R., and Morrison, M. *Biochem. Biophys. Res. Commun.* **40:** 284, 1970.
6. Maddy, A. H. *Biochim. Biophys. Acta* **88:** 390, 1964.
7. Berg, H. C. *Biochim. Biophys. Acta* **183:** 65, 1969.
8. Bretscher, M. S. *J. Mol. Biol.* **58:** 775, 1971.

9. Taylor, R. B., Duffus, P. H., Raff, M. C., and de Petris, S. *Nature* **233:** 225, 1971.
10. Loor, F., Forri, L., and Pernis, B. *Eur. J. Immunol.* **2:** 203, 1972.
11. Kourilsky, F. M., Silvestre, D., Levy, J. B., Dausset, J., Nicolai, M. G., and Senik, A. *J. Immunol.* **106:** 454, 1971.
12. Lin, P. S., Wallach, D. F. H., and Tsai, S. *Proc. Natl. Acad. Sci.* (U.S.) **70:** 2492, 1973.
13. Sundquist, K. G. *Nature* **239:** 147, 1972.
14. Warren, L., Glick, M. C., and Nass, M. K. *J. Cell Physiol.* **68:** 269, 1966.
15. Boone, C. W., Ford, L. E., Bond, H. E., Stuart, D. C., and Lorenz, D. *J. Cell Biol.* **41:** 378, 1969.
16. Rosenberg, M. D. *J. Cell. Biol.* **17:** 289, 1963.
17. Rand, R. P., and Burton, A. C. *Biophys. J.* **4:** 191, 1964.
18. Rand, R. P. *J. Genl. Physiol.* **52:** 173s, 1968.
19. Wallach, D. F. H. In *Specificity of Cell Surfaces,* B. Davis and L. Warren, eds. Englewood Cliffs, N.J.: Prentice-Hall, 1967, p. 129.
20. Hughes, D. E., and Nyborg, W. L. *Science* **138:** 108, 1962.
21. Christiansen, R. O., Lyoter, A., Steensland, H., Saltzgraber, J., and Racker, E. *J. Biol. Chem.* **244:** 4428, 1969.
22. Steck, T. L., Straus, J. H., Weinstein, R. R., and Wallach, D. F. H. *Science* **168:** 255, 1970.
23. Avruch, J., and Wallach, D. F. H. *Biochim. Biophys. Acta* **241:** 24, 1971.
24. Dodge, J. T., Mitchell, C., and Hanahan, D. J. *Arch. Biochem. Biophys.* **160:** 119, 1963.
25. Brakke, M. K. *Advan. Virus Res.* **7:** 193, 1960.
26. Vinograd, J. *Methods Enzymol.* **6:** 854, 1963.
27. Schumaker, V. N., and Rosenbloom, J. *Biochemistry* **6:** 1149, 1967.
28. Steck, T. L., and Wallach, D. F. H. *Methods in Cancer Research* V. H. Busch, ed. New York: Academic Press, 1970, p. 92.
29. Steck, T. L., Straus, J. H., and Wallach, D. F. H. *Biochim. Biophys. Acta* **203:** 385, 1970.
30. Wallach, D. F. H., and Kamat, V. B. *Proc. Natl. Acad. Sci.* (U.S.) **52:** 721, 1964.
31. Anderson, N. G. *Natl. Cancer Inst. Monograph* **21:** 9, 1966.
32. Anderson, N. G. *Quart. Rev. Biophys.* **1:** 217, 1968.
33. El-Aaser, A. A., Fitzsimmons, J. T. R., Hinton, R. H., Reid, E., Klucis, E., and Alexander, P. *Biochim. Biophys. Acta* **127:** 553, 1968.
34. Pfleger, R. C., Anderson, N. G., and Snyder, F. *Biochemistry* **7:** 2826, 1968.
35. Weaver, R. A., and Boyle, W. *Biochim. Biophys. Acta* **183:** 118, 1969.
36. Leighton, F., Poole, B., Beaufay, H., Baudhuin, P., Coffey, J. W., Fowler, S., and De Duve, C. *J. Cell Biol.* **37:** 482, 1968.
37. Mach, O., and Lacko, L. *Anal. Biochem.* **22:** 393, 1968.
38. Wallach, D. F. H. *Anal. Biochem.* **37:** 138, 1970.
39. Wetzel, M. G. W., and Korn, E. D. K. *J. Cell Biol.* **43:** 90, 1969.
40. Greenawalt, J. W., Rossi, C. S., and Lehninger, A. L. *J. Cell. Biol.* **23:** 21, 1964.
41. Leskes, A., Siekevitz, P., and Palade, G. E. *J. Cell Biol.* **49:** 264, 288, 1971.

42. Tsukagoshi, N., and Fox, C. F. *Biochemistry* **10:** 3309, 1971.
43. Wallach, D. F. H., Kranz, B., Ferber, E., and Fisher, H. *FEBS Letters* **21:** 29, 1972.
44. Brunette, D. M., and Till, J. E. *J. Memb. Biol.* **5:** 215, 1971.
45. Albertsson, P. A. A. *Adv. Prot. Chem.* **24:** 309, 1970.
46. Resch, K., Imm, W., Ferber, E., Fischer, H., and Wallach, D. F. H. *Naturwissenschaften* **58:** 220, 1971.
47. Lowry, O. H., Rosebrough, N. J., Farr, A. L., and Randall, R. J. *J. Biol. Chem.* **193:** 265, 1951.
48. Schmidt-Ullrich, Knüfermann, H., and Wallach, D. F. H., *Biochim. Biophys. Acta*, in press.
49. Coleman, R., and Finean, J. B. *Biochim. Biophys. Acta* **125:** 197–206, 1966.
50. Bosmann, H. B., Hagopian, A., and Eylar, E. H. *Arch. Biochem. Biophys.* **128:** 51–59, 1968a.
51. Ferber, E., Imm, W., Resch, K., and Wallach, D. F. H. *Biochim. Biophys. Acta* **266:** 494, 1972.
52. Song, C. S., and Bodansky, O. *J. Biol. Chem.* **242:** 694–699, 1967.
53. Widnell, C., and Unkeless, J. C. *Proc. Natl. Acad. Sci.* (U.S.) **61:** 1050–1057, 1968.
54. Emmelot, P., Bos, C. J., Benedetti, E. L., and Rümke, P. *Biochim. Biophys. Acta* **90:** 126–145, 1964a.
55. Graham, J. M., Higgins, J. A. and Green, C. *Biochim. Biophys. Acta* **150:** 303–305, 1968.
56. Lansing, A. I., Belkhode, M. L., Lynch, W. E., and Lieberman, I. *J. Biol. Chem.* **242:** 1772–1775, 1967.
57. Kamat, V. B., and Wallach, D. F. H. *Science* **148:** 1343–1345, 1965.
58. Lieberman, I., Lansing, A. I., and Lynch, W. E. *J. Biol. Chem.* **242:** 736–739, 1967.
59. Robison, G. A., and Sutherland, E. W. In *Cyclic AMP and Cell Function*, G. A. Robison, G. G. Nahas, and L. Triner, eds. *Ann. N.Y. Acad. Sci.* **185:** 5, 1971.
60. Stahl, W. L., and Trams, E. G. *Biochim. Biophys. Acta* **163:** 459–471, 1968.
61. Hagopian, A., Bosmann, H. B., and Eylar, E. H. *Arch. Biochem. Biophys.* **128:** 387–396, 1968.
62. Wallach, D. F. H., and Hager, E. B. *Nature* **196:** 1004–1005, 1962.
63. Wallach, D. F. H., and Kamat, V. B. *Methods Enzymol.* **8:** 164–172, 1966a.
64. Wallach, D. F. H., and Vlahovic, V. *Nature* **216:** 182, 1967.
65. Ozer, J. H., and Wallach, D. F. H. *Transplantation* **5:** 652–667, 1967.
66. Wallach, D. F. H. *The Plasma Membrane: Dynamic Perspectives, Genetics and Pathology. Heidelberg Science Library* **18:** Springer, 1972, Chapter 6.
67. Zajac, I., and Crowell, R. L. *J. Bacteriol.* **89:** 1097–1100, 1965.
68. McLaren, L. C., Scaletti, J. V., and James, C. G. In *Biological Properties of the Mammalian Surface Membrane*, L. A. Manson, ed. Philadelphia: Wistar Inst. Press, 1968, pp. 123–135.
69. Philipson, L., Lonberg-Holm, K., and Pettersson, U. *J. Virol.* **2:** 1064–1075, 1968.
70. Pardee, A. B., and Watanabe, J. *Bacteriol.* **96:** 1049, 1968.

71. Bender, W. W., Garan, H., and Berg, H. C. *J. Mol. Biol.* **58**: 783, 1971.
72. Schmidt-Ullrich, R., Ferber, E., Knüferman, H., and Wallach, D. F. H., to be published.
73. Bretscher, M. S. *J. Mol. Biol.* **59**: 351, 1971.
74. Bonsall, R. W., and Hunt, S. *Biochim. Biophys. Acta* **249**: 266, 1971.
75. Steck, T. L. In *Membrane Research*, C. F. Fox, ed. New York: Academic Press, 1972, p. 71.
76. Arroti, J. J., and Garvin, J. E. *Biochim. Biophys. Acta* **255**: 79, 1972.
77. Yariv, J. J., Kalb, A. J., Katchalski, E., Goldman, R., and Thomas, E. W. *FEBS Letters* **173**.
78. Fleet, G. W. J., Knowles, J. R., and Porter, R. R. *Nature* **224**: 511, 1969.
79. Kiefer, H., Lindstrom, J., Lennox, E. S., and Singer, S. J. *Proc. Natl. Acad. Sci.* (U.S.) **67**: 1668, 1970.
80. Avruch, J. and Fairbanks, G., *Proc. Natl. Acad. Sci.* (U.S.) **69**: 1216, 1972.
81. Dunham, P. B., and Hoffman, J. F. *Proc. Natl. Acad. Sci.* (U.S.) **66**: 936, 1970.
82. Phillips, D. R., and Morrison, M. *Biochemistry* **10**: 1766, 1971.
83. Phillips, D. R., and Morrison, M. *FEBS Letters* **18**: 95, 1971.
84. Hubbard, A. L. H., and Cohn, Z. C. *J. Cell Biol.* **55**: 390, 1972.
85. Marchalonis, J. J., Cone, R. E., and Santer, V. *Biochem. J.* **124**: 921, 1971.
86. Hubbard, A. L. H., and Cohn, Z. C. Abstracts of papers presented at 12th Annual Meeting of the American Society of Cell Biology.
87. Rifkin, D. B., Compans, R. W., and Reich, E. *J. Biol. Chem.* **247**: 6432, 1972.
88. Himmelspach, K., Westphal, O., and Teichman, B. *Eur. J. Immunol.* **1**: 106, 1971.
89. Suttajitt, M., and Winzler, R. J. *J. Biol. Chem.* **246**: 3398, 1971.
90. Marfey, P. S. In *Red Cell Membranes Function*, G. A. Jamieson and T. J. Greenwalt, eds. Philadelphia: J. B. Lippincott, 1969, p. 309.
91. Weibel, E. R. *Internatl. Review Cytol.* **26**$_6$ 235, 1969.
92. Wallach, D. F. H., Kamat, V. B., and Gail, M. H. *J. Cell Biol.* **30**: 601, 1966.
93. Hoogeven, J. T., Juliano, R., Coleman, J., and Rothstein, A. *J. Memb. Biol.* **3**: 156, 1970.
94. Boyse, E. A., and Old, L. J. *Ann. Rev. Genet.* **3**: 269–290, 1969.
95. Boyse, E. A., Old, L. J., and Stockert, E. *Proc. Natl. Acad. Sci.* (U.S.) **60**: 886–893, 1968.
96. Zwaal, R. F. A., and van Deenen, L. L. M. *Biochim. Biophys. Acta* **150**: 323, 1968.
97. Wallach, D. F. H., and Zahler, P. H. *Biochim. Biophys. Acta* **150**: 186–193, 1968.
98. Takeuchi, M., and Terayama, H. *Exptl. Cell. Res.* **40**: 32, 1965.
99. Hulcher, F. H. *Arch. Biochem. Biophys.* **100**: 237, 1963.
100. Rosenberg, S. A., and Guidotti, G. In *The Red Cell Membrane*, G. A. Jamieson and T. J. Greenwalt, eds. Philadelphia: J. B. Lippincott, 1969, p. 93.
101. Elias, H. *Am. J. Anat.* **85**: 379–456, 1949.
102. Neville, D. M., Jr. *J. Biophys. Biochem. Cytol.* **8**: 413–422, 1960.
103. Coleman, R., Mitchell, R. H., Finean, J. B., and Hawthorne, J. N. *Biochim. Biophys. Acta* **135**: 573–579, 1967.
104. Goodenough, D. A. G., and Stoeckenius, W. S. *J. Cell. Biol.* **54**: 646, 1972.

105. Holland, J. J. *Virology* **16**: 163–176, 1962.
106. Holland, J. J., and Hoyer, B. H. *Cold Spring Harbor Symp. Quant. Biol.* **27**: 101–112, 1962.
107. Hays, R. M., and Barland, P. *J. Cell Biol.* **31**: 209–214, 1966.
108. Emmelot, P., and Benedetti, E. L. In *Carcinogenesis: A Broad Critique.* Baltimore: Williams & Wilkins, 1967, pp. 471–533.
109. Benedetti, E. L., and Emmelot, P. *J. Cell Sci.* **2**: 499–512, 1967.
110. Emmelot, P., and Bos, C. J. *Biochim. Biophys. Acta* **150**: 354–363, 1968.
111. Emmelot, P., Visser, A., and Benedetti, E. L. *Biochim. Biophys. Acta* **150**: 364, 1968.
112. Kaulen, H. D., Henning, R., and Stoffel, W. *Hoppe Seyler Ztschr. für Physiol. Chem.* **351**: 1555, 1970.
113. Perdue, J. F., and Sneider, J. *Biochim. Biophys. Acta* **196**: 125, 1970.
114. Wheeldon, L. W. W., and Gan, K. G. *Biochim. Biophys. Acta* **233**: 37–48. 1971.
115. Korn, E. D. In *Theoretical and Experimental Biophysics*, A. Cole, ed. New York: M. Dekker, 1967, p. 2.
116. Fleischer, B., Fleischer, S., and Ozawa, H. *J. Cell. Biol.* **43**: 59, 1969.
117. Morré, D. J., Hamilton, R. L., Mollenhauer, H. H. Mahley, R. W., Cunningham, W. P., Cheetham, R. D., and Leguire, V. S. *J. Cell. Biol.* **44**: 484, 1970.
118. Kleinig, H. *J. Cell. Biol.* **46**: 396, 1970.
119. Romeo, D. R., and De Bernard, D. *Nature* **212**: 1491, 1966.
120. Sawant, P. L., Desai, I., and Tappel, A. L. *Arch. Biochem. Biophys.* **105**: 247, 1964.
121. Schuel, H., Berger, E. R., Wilson, J. R., and Schuel, R. *J. Cell Biol.* **43**: 125a, 1969.
122. Swick, R. W., Tollaksen, S. L., Nance, S. L., and Thomason, J. F. *Arch. Biochem. Biophys.* **136**: 212, 1970.
123. Allmann, D. W., Bachman, E., Orme-Johnson, N., Tais, W. C., and Green, D. E. *Arch. Biochem. Biophys.* **125**: 981, 1968.
124. Ernster, L., and Kuylenstiarna, B. In *Membranes of Mitochondria and Chloroplasts*, E. Racker, ed. New York: Van Nostrand, **81**: 1969, p. 172.
125. Beattie, D. S. *Biochem. Biophys. Res. Commun.* **31**: 901, 1968.
126. Schnaitman, C., and Greenawalt, J. W. *J. Cell Biol.* **38**: 158, 1968.
127. Sjöstrand, F. In *Structural and Functional Aspects of Lipoproteins in Living Systems*, E. Tria and A. M. Scanu, eds. New York: Academic Press, 1969, p. 79.
128. Mokrash, L. C. M. In *Methods of Neurochemistry*, F. Fried and L. Rainer, eds. New York: M. Dekker, 1971, p. 1.
129. Autilio, L. A. A., Norton, W. T., and Terry, R. D. *J. Neurochem.* **11**: 17, 1969.
130. O'Brien, J. S., and Sampson, E. L. *J. Lipid Res.* **6**: 537, 545, 1965.
131. Stevens, B. J., and Andre, J. In *Handbook of Molecular Cytology*, A. Lima-de-Feria, ed. Amsterdam: North Holland Publ. Co., 1969, p. 837.
132. Franke, W., Deumling, B., Ermen, B., Jarasch, E. D., and Kleinig, H. *J. Cell Biol.* **46**: 379, 1970.

CHAPTER 2

MEMBRANE "MACROMOLECULES"

Introduction

All known biomembranes contain protein, lipid, and carbohydrate. RNA is thought to be a plasma membrane component of some cells, and some DNA is intimately associated with nuclear membranes. As far as defined macromolecules are concerned, the data on RNA and DNA remain fragmentary, and membrane lipids are not macromolecules, but aggregate into supramolecular complexes in aqueous solvents; thus, membrane proteins constitute the major macromolecular components of biomembranes. Except for myelin, they comprise 60–70% of the membrane mass. Membrane carbohydrate is primarily associated with glycoproteins and to a lesser extent with glycolipids.

Here we shall concentrate our attention primarily on membrane proteins and glycoproteins. However, we shall first comment briefly on membrane lipids, since these have been and continue to be the focus of much membrane research, including many refined spectroscopic approaches.

Lipids

General

The biochemistry of membrane lipids has been often and well reviewed (1–4). The techniques used in lipids biochemistry are extensively treated by Johnson (5) and Davenport (6).

The lipids found in plasma membranes consist of three major classes: glycerophosphatides (Fig. 2.1), sphingo- and glycolipids, and steroids. Most abundant in membranes are glycerol phosphatides, such as phosphatidyl choline, phosphatidyl serine, phosphatidyl ethanolamine, and phosphatidyl inositol.

Glycerophosphatides comprise derivatives of glycerol-3-phosphoric acid. The hydrocarbon side chains are highly soluble in lipid solvents (and in

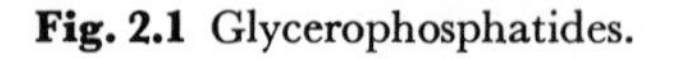

Fig. 2.1 Glycerophosphatides.

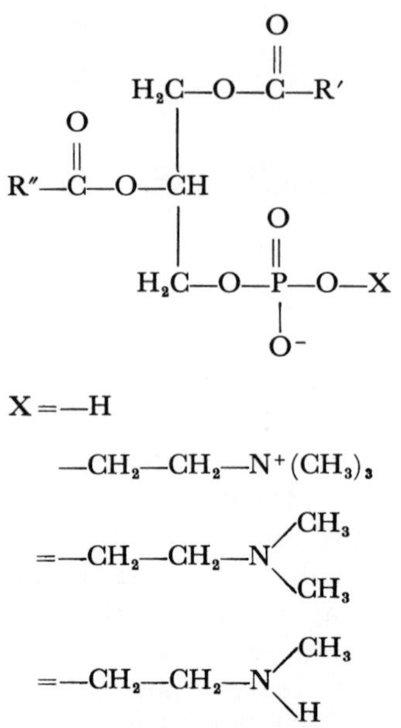

General formula for glycerophosphatides*

$X = —H$ Phosphatidic acid

$—CH_2—CH_2—N^+(CH_3)_3$ Phosphatidyl choline or lecithin

$= —CH_2—CH_2—N(CH_3)_2$ Phosphatidyl (N-dimethyl)-ethanolamine

$= —CH_2—CH_2—N(CH_3)H$ Phosphatidyl (N-methyl)-ethanolamine

$= —CH_2—CH_2—NH_2$ Phosphatidyl ethanolamine

$= —CH_2—CH(NH_2)—COOH$ Phosphatidyl serine

(*continued*)

---

*R′ = a saturated fatty acid commonly
R″ = unsaturated commonly

The chain length of R′ and R″ is usually 10–20. In *inositol phosphatides* $X = —C_6H_6(OH)_5$, but the 4-, and/or 5-positions of the inositol may also be phosphorylated; *acetal phosphatides* (plasmalogens) consist of glycerylphosphorylcholine (or —ethanolamine, or —serine) with one esterified fatty acid and one fatty acid in enol—ether linkage, as shown below:

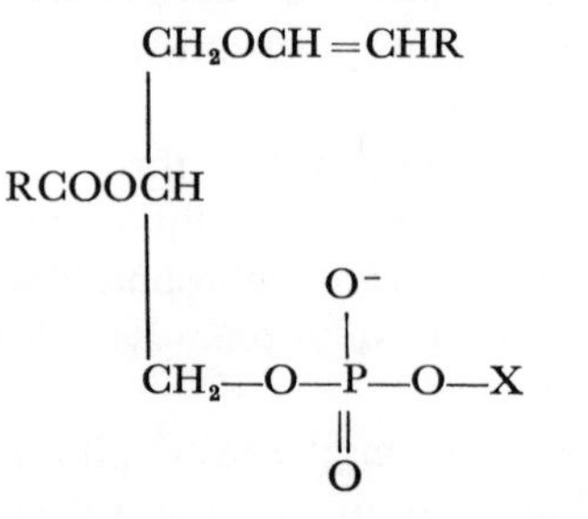

Ethanolamine plasmalogen

Fig. 2.1 (*concluded*).

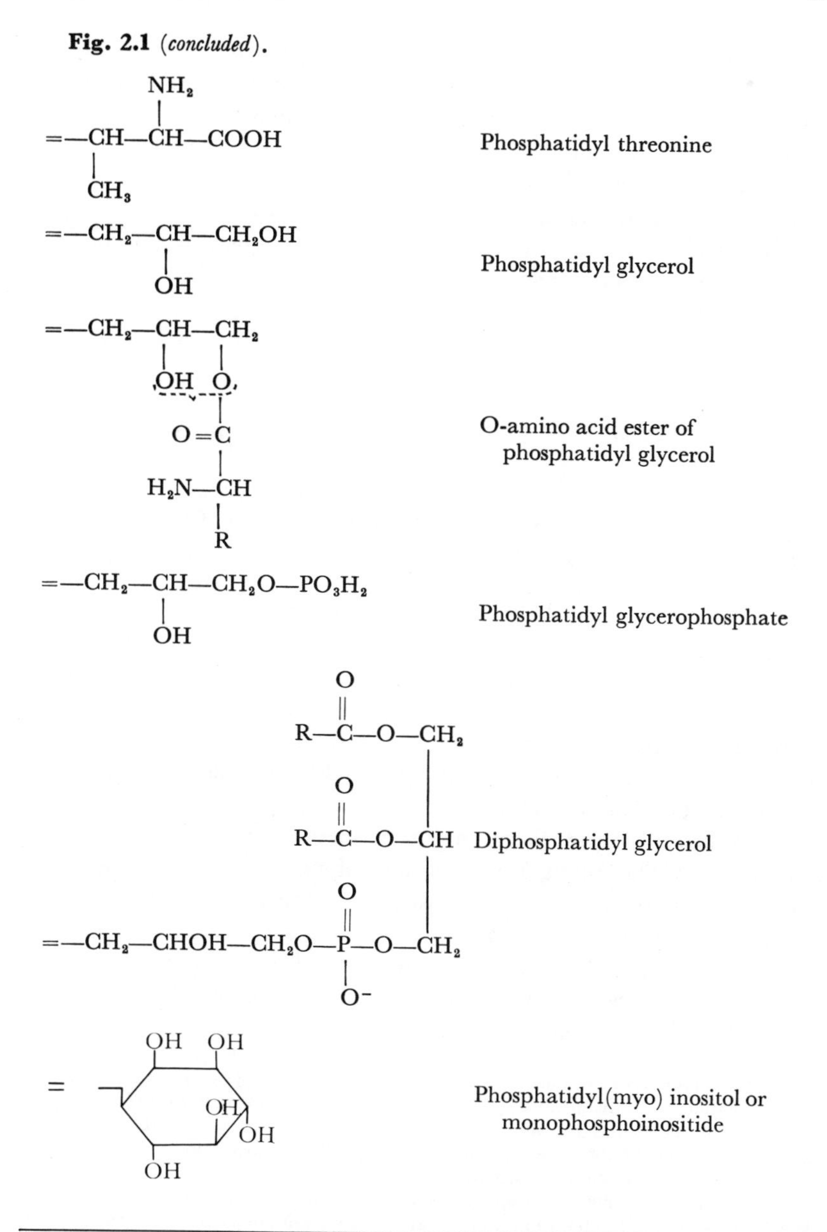

each other), whereas the polar "head groups" containing the glycerol moiety, the phosphate group and the choline, serine, ethanolamine or inositol are water soluble. Therefore, such molecules will orient at an oil-water interface in such a way that the side chains are in the oil and the polar groups are in the water.* At physiologic pH, the polar group may lack a net change as in phosphatidyl choline (lecithin) and be slightly anionic as in phosphatidyl ethanolamine or exhibit a distinct net negative charge as in phosphatidyl serine and in phosphatidyl inositol.

Closely related to the common glycerophosphatides is the more complex lipid phosphatidyl glycerol, in which two molecules of phosphatidic acid are esterified with a third molecule of glycerol. Phosphatidyl glycerol (cardiolipin) carries a negative charge due to the two phosphodiester bonds.

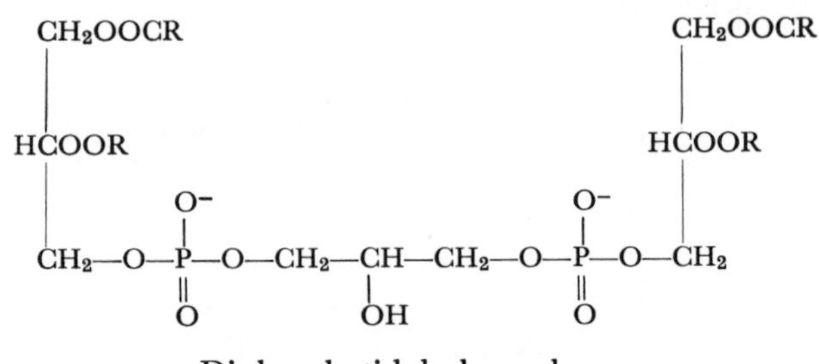

Diphosphatidyl glycerol

This lipid abounds in mitochondria.

Plasmalogens differ from the phosphoglycerides in that one of the hydrocarbon chains is an γ,β unsaturated ether linkage with the glycerol moiety. On hydrolysis, the plasmalogens yield one molecule of fatty acid and one of long-chain fatty aldehyde.

A *second* class of lipids found in plasma membranes are the *sphingolipids and glycolipids*. The sphingolipids contain sphingosine (or dihydrosphingosine) with a fatty acid in amide linkage with the sphingosine. Fatty acid amides of sphingosine are known as ceramides (Fig. 2.2). In sphingomyelin (Fig. 2.2), a phosphate is in diester linkage between choline and the hydroxyl on the terminal carbon of ceramide. Like lecithin, sphingomyelin bears no net charge at neutral pH.

* Potential energy calculations for noncovalently bonded systems have been applied to phosphatides to indicate their energetically preferred conformations and interactions (J. McAlister, N. Yathindra, and M. Sundaraligam, *Biochemistry* **12:** 1189, 1973). The computations suggest that intramolecular chain packing restricts the conformations of phosphatides to a few stable possibilities.

The glycerol moiety of phosphatides can exist in two orientations, in which the β-chain folds unto the γ chain to initiate chain packing. Although the hydrocarbon chain axes lie parallel in both cases, the C-atoms distal to the two ester linkages intersect at dihedral angles of 72° or 57°, depending on the glycerol conformation. The tilt of the fatty acid chain planes optimizes contacts between the fatty acid termini. In contrast to the β- and γ-residues, the polar residue at the α-position of glycerol can assume multiple conformations because of several energetically favored orientations of the choline and phosphodiester residues. These conformations might also depend markedly upon diverse molecules which associate with the headgroups.

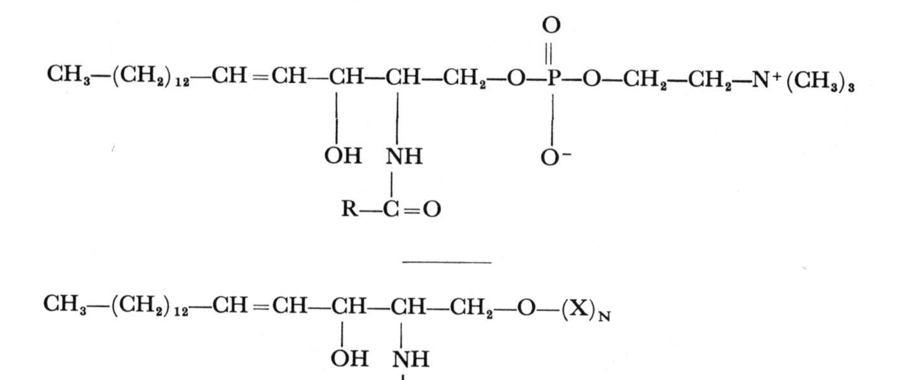

Fig. 2.2 The structure of sphingomyelin, above, and glycolipids, below. In *ceramide monohexosides* or *cerebrosides* X = *galactose* or *glucose* and N = 1. In *ceramide oligohexosides* N = 1, but there are equivalent molar amounts of sphingosine, hexose and fatty acid (R); the only nitrogen is in the sphingosine. In *globosides* (e.g., from human erythrocytes), additional nitrogens occur in the sugar moiety, when X=hexose(s)+hexosamine(s), or in *mucolipids* containing *neuraminic* acids (e.g., in horse erythrocytes). Then X = hexose(s) or hexosamine(s) +neuraminic acid.

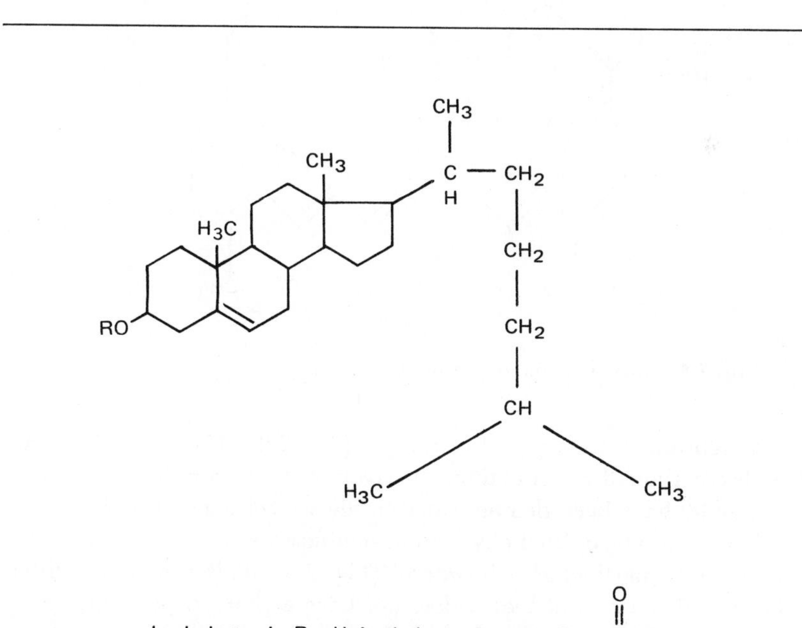

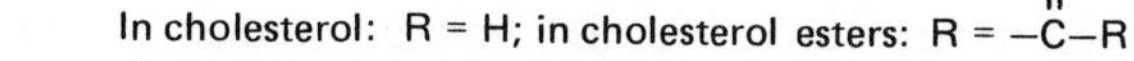

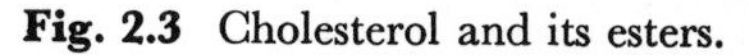
Fig. 2.3 Cholesterol and its esters.

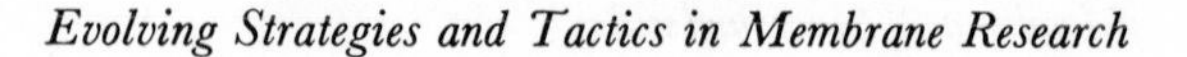

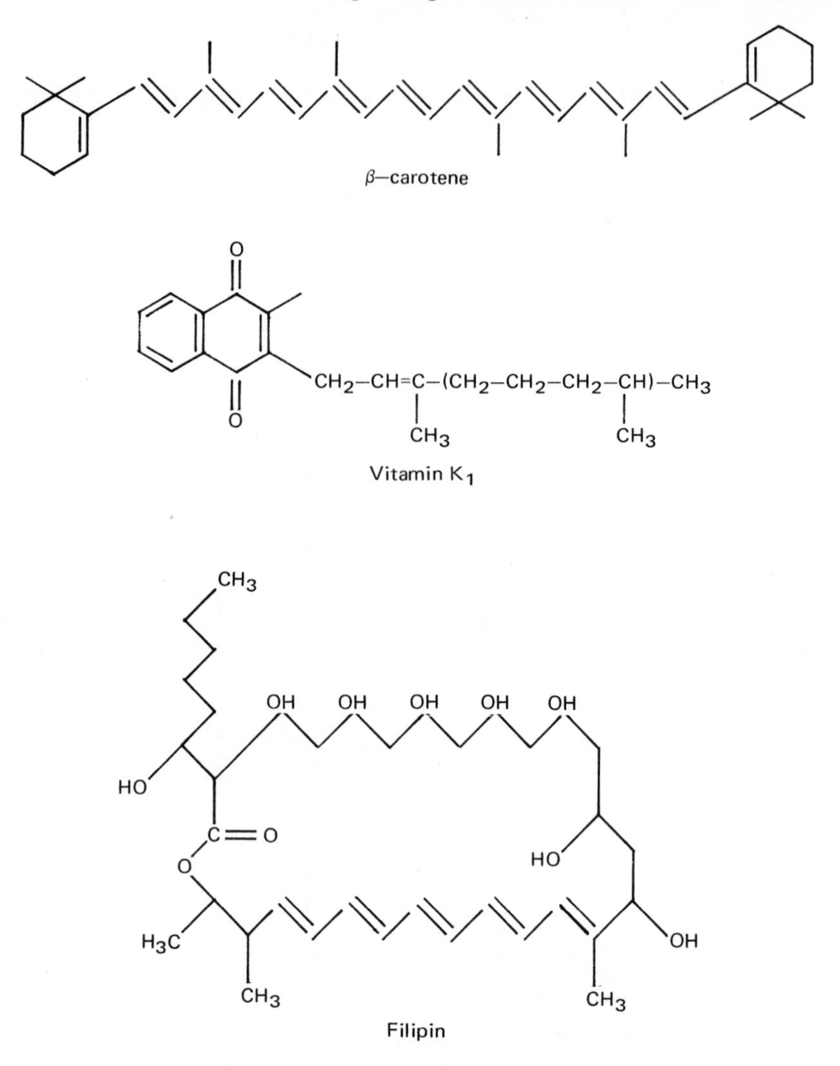

Fig. 2.4 Important minor membrane lipids.

Cerebrosides are glycosyl ceramides (Fig. 2.2). The gangliosides are complex carbohydrate-containing derivatives of ceramide. Many such gangliosides have been demonstrated in diverse cell membranes.

The *third* type of lipid observed in significant amounts in membranes is the *steroids*, particularly cholesterol (Fig. 2.3). This compound differs from the other lipids in that it does not have a charged polar group. As will be discussed later, cholesterol is probably found in the interior hydrophobic region of membranes.

Certain quantitatively minor, but biologically important membrane lipids are given in Fig. 2.4. Of particular interest are β-carotene and its derivatives, vitamin K, and polyene antibiotics such as filipin, which form tight, disruptive complexes with cholesterol in membranes containing this steroid.

We shall now turn our attention to selected aspects of lipid technology that are particularly pertinent to membrane biology in general and those aspects of the field treated elsewhere in this volume.

Use of Phospholipases

These enzymes are used increasingly to modify and extract membrane phospholipids and their components, as well as to investigate the interactions of phosphatides with other membrane components *in situ.* We shall return to the last in later chapters.

The actions of diverse phospholipases are illustrated in Fig. 2.5, and the properties of these enzymes are summarized in Table 2.1. The A_2 group enzymes have been most widely purified. They are all resistant to heat and acid denaturation. The snake venom enzymes are peptides of mol wt ~5000. The pancreatic enzyme comprises a single 16,000 dalton peptide, stabilized by six internal S—S bonds.

Known phospholipases will act on intact membranes under suitable conditions, which can depend very much on membrane state. Zwaal et al. (7, 8) show a correlation in the capacities of phospholipases A_2 and C to lyse intact erythrocytes, depending upon the state of osmotic "expansion" of the erythrocytes. Chemical analyses indicate that no more than 5% of the membrane phosphatides require cleavage for cell lysis to occur. The authors point out that hypoosmotic conditions

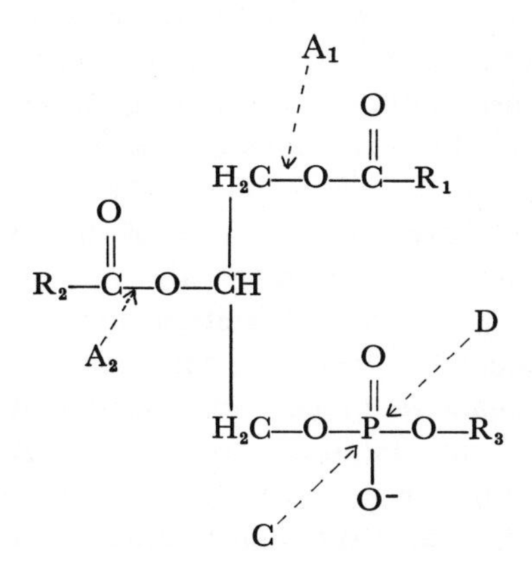

Fig. 2.5 Action of phospholipases.

Table 2.1 Properties of Some Phospholipases

Enzyme	Substrate	Cofactor	Source	Purity
A_1	Phospholipids glycerides	None	Widely distributed	Partially purified from brain
A_2	Phospholipids	Ca^{++}	Snake and bee venom; mammalian pancreas	Homogeneous; crystalline
C	Phospholipids sphingolipids	Ca^{++}	Bacteria	Homogeneous (B. cereus)
D	Phospholipids	Ca^{++}	Cabbage	Slightly purified

make erythrocytes (or ghosts) relatively more susceptible to the action of phospholipase C (B cereus) than to pancreatic phospholipase A_2, i.e., the polar head groups are more vulnerable than the fatty ester linkages. The effect is *not* fully reversed by resealing hemoglobin-deprived ghosts, nor is it modified by pretreatment of intact cells with diverse proteases. The authors carefully suggest that the effect may derive from either (a) mechanical expansion of the membrane and/or (b) architectural changes in lipid-associated membrane proteins. Clearly, such data indicate caution in terms of the degree to which erythrocyte ghosts represent the membranes of their native cells.

Phospholipase A_2 has been used in a novel way to extract glycerophosphatides (as their split products, fatty acids + lysophosphatides) from diverse membranes (9–11). In these experiments, the phospholipase-treated membranes are equilibrated with defatted serum albumin (10–40 mg/ml, 25°C, 10 min). In their studies on sarcoplasmic reticulum, Fiehn and Hasselbach (9) observe ~70% phosphatide cleavage and find that 90% of the liberated fatty acid and 65% of the lysolecithin could be removed by albumin. Weidekamm et al. (11), studying erythrocyte ghosts, found 50–70% cleavage of lecithin and 70–95% hydrolysis of phosphatidyl ethanolamine. However, they monitored the fate of the albumin, using ^{125}I-labeled protein and found that about 2% of the ghost protein after albumin treatment could be accounted for in this substance, even after multiple washing. Intact erythrocytes, in contrast, were easily freed of ^{125}I-labeled albumin. They also noted that albumin laden with ^{14}C-oleate readily exchanged this fatty acid with erythrocyte ghosts.

In contrast to phospholipase A, phospholipase C does not effect release of the lipid moieties of cleaved phosphatides (diglycerides, ceramides) by centrifugation or treatment with fat-free albumin or immiscible organic solvents, although the phosphoryl bases go into solution (9).

The membrane biologist most commonly uses phospholipases *extrinsically* as aids in the study of membrane organization. However, many biomembranes contain *intrinsic* phospholipases of the A type, which allow some insight into membrane phospholipid dynamics. These enzymes form part of what would appear to constitute an important cycle, allowing replacement or repair of phosphatide fatty acids in situ (Fig. 2.6).

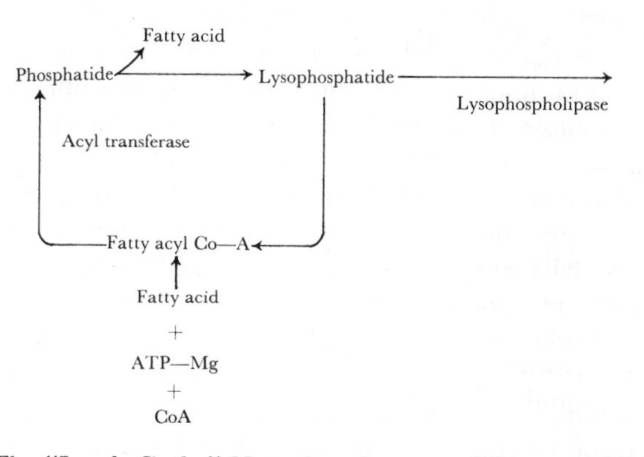

Fig. 2.6 The "Lands Cycle." Note that there are different acyl transferases for saturated and unsaturated fatty acids.

This system, first discovered by Lands (12) and recently reviewed by Ferber (13), interests us here primarily because it provides a means for *labeling* the phosphatides of isolated membranes. However, the cycle represents a system of intrinsic interest, serving not only in the in situ exchange and/or repair of phosphatide fatty acids, but, together with lysophospholipase, prevents accumulation of lytic levels of lysophosphatides.

Use of Organic Solvents

The most widely used approach to the study of membrane lipids involves their extraction by organic solvents, e.g., alcohols, ethers, chloroform, 2-chloroethanol, and diverse solvent mixtures. This matter is not as simple as it may seem, however, and the following considerations are essential.

Many membranes contain lipids which are (a) *loosely bound* and (b) *tightly bound* (14), i.e., extractable by (a) solvents that leave membrane proteins relatively intact or (b) only by denaturing solvents. Cholesterol is always loosely bound and can, indeed, be removed from erythrocyte membranes by equilibrating them with phosphatide dispersions lacking the steroid; it can also be added to normally cholesterol-deficient mitochondria by exposing them to cholesterol-enriched liposomes. Some phosphatides, particularly lysophosphatides, are loosely bound. However, those phosphatides, which define the genetic lipid specificity of erythrocyte membranes, are "tightly bound."*

In many cases, most membrane phosphatides and all the cholesterol are extracted by organic solvents such as 9:1 *acetone:water*, in which the lipids are *not soluble*, but form supramolecular aggregates (15).

Chloroform:methanol (2:1; v/v) is the lipid extractant of choice. However, many sugars, amino acids, small peptides, and some larger peptides are readily soluble in this mixture. These contaminants can be removed to a large extent by effecting a phase separation through addition of aqueous KCl, NaCl, or CaCl. Then phosphatides and cholesterol accumulate in the dense, chloroform:methanol phase and more polar molecules in the upper water:methanol layer. Repeated washing with "artificial upper phase" and phase separations are usually required (16), but even then, some highly polar lipids (glycolipids) may go into the aqueous phase, whereas some proteins and/or peptides with apolar characteristics (e.g., proteolipids) accumulate in the "organic phase." This is well recognized for myelin (17), but is also true for other membranes, such as erythrocyte ghosts (18, 19).

Membrane proteins and membrane lipids (+ small lipophilic molecules) can be effectively separated by gel permeation chromatography on lipophilic Sephadex (LH-20) as first shown by Zahler and Wallach (20). This procedure involves the use of a unique solvent—*2-chloroethanol*—which dissolves membrane proteins, lipids, and most small contaminants. These can then be separated on Sephadex (LH-20), the proteins being excluded and the smaller molecules retarded. 2-chloroethanol is somewhat unstable, tending to release some HCl. It has a dielectric constant of 25.8, a boiling point of 128.6°C, melting point of −67.5°C, a density of 1.202, a refractive index of 1.444, and a viscosity of 3.913. None of these properties explain the unique solvation properties of 2-chloroethanol, but some pertinent aspects of the subject are discussed by Singer (21). Very important in this respect is the fact that 2-chloroethanol tends to promote

* Further evidence demonstrating that membrane proteins can exert selective affinities for phosphatides containing certain unsaturated fatty acids, exists for mitochondrial components (F. D. Collins, G. G. de Pury, M. Havlicek, and C.-S. Lim, *Chem. Phys. Lipids* **7**: 144, 1971). Lecithins containing mixtures of oleic and lineolic acids, or 5, 8, 11-licosatrienoic acids react more effectively with mitochondrial proteins than those containing arachidonic acid (cf. ref. 70).

helical folding, rather than the unwinding of the peptide chain induced by most agents used in membrane studies (e.g., sodium dodecyl sulfate, urea, guanidine HCl). We shall return to this.

Fluorinated hydrocarbons, e.g., trifluoroethanol and hexafluoroacetone, do not substitute for 2-chloroethanol in our experience. However, we find that erythrocyte and other membranes dissolve (i.e., nonsedimentable at 5×10^6 g-min) in 8:1 2-chloroethanol: water mixtures, buffered at pH *7.4* with 20 mM phosphate. This suggests that 2-chloroethanol:water systems, appropriately buffered, may add to our impoverished armamentarium for the fractionation of membrane components.

The techniques required to effect this strategy will be discussed further on. Suffice it to say here that gel permeation chromatography in 2-chloroethanol provides an unusually effective means of separating small (<1000 daltons) from large membrane molecules.

A major difficulty in applying molecular sieving techniques to membrane components until recently has been the lack of suitable sieving media. Thus, the lipophilic Sephadexes are suitable for separating small and large membrane components, but not for fractionating these classes. This situation may have been resolved by the availability of *controlled-pore* glass beads.

These consist of granules of high-silica glass, penetrated by channels of uniform and controlled diameter, from 7.5–200 nm, and independent of solvent characteristic, flow, or pressure. Unlike other molecular sieving substrates, the pore diameters of these glasses are unrelated to the pore volumes; the latter vary slightly with manufacturing conditions. Typical conditions are given in Table 2.2.

An important aspect is using controlled-pore glasses in the surface

Table 2.2 Critical Characteristics of Some Controlled-Pore Glasses

Mean Pore Diameter (nm)	Surface Area m^2/g at Specific Pore Volume of 1.00 cm^3/g
7.5	340
12.0	210
17.0	150
24.0	110
35.0	75
70.0	36
140.0	18
200.0	13

activity of the reacting sites. The glasses consist primarily of silica with traces of B_2O_3 and Na_2O. The pore surfaces accordingly exhibit a slight negative charge with aqueous solutions near neutral pH. Moreover, these glasses tend to adsorb nonpolar substances. Accordingly, for fractionation of substances in organic solvents, the beads should be treated with alkyl silanes (22). Although we have stressed this new technology, we must also emphasize that it has yet to be applied to the membrane field.

Exchange of Lipids between Diverse Membranes

It is well established that cholesterol can exchange freely between erythrocyte membranes and liposomes or lipoproteins and that the rate of exchange is similar for the internal and external surfaces of erythrocyte ghosts (23). This allows the investigator to manipulate membrane cholesterol over a wide range.

Other lipids can also be exchanged between membranes and/or membranes and other lipophilic systems. In general, phosphatides do not exchange readily, but in Chapters 7 and 8, we shall discuss in some detail how fluorescent and spin-labeled phosphatides, as well as other lipids can be introduced into biomembranes and model membranes. Moreover, in Chapter 1 we have alluded to the efforts of Fox and associates (24) to utilize bromostearic acid as a membrane density perturbant (p. 15).

Not all lipid exchange is passive, however. We have already described the "Lands cycle" (p. 65), allowing fatty acid exchange in membrane-located phospholipids through the aid of a series of enzymes and cofactors. Equally important is the discovery by Wirtz and Zilbersmit (25) that a soluble cytoplasmic protein in rat liver promotes the rapid exchange of phosphatides, particularly phosphatidyl ethanolamine and phosphatidyl-choline between microsomal and mitochondrial membranes.

We cite these two systems as examples of hitherto unexplored avenues for the efficient integration of biochemical and biophysical approaches to membrane biology.

Epilogue

For decades the membrane biologist has focused his attention on membrane lipids, even though they are not the predominant components of most biomembranes and do not endow them with their genetically determined functions. As we shall see subsequently, much current effort, employing elegant and sophisticated techniques, continues to deal with the lipids of membranes in preference to their proteins. These trends are fully understandable: Lipids are small molecules, rather easily extracted and purified; many methods are available for their precise fractionation and structural characterization; they lend themselves readily to the formation of diverse derivatives and readily accessible model systems.

However, we feel that many questions remain to be resolved even in the biochemistry and biophysics of membrane lipids and that the *burning issues of membrane biology intimately concern membrane proteins.*

Proteins

General

An important recurrent theme concerns the questions: (a) *Which are membrane proteins?* and (b) *which are possible contaminants?* We shall return to the matter of true contamination shortly, but believe, as conceptualized in Fig. 2.7, that no simple definition of a membrane protein exists and that a large number of variables must be considered even for a well-defined membrane domain, in a given cell of defined developmental and physiologic state.

We shall concentrate here on "core proteins," realizing that these can be defined only operationally. These comprise a class of proteins that remain membrane associated unless drastic extraction methods are employed, and/or that serve distinct membrane functions. However, we lack assurance that such entities may not exist elsewhere in a different configuration and/or functional capacity, i.e., in a guise that may prevent recognition of the fact that they are the same gene product as a given "membrane protein." However, we do know that well-recognized soluble proteins, such as *immunoglobulins*, can also serve as functionally important membrane proteins in *lymphocytes* (26). We also recognize that numerous proteins (e.g., cytochrome c) exert their role preferentially or differently when membrane associated, although they function recognizably in solution.

In addition, some proteins are present, particularly in the intracellular spaces, in concentrations sufficient to alter the membrane microenvironment. *Hemoglobin* represents a prominent example in the erythrocyte, especially since it undergoes large architectural changes during physiological oxygenation and deoxygenation.

Another category to be considered here is *regulatory proteins*, e.g., hormones with very high membrane affinities. These would not generally fall into the membrane protein category. Finally, many proteins indubitably undergo rather trivial contacts with diverse biomembranes.

As noted before, we shall deal here with strategies and techniques attempting to handle those proteins that are strongly and permanently associated with membranes. But, first, we must reemphasize a matter raised in Chapter 1, i.e., the problems arising in membrane isolations.

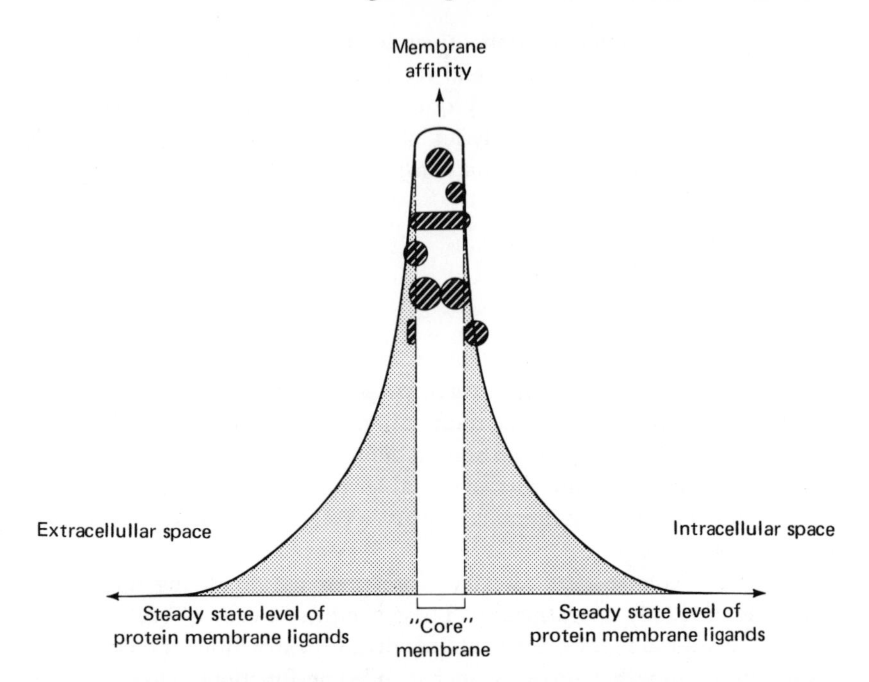

Fig. 2.7 Heuristic depiction of the spatial disposition of "membrane" proteins. We envisage membranes to comprise a "core" (stippled) containing all the lipid and those proteins (black symbols) that can be separated from native membranes only by disrupting their basic structure. Possible but hypothetical dispositions of such "core" proteins are suggested.

The ordinate represents "membrane affinity," the abscissa the proportion of diverse noncore proteins in the intra- and extracellular spaces, respectively. For illustration only, the "core proteins" are depicted showing lesser affinities with decreasing penetration; this need not be the case.

The distribution curves enclosing the hatched areas are drawn symmetrically, but there is no information that this should be so necessarily. They are intended to show that diverse proteins will exhibit various membrane affinities, from high to trivial, depending upon circumstances, and that the definition of a membrane protein is at best an operational one. The diagram shows not all core proteins to have a higher affinity than other possible membrane-associated proteins (ligands). This may or may not be so, but should be tested. The distribution curve is intended to include labile proteins (e.g., hormones), with high membrane affinity, but which are, for specific reasons, only transitorily membrane associated.

We want to stress also that the core composition and the extra- and intracellular distributions may vary during development, between cells, between different cell membranes, between different regions of the same membrane and with physiologic state.

(a) One must distinguish between membrane and organelle. In mitochondria, for example, most proteins are not membrane associated (27).

(b) Many membrane isolates are contaminated by adsorbed proteins, e.g., nucleohistones, or contain soluble proteins trapped when membrane vesicles form during cell disruption. Avoidance of such artifacts has been discussed in (28) and (29). Ionic adsorption can be avoided or reversed by choice of suitable pHs and ionic strengths, and trapped components may be released osmotically, by the use of sonication or by surfactants.

(c) Most cells contain lysomal proteases, and some membranes appear to bear endogenous peptidases. Both could introduce artifact during the preparation of a membrane or its proteins.

(d) Certain proteins that exhibit high membrane affinity in intact cells lose this during membrane-isolation procedures and elute from the membranes during some isolation steps. The acetylcholinesterase of erythrocytes represents a good example of this situation: Almost certainly located on the exterior of intact cells, it cannot be eluted therefrom at an ionic strength >0.5, but these conditions easily extract the protein from erythrocyte ghosts. Also, an electrophoretically defined peptide can be eluted at physiological salt concentrations from erthrocyte *ghosts* but not intact cells.

(e) The ambiguities due to inhomogeneities in membrane topology, discussed in Chapter 1, clearly impinge upon membrane protein analyses.

(f) Similarly, membrane variations due to physiologic state, cell cycle, age, etc., will assuredly reflect in their protein composition.

Strategies Employed in Studies of Membrane Proteins

The organization of proteins in membranes, their interaction with each other, with membrane lipids, and with regulatory agents, i.e., all topics dealing with the role of proteins in membrane architecture, require their examination in situ. The extensive efforts in this direction will be treated in Chapters 3 to 8.

However, we must also define the following:

(a) What protein species populate a given membrane domain?
(b) How are these disposed in the plane of the membrane and parallel to it?
(c) Are they oligomeric or monomeric?
(d) What is their amino acid composition?
(e) What is their amino acid sequence?
(f) What is their carbohydrate composition?
(g) How are they synthesized, deposited in the membrane, and degraded?
(h) What distinguishes the proteins of different domains of one membrane or of different membranes?

(i) What, if any, unique primary structural principles characterize membrane proteins?

These and related questions require primarily chemical approaches and will be treated here. We shall first address techniques utilizing intact cells or "native" membrane isolates before turning to various methods and stratagems designed to bring membrane proteins into solution.

Disaggregation, Solubilization, Monomerization

In general, the lipids and proteins of membranes can be separated under mild conditions, e.g., by use of organic solvents and detergents. They must, therefore, associate through weak interactions, i.e., coulombic interactions, hydrogen bonding, London–Van der Waals forces, and "hydrophobic bonding," rather than by covalent linkages.

We reason, as before (30), that the formation and dissociation of biomembranes involves a complex set of interacting equilibria, which can be heuristically represented as follows:

$$M \underset{1}{\rightleftharpoons} m \underset{2}{\rightleftharpoons} \sum_{i=1}^{n} L_{x(i)} P_{y(i)}^{(i)} \underset{3}{\rightleftharpoons} \left[\sum_{i=1}^{n} x_{(i)} L\right] + \left[\sum_{i=1}^{n} y_{(i)} P^{(i)}\right]$$

$$\overset{4}{\rightleftharpoons} \sum_{i=1}^{r} L_{A(i)} \qquad \overset{5}{\rightleftharpoons} \sum_{i=1}^{n} y_{(i)} P_{sol}^{(i)} \qquad \overset{6}{\rightleftharpoons} \sum_{k=1}^{t} P_{w(k)}^{(k)}$$

Here M represents the intact membrane and m small fragments thereof;

$$\sum_{i=1}^{n} L_{x(i)} P_{y(i)}^{(i)}$$

represents a summation over the n lipoprotein subunits in the small fragments; $x_{(i)}$ and $y_{(i)}$ represent the number of molecules of lipid and protein, respectively, in the ith subunit. L does not refer to one particular lipid, but to all the classes of lipid in the membrane, some of which (e.g., cholesterol) are loosely held, while others (e.g., most phosphatides) are strongly bound. Also the $y_{(i)}$ protein molecules within a subunit may be different molecules.

$$\sum_{i=1}^{r} L_{A(i)}$$

represents r different aggregated states of lipid with $A_{(i)}$ molecules per aggregate (micelle). Similarly,

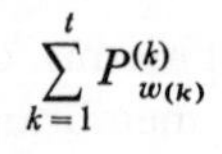

$$\sum_{k=1}^{t} P_{w(k)}^{(k)}$$

represents t aggregates of protein containing $w_{(k)}$ molecules of protein per aggregate, and

$$\sum_{i=1}^{n} y_{(i)} P_{(sol)}^{(i)}$$

represents the n protein species in states permitting solubility in water. We suspect that such states for true membrane proteins would involve different quaternary, tertiary, and even secondary structures than apply to their state *in situ*. Thus, first principles indicate that membrane proteins in their peculiar, partly apolar, environment assume a structure providing a minimal free energy state. We cannot visualize how the same structure can provide a minimal free energy state in water. This means that the isolation of membrane proteins in their native state faces formidable obstacles. However, one should succeed in purifying diverse membrane proteins in a soluble state, then devising means to return them to their native architecture. This hope rests on extensive experience over the past decade, indicating that structural changes in proteins are generally reversible.*†

Step 1 represents the disruption of large membrane sheets, e.g., intact plasma membranes or endoplasmic reticulum, into small fragments, such as occurs with liquid shear or electrostatic disaggregation. The dissociation of membrane fragments into their component lipoproteins (Step 2) does not proceed readily except in the unusual case of *halobacterium halobium*, where it occurs as the ionic strength is dropped below 3 in media lacking divalent cations. The action of various surface-active agents on membranes does not produce native lipoprotein subunits, but a variety of complexes into which the added detergent is incorporated.

* The role of micellar state and "hydrophobic free energy" clearly relates to the association of proteins with amphiphiles, such as occurs in membranes. This matter has been addressed by Tanford and associates (C. Tanford, *J. Mol. Biol.* **67:** 59, 1972; R. Smith and C. Tanford, *Proc. Natl. Acad. Sci.* **70:** 289, 1973). They find that under ideal conditions the energy of transfer between long chain, n-alkyl carboxylic acids between heptane and water varies linearly with alkyl chain length to C_{21}. However, in the binding of such molecules to proteins this pattern extends to short alkyl chains only, pointing up the limited regions in proteins free for apolar associations. Moreover, some proteins appear to bind amphiphiles by both polar and apolar associations. The authors suggest that both of the latter reactions participate in membrane structure and that all proteins endowed with apolar binding sites compete for amphiphiles, such as phosphatides, with supramolecular aggregates, such as artificial and biological membranes.

† In evaluating possible mechanisms of lipid-protein interactions, one should consider the importance of leucine residues in determining protein structure (P. Y. Chou and G. D. Fasman, *J. Mol. Biol.* **74:** 263, 1973). This analysis of fifteen proteins with known amino acid sequences and structures determined by X-ray crystallography suggests that leucine comprises the strongest structure-forming residue. Its common occurrence in the inner helical cores of proteins and its experimentally documented helix-forming tendency suggest that leucine may play a major role in initiating and organizing helical protein segments.

Step 3—subunit dissociation into lipid and protein—is critically important to us here, but cannot readily be separated from the preceding reaction. We suspect that it involves unstable intermediates indicated by [ ]. This is to be expected from the known behavior of membrane lipids and apolar proteins in aqueous media. Thus, phosphatides with long acyl side chains form micelles in water when present at concentrations of more than 10^{-5} M. Therefore, one would expect that any released lipid would tend to aggregate into micelles (Step 4). Also, if the structure of the proteins in the membrane is such as to expose large hydrophobic regions, e.g., murein-lipoprotein (31), they are not likely to exist as stable monomers in aqueous media. However, as shown in the equation, one must consider two possibilities: (a) The proteins aggregate through apolar associations; (b) they rearrange into structures that are thermodynamically stable in aqueous media, i.e., with hydrophobic residues concentrated in the interior and polar groups at the surface of the molecules.

The equilibrium of Step 3 obviously lies far to the left. It has not been examined in any detail, but has been shown to exist in erythrocytes, which readily exchange cholesterol and certain phosphatides with plasma lipoproteins. Many organic solvents and also ionic detergents clearly displace the equilibrium in Step 3 to the right. Ionic detergents yield lipid-detergent and protein-detergent complexes, which are usually separable. Organic solvents generally bring only the lipids (and proteolipids) into solution. However, there are some exceptions. For example, most membranes are soluble in 2-chloroethanol, permitting separation of their lipid and protein components by dialysis or molecular sieving. Also, under certain conditions, treatment of erythrocyte ghosts with *n*-butanol extracts most of the lipid into the organic phase, leaving most of the soluble protein in aqueous phase. We shall later return to the specific applications of detergents and organic solvents in the fractionation of membrane proteins.

In our equation, we show all steps on the path to true molecule solution to be reversible. But in as complex a system as a membrane, nonspecific reaggregation ("scrambling") may occur, e.g., by removal of a solubilizing agent, unfavorable ionic conditions (ionic strength, polyvalent cations, pH$\sim$ apparent isoelectric point), excessive concentration (e.g., during freezing), oxidation of SH groups to interchain S—S and "cross beta" structures.

What is clear from our considerations here is that membranes usually do not pass from their supramolecular state into fundamental molecular dispersions of their components. The term *solubilization* so commonly used is thus ambiguous and usually defined only operationally. Many investigators consider material that does not sediment at 10^6 g-min as soluble, although "microsomal membrane" vesicles sediment incompletely under these conditions. Few investigators set their criterion higher than "nonsedimentable" at 10^7 g-min, but this is not adequate to fully sediment 70S ribosomes, which are certainly complex, supramacromolecular aggregates.

In addition, few investigators sufficiently consider the densities and viscosities of the centrifugation media in such experimentation. Thus, 6 M guanidine-HCl has a density of about 1.15 g/cm³ and a relative viscosity of 1.61, while the values for 8 M urea are approximately 1.16 g/cm³ and 1.66, respectively. Since the sedimentation coefficient s_p of a particle is given by

$$s_p = 2r^2 (\rho_p - \rho_s)\ 9\eta \tag{2.2}$$

where ρ_p and ρ_s are the particle and solvent densities and η the viscosity, one can show that a particle of $\rho_p = 1.36$ will sediment at $\frac{1}{3}$ the rate in 6 M guanidine-HCl, as in water, in a given rotor.

Ionic Manipulations

Membrane biologists have long been imbued with the concept that ionic interactions play a major role in lipid protein interactions. Indeed, polar interactions were the only ones invoked in the original Danielli-Davson model (32). Wallach and Zahler (33) pointed out that because of the high *amidation* of membrane glutamic and aspartic acids, salt bridges—via Ca^{++} and Mg^{++}—to negatively charged phosphatides cannot occur very frequently. Also, we now know that apolar associations play at least as great a role.

Several authors attempt to explain the evidence for apolar lipid protein interactions by suggesting that apolar amino acid residues of proteins located on the faces of a hypothesized lipid bilayer penetrate into the hydrophobic core of the latter. However, Haydon and Taylor (34) show that such an arrangement is spatially and energetically impossible and that "penetration" would at best extend only to the rather polar glycerol-ester region. This point is illustrated in Fig. 2.8.

One must, however, appreciate that apolar interactions are driven entropically, that is, by the fact that the presence of an apolar residue in water tends to structure the solvent. To minimize this effect, such residues tend to be excluded from water. Such "entropically driven," hydrophobic effects naturally depend on the state of the solvent. Thus, water exhibits different behavior at varying pHs and $\Gamma/2$s and is much less structured the higher the temperature. Moreover, D_2O behaves other than H_2O. Divalent ions would be expected to exert different effects from monovalent species and inorganic ions from organic ones.

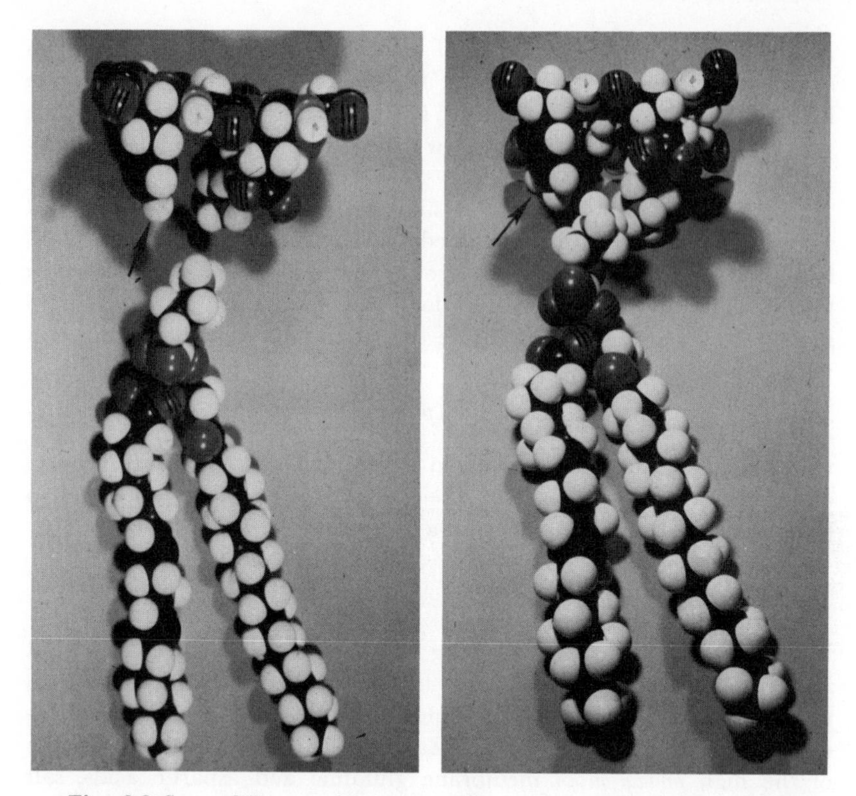

Fig. 2.8 Space-filling models showing the steric improbability of hydrophobic interactions between the apolar groups of a phosphatide bilayer and apolar residues of hypothetical, surface-located polypeptide.

Both panels show a polypeptide segment in the β-conformation, extended parallel to the surface of a hypothetical phosphatide bilayer. The polypeptide has a large *apolar* residue, Phe, and an Asp residue oriented toward the bilayer plane. A molecule of phosphatidyl-ethanolamine lies below. (a) On the *left* the two moieties are separated; (b) on the right a salt bridge has formed between the lipid (NH_3^+) and the Asp (COO^-). Note that the *Phe residue* (arrow) cannot extend into the apolar lipid domain. Other possible head group configurations would not permit location of the Phe deeper than in the glycerol region.

We shall separately discuss so-called chaotropic salts and ions. Another point of importance is the fact that apolar surfaces tend to adsorb many inorganic anions more than their companion cations (e.g., Cl^-), tending to give such domains a net negative charge in a neutral salt solution (35). Thus, a triglyceride droplet in NaCl solution will electrophorese toward the anode.

The possible ionic and other polar interactions between membrane proteins and lipids, as well as between membrane proteins themselves, thus comprise a complex and bewildering field. Membrane biologists accordingly deal with the problem rather empirically.

pH manipulations

Erythrocyte membrane proteins resist solubilization maximally near pH 4, but exhibit greater solubility above and below this region. This behavior is common to many soluble proteins, but it is interesting in erythrocyte membranes, whose external surface does not exhibit an isoelectric point above pH 2. As we shall note further, concentrated acetic and formic acids serve as membrane solvents, but dilute acetic acid appears to effect release of certain proteins from erythrocyte ghosts in a rather selective fashion (36).

Several investigators employ alkaline pH to release certain membrane proteins. Thus, Neville (37) treats rat liver bile fronts (q.v., Chap. 1) with 0.05 M K_2CO_3 at 25°C for 30 min to release about 70% of the membrane protein in what is most likely an aggregated state, but not sedimentable under the mild centrifugation conditions of 6×10^5 g-min. However, one of the eluted components ($\sim$10% of the protein) appears truly soluble. The protein appears limited to liver bile fronts and exhibits an unusual optical activity, which suggests that it extracts as a highly helical 2–3 chain rod (38, 39). Neville finds that this protein can also be eluted by dialysis against 1 mM ethylene-diamine tetraacetate (EDTA) at neutral pH.

Low ionic strength

Mazia and Ruby (40) introduced a method of extreme deionization for the solubilization of bovine erythrocyte membranes. This involved prolonged dialysis against water brought to pH 9 with ammonia and containing a "mixed bed" deionizing resin. They claim considerable success, but their data suggest that they have discarded much of their material. Moreover, they include 0.1% Triton X-100 in their extraction, making the procedure even more difficult to evaluate. Finally, Hamaguchi and Cleve (41), in testing this procedure with human erythrocyte ghosts, find that the protein release requires the detergent and that it is at best incomplete and selective. Harris (42, 43) uses prolonged dialysis against distilled water to extract a protein, which exhibits a striking association pattern by negative staining electron microscopy; i.e., rings of 10 similar subunits often stacked to form hollow cylinders.

One consequence of exposing erythrocyte ghosts to low $\Gamma/2$ (and low divalent cations) is the induction of endocytosis with formation of inside-out vesicles. In this process, several high-molecular-weight ghost components go into solution.

Removal of inorganic ions

Reynolds and Trayer (44) and Reynolds (45) argue that the divalent and also some monovalent cations in large part determine the association of membrane components. Ternary complexes between proteins, cations, and lipid head groups are invoked. They accordingly treat ghosts with 5–100 mM ethylene-diamine tetraacetate (EDTA) and/or 100 mM tetramethyl ammonium bromide for 4–5 days and finds this procedure makes about 90% of the protein and much of the phosphatide nonsedimentable at 3×10^6 g-min. However, the released material is by no means *soluble.*

Removal of divalent cations

As first shown by Marchesi and associates (46), reduced ionic strength, slightly alkaline pH and sequestration of divalent cation, permits elution of a rather high-molecular-weight protein, "spectrin," which may account for as much as 25% of human-ghost protein (47). As noted, removal of divalent cations is also essential in the Reynolds protocol (45) and can achieve elution of Neville's liver-specific protein (38, 39).

In a number of instances, certain ATPases can be brought into solution by divalent cation removal. Two striking examples are (a) inner mitochondrial (F_1) ATPase (48), (b) a major protein of thylakoid membranes (49). Both proteins contain the excess of acidic over basic amino acid residues, necessary for ion triplet formation. However, one cannot tell from available data to what extent the aspartate and glutamate residues are amidated.

High salt extractions

We shall separately discuss the solubilizing action of so-called chaotropic ions and here limit ourselves to efforts concerned with solubilizing membrane components by interfering with ionic attractions by raising ionic strength. There are models for this, e.g., the elution of mitochondrial cytochrome c (50), but we suspect that these are exceptions rather than the rule.

This approach was used by Mitchell and Hanahan (51), who found that overnight extraction of erythrocyte membranes in cold 0.6–1.2 M NaCl released considerable membrane material, including acetylcholinesterase from the ghosts. However, they did not employ very stringent solubility criteria (solubility defined as nonsedimentable at 1.65×10^5 g-min!).

A more defined example has been demonstrated by Fairbanks et al. (47), who could selectively elute an electrophoretically distinct single protein/peptide by extracting human erythrocyte *ghosts* overnight with cold saline at *physiologic ionic strength.*

Chaotropic Agents

The major action of the following substances resides in their ability to "destructure" water, thereby reducing apolar attractions in aqueous media. One should note that such agents, or *any agent that effectively disrupts a membrane*, will make the lipid components more susceptible to autooxidation.

Guanidine-HCl and urea

These agents tend to unfold soluble proteins presumably by two mechanisms (52, 53): (a) interference with entropically driven, structural interactions and (b) hydrogen-bond rupture. Both tend to generate random structures, but neither agent alone can solubilize and monomerize all membrane proteins (54).

Guanidine-HCl

6 M guanidine-HCl generally denatures globular proteins into random polypeptide chains, but some oligomeric proteins, e.g., horse hemoglobin, denature at a rather slow rate. Gwynne and Tanford (55) use this reagent to solubilize the proteins of human erythrocyte ghosts and find that it makes about 60% of the ghost protein nonsedimentable (according to their criteria). This fits observations on the relative effects of guanidine-HCl and urea vs. Li-diiodosalicylate (56), showing that *glycoproteins* are poorly extracted by the former reagents. However, guanidine thiocyanate appears to be a more general solubilizing agent.

Triplett et al. (54) also find that guanidine-HCl does not effectively disaggregate membrane proteins. They put ^{3}H-acetylated ghosts into 2% cetyl trimethyl ammonium bromide (CTAB) (see below), mix, bring the material to ~ 6 M in guanidine-HCl (pH 8.5, 1% mercaptoethanol), and equilibrate overnight at room temperature under N_2, before chromatography on 4% Biogel A-15 M and P-100 in 5.5 M guanidine-HCl, pH 6.5, 0.01% mercaptoethanol. No separation of lipid from protein is achieved on either agarose or polyacrylamide, which indicates that this system yields lipid-protein aggregates of high particle weight. Removal of CTAB and guanidine-HCl by dialysis allows separation of protein and lipid components on P-100 polyacrylamide *using 1% sodium dodecyl sulfate as eluant.* These results indicate that *combination* of detergents with reagents such as guanidine-HCl could actually vitiate the primary aims of this type of experiment, i.e., separation of protein from lipid and fractionation of the protein.

Maddy and Kelly (36) also explore the use of guanidine-HCl in membrane solubilization and find that the material that is not solubilized by guanidine-HCl is enriched in lipid and glycoprotein. This is concordant with the work of Marchesi and Andrews (56). Maddy and Kelly also observe that 6 M guanidine-HCl does not dissociate protein aggregates formed during other procedures for membrane purification. However, one should note that membrane proteins dissolved in 2-chloroethanol can be transferred to 6 M guanidine-HCl without solubility loss (20).*

Inorganic chaotropic ions

Inorganic chaotropic ions behave as urea and guanidine-HCl in that they favor the transfer of apolar groups to water. Some had been noted by Zahler and Wallach (20) to maintain membrane proteins dissolved by 2-chloroethanol in solution after removal of the solvent. However, Hatefi and Hanstein (57) have explored the utility of these ions to membrane solubilization in a more systematic fashion.

In their studies, they define a protein as solubilized when non-sedimentable at 9–12 × 10^6 g-min. The effectiveness of diverse anions in raising the solubility of apolar groups in water increases in the order

$$Cl^- \; Br^- \; I^- \; NO_3^- \; ClO_4^- \; SCN^-$$

which is also the sequence of solubilization effectiveness for membranes. However, not more than 28% of erythrocyte ghost protein can be extracted under optimal conditions. Interestingly, the more chaotropic the ion the more it labilizes unsaturated membrane lipids to *peroxidation.* This indicates that these compounds not only solubilize some membrane proteins but also perturb (loosen) membrane lipid organization.

Organic Solvents

Many organic solvent systems denature and aggregate membrane proteins, allowing extraction of the membrane lipids. In our comments on membrane lipids, we have already pointed out that this is a complex matter, some lipids being tightly associated with membrane proteins, others not. In general, one can relate the role of organic solvents to the lowering of medium dielectric constant, allowing a concomitant increase in coulombic attraction between protein entities (52, 53), but the matter

* Guanidine-HCl (6 M) extracts essentially the same set of electrophoretically-defined erythrocyte membrane polypeptides as do 0.1 N NaOH, 5–20 mM lithium diiodo salicylate, 1mM parachloromercuribenzoate (or parachloromercuribenzene sulfonate) and acid anhydrides (T. L. Steck and J. Yu, *J. Supramolec. Structure*, **1:** 220, 1973).

is indubitably more complicated, because some organic solvents (e.g., 2-chloroethanol) are helicogenic, while others (e.g., trifluoracetic), tend to generate extended and/or disordered structures.

As a general rule, solvents that mix poorly with water tend to extract most lipids into the organic phase, leaving the rest and proteins in the aqueous compartment. However, proteins strongly associated with lipids may partition from water into a predominantly chloroform phase (58). Ionic complexes of proteins and charged phosphatides typically behave in this fashion (50).

Let us now consider some specific organic solvent systems used for the fractionation of membrane proteins, keeping in mind that some promising approaches, e.g., the use of controlled-pore glasses, have at this writing not been applied to the biomembrane field.

Alcohols

BUTANOL

When deionized erythrocyte membrane suspensions are rigorously extracted at 0–4°C with *n*-butanol, most of the phospholipid and all of the cholesterol partition into the predominantly organic phase, most of the proteins remaining in the more aqueous phase, with some protein accumulating at the interface (59, 60). The method is extremely difficult to reproduce, depending critically on temperatures, mixing rates, etc. Even under optimal conditions, removal of butanol, or increase of ionic strength to less than minimal levels, precipitates the protein (60).

Triplett et al. (54) examine the solubilization characteristics of various alcohol-water systems more critically than has been the general practice. For this, they react 5 ml packed erythrocyte ghosts in isotonic phosphate, pH 7.4, with ^{3}H-acetic anhydride (2 mCi) for 30 min at room temperature, remove excess label by overnight dialysis against cold 0.02% sodium azide and collect the acetylated ghosts at 2.4×10^{6} g-min. They then solubilize the ghosts in butanol:acetic acid:water 1:1:1 (v/v/v) and chromatograph on Sephadex (LH-20) in the same solvent system. They achieve convincing separation of protein and lipid, but no significant fractionation of either.

CHLOROFORM-METHANOL

Hamaguchi and Cleve (61) show that the major glycoproteins in the erythrocyte ghosts of diverse species can be extracted in a facile fashion as follows:

To one volume of ghosts in 0.01 M tris-EDTA-HCl, pH 7.4, 9 volumes of 2.1 $CHCl_3$:CH_3OH are added, the mixture stirred 30 min at room temperature and then allowed to separate into two phases. The upper, aqueous layer is concentrated 10x in vacuo and cleared by centrifugation at 6×10^6 g-min. The major glycoproteins remain in the supernatant, but their state of dispersion there is not given.

Hamaguchi and Cleve (62) have extended their method recently, precipitating the aqueous phase components by addition of 9 volumes of ethanol/volume upper phase followed by gel-permeation chromatography on Sephadex (G-100) in 1% sodium dodecyl sulfate.

ETHANOL

Hot 75% ethanol can be utilized to extract the major glycoproteins from the erythrocytes of diverse species (63, 64). The method is drastic—erythrocyte ghosts are extracted first with hot acetone and 100% hot alcohol. The residue is then extracted with hot 75% ethanol and the soluble material concentrated in vacuo, dialyzed against dH_2O and lyophilized, yielding 2–4% of the original ghost mass and all the MN antigenic activity (in human material). In aqueous solutions, the glycoproteins exist as aggregates with particle weights >200,000.

CHLOROETHANOL

In our comments on membrane lipids (p. 66), we have discussed some of the properties of 2-chloroethanol, including its effectiveness in dissolving both membrane lipids and membrane proteins, as well as its unusual peptide-structuring tendency. More than any other organic solvent, 2-chloroethanol promotes the formation of α-helix. This is true also for membrane proteins (65, 66) and bears special significance for membrane protein fractionations because (a) disordering agents such as trifluoro-acetic acid and anonic detergents tend to produce *entanglement* of peptide chains, especially when these agents are removed (e.g., for membrane reconstitution); (b) agents promoting β-structuring, e.g., many acid solvents, also tend to generate cross-linking between peptide chains, which hinders fractionation of monomers (we note here that sodium dodecyl sulfate generates β-structured aggregates with basic polypeptides) (67); (c) promotion of α-helix, as by 2-chloroethanol, implies formation of more (or maximal) *intra*chain hydrogen bonds, i.e., *minimal scrambling*. We consider this a most desirable goal for the fractionation of polypeptide mixtures, such as exist in membranes, and the validity of our reasoning is attested to by the fact that many enzymes can be put through a chloroethanol cycle without loss of activity.

It was originally thought that an acid pH is necessary for the solvent action of 2-chloroethanol, but recent studies (68, 69) show that 80%

2-chloroethanol, 20% 50 mM phosphate, pH 7.4 has approximately ½ the solubilization power for erythrocyte ghosts as 9:1 2-chloroethanol: water, pH 2. Moreover, Zahler and associates (70) have demonstrated that erythrocyte ghost proteins solubilized in 2-chloroethanol retain their species-specific phospholipid-binding properties.

Zahler and Wallach (20) developed a facile technique for the separation of membrane proteins and membrane lipids in 9:1 2-chloroethanol, using organophilic Sephadex (LH-20). The proteins remained nonsedimentable after dialysis into concentrated urea, guanidine-HCl, and KI. They could not achieve fractionation of the excluded protein moiety for want of suitable molecular sieves. However, the controlled-pore glasses discussed earlier (p. 67) may resolve this difficulty. Mono- and trifluoroethanol, as well as hexafluoroacetone, do not substitute for 2-chloroethanol, nor does N, N-dimethyl-formamide (71).

2-chloroethanol received by various manufacturers must generally be redistilled and further purified by filtration through active charcoal. It should then be stored at $\sim -20°C$ in the dark. For membrane use, the solvent should be transparent to 230 nm. The solvent is hepatotoxic and must be handled with care. Some evidence indicates that 2-chloroethanol may form carboxylic acid esters at low pH.

N-PENTANOL

Zwaal and van Deenen (72) show that *n*-pentanol can solubilize erythrocyte ghosts, but that unlike *n*-butanol, this solvent causes 80–85% of the membrane proteins and more than 80% of the membrane lipids (phosphatides and cholesterol) to partition into the aqueous phase as lipid-protein aggregates. These components continue in association even upon differential centrifugation in sucrose density gradients. Using various butanol-pentanol mixtures, the authors find that the recovery of lipids in the aqueous phase depends markedly upon the proportion of butanol.

PHENOL

Phenol has long been employed to solubilize glycoproteins. For example, the major glycoprotein of human erythrocyte ghosts can be effectively extracted into the aqueous layer after treatment with 1:1 phenol:water at 65°C (73). Acidified phenol water mixtures, such as phenol/formic acid/water (74) and phenol/acetic acid/water (75), have also been explored for the purpose of membrane solubilization, but offer no distinct promise.

In their recent careful studies, Triplett et al. (54) also study the solubilization potency of a phenol water mixture. They acetylate their erythrocyte ghosts as described above (p. 81), solubilize these in phenol: acetic acid-water, 1:1:1 (w/v/v) and chromatograph on biogel P-100 with the same solvent, using cytochrome c and ^{14}C cholesterol as column markers and measuring elution of protein and phospholipid. The system yields essentially full separation of protein, which elutes in the void volume, and lipid, which is retarded in low-molecular-weight form. Presumably because of aggregation, no protein fractionation is achieved.

AQUEOUS PYRIDINE

Blumenfeld and associates (76) utilize 33% aqueous pyridine to solubilize erythrocyte membrane proteins. After treating ghosts with this reagent, excess pyridine is removed by dialysis against d H_2O for 16 hr at 4°C. Longer dialysis leads to extensive protein precipitation, as does the presence of diverse salts or freezing of the membranes prior to extraction. After dialysis, the extract is centrifuged at 4×10^6 g-min, leaving 40% of the ghost protein in the supernatant. Most of this is still aggregated as revealed by gel permeation chromatography in 33% aqueous pyridine. Importantly, however, the major membrane glycoproteins appear to be extracted in soluble form by this procedure.

Organic acids

ACETIC AND FORMIC ACIDS

Concentrated acetic and formic acids readily dissolve many biomembranes, but produce drastic conformational and, in the case of formic acid, chemical changes. Both compounds, and low pH in general (33), induce β-structures in membrane peptides. Concentrated formic acid, in addition, leads to the formation of orthoformates and other adducts, which are readily apparent upon infrared spectroscopy (77) (Fig. 2.9). In view of the extensive parallel-β-structuring, one must suspect some cross-linking between peptide chains; solubilization may thus not indicate monomerization.

TRIFLUOROACETIC ACID

Trifluoroacetic acid (TFA) constitutes an excellent protein solvent, which tends to unfold peptide chains. Drey et al. (78) report on its use in membrane protein fractionation by gel permeation on polyacrylamide (PA). Although fragile in 98% TFA, polyacrylamide can be used for this purpose, since it is stable between pH 1–10. The authors employ Biogel P-150 (operating range 15,000–150,000 daltons) and Biogel P-300

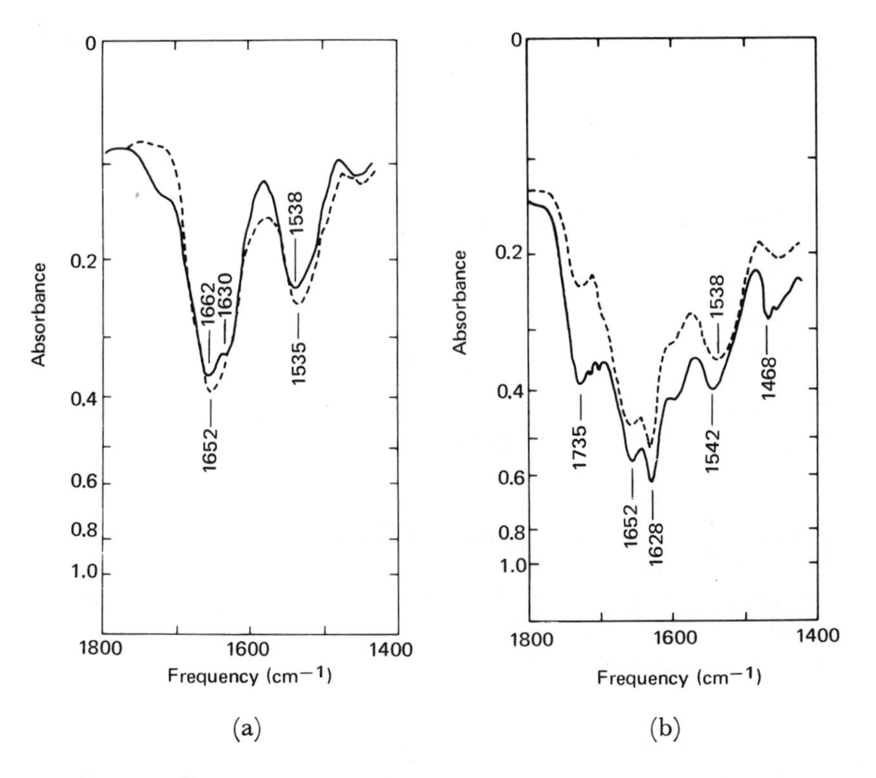

Fig. 2.9 Effect of acid pH and formic acid on the amide I and II spectra of plasma membranes. (a) Role of acid pH: ----------- untreated; ——— exposed to pH 3. (b) Membranes dissolved in 90% formic acid: ——— untreated; ------------ lipid extracted. The spectra in (a) were obtained on films dried from aqueous suspension; those in (b) were obtained from films dried from 90% formic acid. All films were cast unto AgCl. *Note*: the absorption maximum at 1628 cm^{-1} in the case of formic acid corresponds to the shoulder in the pH 3 spectrum. Both indicate β-structuring (Chap. 4).

(operating range 70,000–300,000 daltons), calibrating their columns with bovine serum albumin, cytochrome c, and dinitrophenylarginine. Their data indicate that serum albumin (69,000 daltons) is excluded by Biogel P-300 and cytochrome c (124,000 daltons) by Biogel P-150. The data imply that (a) the effective pore size of the gel becomes reduced in TFA; (b) the standards unfold and develop larger effective diameters than in the native state, or else they aggregate. Indeed, no information as to the degree of dispersion of protein in TFA is given. The authors find that nearly all membrane protein is excluded by both gels, whereas lipid is not. They use this finding as an argument against the presence of low-molecular-weight peptides in membranes, but the system does not appear adequate for this purpose. A more successful application of TFA as solvent would be in conjunction with the controlled-pore glasses described on p. 81.

Surfactants (Detergents)

General

This class of amphiphilic compounds at present includes the most favored membrane solubilizing agents. The reasons for this are empirical. Less is known about the role of detergents than about the actions of chaotropic agents or organic solvents; they are difficult to separate from the protein and/or lipid complexes they form so readily, commonly denature proteins, and are not all easily purified. The last objection is partly offset by the fact that detergents exhibit high affinities for membrane lipids and most membrane proteins, and act effectively at levels 10–100 fold lower than chaotropic agents and organic solvents. Diverse detergents act differently; some (e.g., lysolecithin) generate small, lipid-protein complexes; others (e.g., sodium dodecyl sulfate) usually monomerize oligomeric proteins and separate proteins from lipids.

Nonionic and zwitterionic detergents

Many nonionic detergents are used in membrane studies. The *Triton* series, Triton X-100, enjoys particular favor. These compounds have the following general structure:

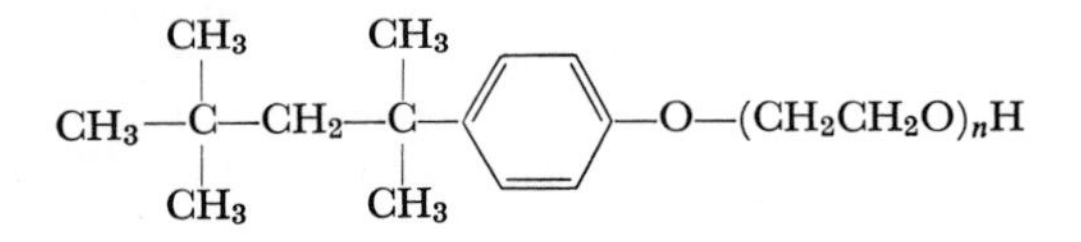

In Triton X-100, $n = 9$–10. Triton X-114 and X-45, also commonly employed, have n values of 7–8 and 5, respectively.

The *Tweens* are also used widely. They comprise polyoxyethylene ethers of mixed, partial fatty acid esters of sorbitol anhydrides:

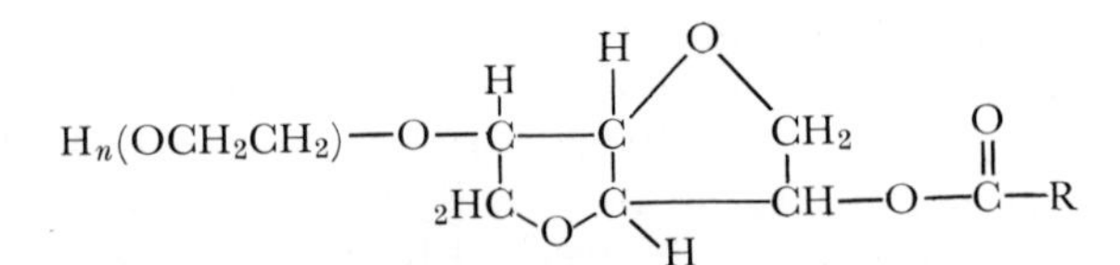

Tween series

In *Tween*-40, $n = 20$ and R = palmitate. In *Tween*-80, $n = 20$, but R = oleate.

Tergitol TMN, another useful nonionic detergent, is polyoxyethylenated trimethyl nonyl alcohol, containing 6-oxyethylene residues. The *Brij* series comprises polyoxyethylenated cetyl, lauryl, oleyl, or stearyl alcohols. *Lubrols* are also mixtures of polyoxyethylenated long-chain alcohols.

The only zwitterionic detergent that has found use in membrane studies is natural or synthetic *lyso*phosphatidyl choline.

Available evidence indicates that this class of detergent disperses membranes into large lipid-protein aggregates, which do not, however, pellet at 10^7 g-min. Interestingly, some membranes resist the action of nonionic detergents. Thus, the in vivo accumulation of Triton WR-1339

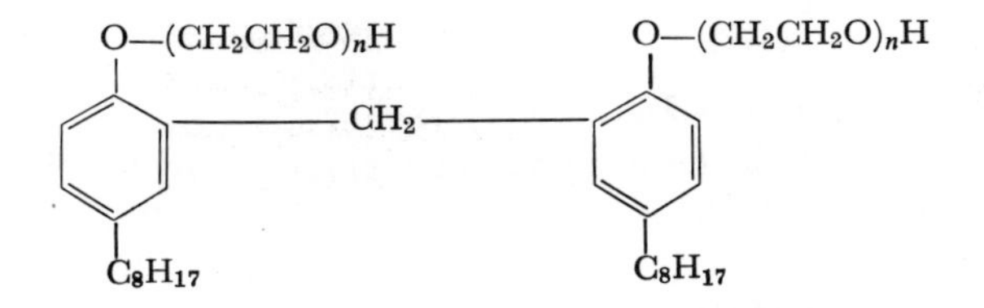

in lysosomes forms the basis for the purification of these organelles by "density perturbation" (see Chap. 1).

The solubilization power of a detergent is not simply related to its surface activity, and some highly surface-active compounds are quite ineffective in bringing membrane components into solution.

The increasing attraction of nonionic detergents derives from the following:

(a) They are less apt to alter protein structure than other solubilizing agents, permitting recovery of enzymatic immunologic and other activities.

(b) Their lack of charge permits purification of proteins according to their ionic character, either by electrophoresis or ion-exchange chromatography.

(c) They often stimulate the activities of certain membrane enzymes, either through a structural change in the enzyme or by removing permeation barriers to its substrate.

(d) They permit elution of some weakly bound proteins from membranes in a relatively selective fashion.*

* Treatment of erythrocyte ghosts with Triton X-100 and related surfactants preferentially solubilizes membrane glycoproteins at ionic strength near 0.04 (J. Yu, D. A. Fischman, and T. L. Steck, *J. Supramolec. Structure* **1:** 233, 1973). At lower ionic strength some non-glycosylated proteins are extracted also.

The disadvantages of this class of compounds includes
(a) their polydispersity and often impurity;
(b) their generally high UV absorbance;
(c) the resistance of most membrane proteins to their solubilization action.

Another nonionic detergent with highly selective properties should be included here. This is digitonin.

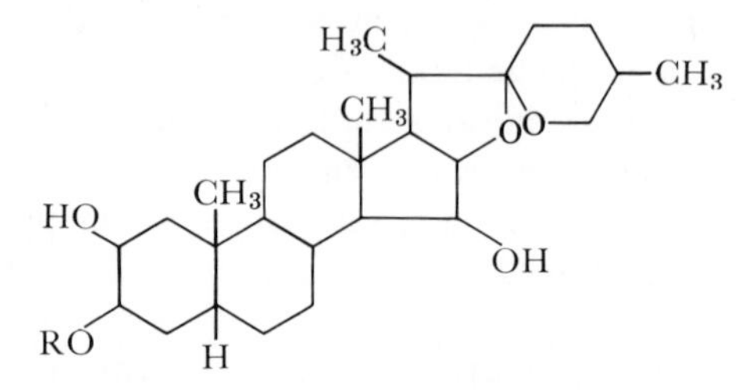

where R = 2-galactose + 2-glucose + 1-xylose.

This disrupts cholesterol-containing membranes by forming tight 1:1 molar complexes with cholesterol.

Anionic detergents

BILE SALTS

Salts of bile acids, e.g., sodium cholate and sodium deoxycholate (DOC), have long found use in the solubilization of diverse membranes, particularly those of mitochondria (48).

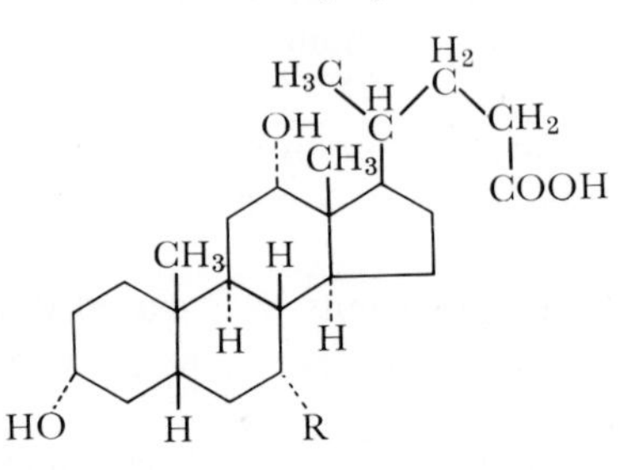

R = –OH in cholic acid and –H in deoxycholic acid.

Bile salts behave quite differently with various membranes; in some cases they may simply fragment membranes into smaller lipid-protein aggregates (79), but they can also effect the separation of membrane proteins from membrane lipids compiexed with detergent. Thus, Salton and Schmitt (80) show that DOC allows complete electrophoretic separation of radioactively marked phosphatides, cholesterol, and fatty acids from the proteins of *Micrococcus Leisodeikticus*.

The situation may be more complicated in animal cells as indicated by the careful study of Philipott (81). His data show that the solubilization of erythrocyte membrane proteins increases with DOC concentration, but is never complete, although proceeding twice as rapidly as phospholipid solubilization. (The solubility criterion was "not sedimentable" at 2.4×10^6 g-min.) In contrast, the solubilization of phospholipid depends upon the amount of DOC and becomes complete at the ratio of 1 mole phosphatide per mole of DOC.

Chromatography on Sepharose 6B indicates that the protein remains aggregated while the phospholipid migrates primarily as a complex, in which 10 molecules of phosphatide are associated with 130 molecules of DOC. *However*, an appreciable proportion of phosphatide appears with the protein in the exclusion volume. It is not clear whether this lipid is in the form of large aggregates or still protein associated.

An attractive feature of the bile salts is the fact that many enzyme activities can be recovered after their removal (48). A good example is erythrocyte membrane $Na^+ - K^+$-activated ATPase (82).

LITHIUM DIIODOSALICYLATE (LIS)

Robinson and Jencks (83) have observed that diiodosalicylic acid, lithium salt

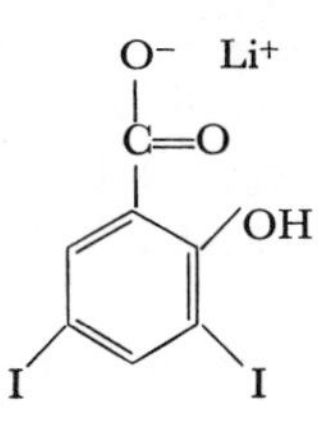

effectively disaggregates synthetic polypeptides. Accordingly, Marchesi and associates (56, 84) have used this reagent in their efforts to purify membrane proteins. They find (a) that LIS disrupts membranes at concentration of 10–25 mM and, importantly, (b) suggest that at concentrations of 0.1–0.3 M, LIS offers unique utility in the solubilization of membrane *glycoproteins*. However, the published procedures suggest that LIS extracts membrane proteins nonspecifically, perhaps like sodium dodecyl sulfate, but allows their purification in a more facile manner than other procedures.*

* Low concentrations of lithium diiodosalicylate (5–20 mM) elute the same types of erythrocyte ghost polypeptides as 6 M guanidine HCl (T. L. Steck and J. Yu, *J. Supramolec. Structure* **1:** 220, 1973).

SODIUM DODECYL SULFATE (SDS)

More widely used than any other detergent, SDS efficiently dissolves membrane proteins and lipids and is generally thought to separate these two components as soluble SDS complexes.

SDS, as provided by various manufacturers, varies considerably in purity. This commonly manifests itself by annoying absorption in the aromatic region of the UV spectrum, or even more so by a tryptophan-like fluorescence. We have found repeated recrystallizations from ethanol methanol and water, as well as treatment with activated charcoal to be relatively ineffective in eliminating fluorescent impunities. However, fluorescence can be obviated by prolonged irradiation of SDS at 280–290 nm.

Few quantitative data exist as to the degree to which SDS-peptide and SDS-lipid complexes separate. Thus, Triplett et al. (54) find that 10–20% of membrane phosphalipid may be excluded, together with protein, upon gel-permeation chromatography in 1% SDS using P-100 columns. Our experience also indicates less than complete separation of phosphatide from peptide during electrophoresis in SDS-laden polyacrylamide gels (69).

In most cases, the interaction between SDS and peptides is principally apolar. However, this cannot be generalized to all situations. Thus, the complexes between SDS and poly-L-lysine indubitably involve a large electrostatic element (67). Because of its high affinity for proteins and phosphatides, SDS cannot be easily removed from these membrane components. This problem is compounded by the fact that at the concentrations usually employed, i.e., $>0.2\%$, SDS is principally in micellar rather than monomeric form, with particle weights of $\sim 10^3$–10^4. The efficacy of SDS removal is best determined using ^{35}S-labeled detergent.

The interaction of increasing amounts of SDS with biomembranes appears to proceed in steps, with diverse functional and structural consequences. At very low detergent concentrations (0.0005–0.001%), enzymes such as Na^+, K^+, ATPase are activated (85). At higher levels (e.g., 0.1%), these enzymes are fully inhibited, and optical activity studies suggest considerable peptide disordering (85). However in some membrane systems, 0.1% SDS stimulates adenylate cyclase (86).

It is generally assumed that nearly all proteins bind about 1.4 g SDS as monomer per gram protein at saturation (87), but we now recognize that there are many exceptions to this rule. Thus, Dunker and Rueckert (88) prepared three mixed disulfide derivatives of lysozyme: (a) β-carboxymethyl lysozyme, (b) β-hydroxymethyl lysozyme, and (c) β-aminomethyl lysozyme, to replace existing SH groups with anionic, neutral, and cationic substituents of equivalent size. They find the negative lysozyme derivative to bind less SDS than the neutral species, which, in turn, binds less detergent than the cationic derivative.

Concordantly, Segrest et al. (89) clearly show that soluble and membrane glycoproteins bind only about ½ as much SDS as serum albumin and ovalbumin.

Most critical, however, is the recent study of Nelson (90). Using equilibrium SDS concentrations of 1–2 g%, he made the following important observations.

(a) Various soluble proteins differ drastically in their SDS binding capacity in 0.1 M NaCl (Table 2.3). Some (e.g., pepsin) bind virtually no detergent, while others (e.g., chymotrypsinogen) bind more than 1.4 g/g protein.

Table 2.3 SDS Binding by Proteins

Protein	Binding at Ionic Strength of 0.1	Rate of Binding	Native Protein Mol Wt (no. of subunits)
	g SDS/g protein		
Papain	0.2		21,000
Pepsin	0.2		36,000
Glucose oxidase	0.2		150,000 (2)
Chymotrypsinogen	2.2	Fast	26,000
Carboxypeptidase	1.4	Fast	34,000
Pepsinogen	1.7	Fast	40,000
Ovalbumin	1.3	Fast	45,000
Fab (immunoglobulins)	1.1	Slow	52,000
Bovine serum albumin	1.4	Fast	69,000
Succinyl BSA no. 1	0.7	Fast	70,000
Succinyl BSA no. 2	1.1		70,000
Carbamyl BSA	1.1	Fast	70,000
Glyoxal modified BSA	1.0	Fast	70,000
Apotransferrin	1.3	Fast	77,000
Alcohol dehydrogenase	1.3	Slow	150,000 (4)
Aldolase	1.7	Fast	160,000 (4)
Catalase	1.4	Slow	232,000 (4)
Thyroglobulin	1.6	Fast	669,000 (2)

(b) SDS does not necessarily dissociate the subunits of oligomeric proteins, e.g., glucose oxidase (Table 2.3).

(c) When SDS does monomerize oligomeric proteins, this can proceed *very slowly*. Thus, after equilibration of 0.66 g catalase/100 ml with 2 g SDS/100 ml at ionic strength = 0.2, pH 8.1, 42% native tetramer remains after 24 hr at 25°C and even after 72 hr, 29% of the enzyme is still in tetrameric form! We shall later comment on the profound significance of these findings on the widespread use of electrophoresis in SDS-laden polyacrylamide for molecular-weight determinations. In this connection, Wallach and Weidekamm (91) point out that 1% SDS 0.04 M dithiothreitol does not fully reduce human hemoglobin to the monomeric state, nor extract all the noncovalently bound heme even from those monomers that are formed; i.e., this protein also does not monomerize and unfold in the manner of most proteins.

(d) SDS binding can vary by 20–50%, depending upon protein charge and may increase by 35–100% as ionic strength is raised from 0.1 to 0.4.

The recent work of Katzman (92) is also extremely pertinent. He finds that 1% SDS and 0.01% mercaptoethanol fully solubilizes dehydrated, lipid extracted, bovine brain residue (principally myelin) within 24 hr at room temperature (nonsedimentable at 2.4×10^7 g-min). However, chromatography of the solubilized material on sepharose 4B yielded three peaks each with identical amino acid composition and indistinguishable electropherograms (in SDS-laden polyacrylamide). Rechromatography of the column peaks again yielded three bands, also with the same electrophoretic characteristics. Katzman argues that, while SDS may *solubilize membrane proteins, it cannot be assumed to monomerize them.* Reynolds and Tanford (93) and Fish et al. (94) suggest that when a peptide becomes saturated with SDS, it folds into a helical rod doubled over in hairpin fashion near its middle, surrounded by an SDS shell. If such an arrangement occurred with all peptides, these should show a regular relationship between electrophoretic mobility in SDS and molecular weight. Indeed, such a relationship is generally assumed to be valid, despite increasing evidence to the contrary. We shall address this matter in our comments on electrophoresis (p. 110). Note that 16-, 18- and 20-carbon homologues of SDS show no particular advantages to SDS in membrane solubilization (95).

SODIUM N-LAURYL SARCOSINATE

This detergent

$$CH_3(CH_2)_{11}—\overset{\displaystyle CH_3}{\overset{|}{N}}—CH_2—COO^-Na^+$$

does not solubilize membranes in a drastic fashion, but allows recovery of certain specialized membrane entities. Specific applications include isolation of bacterial DNA-membrane complexes (96) and purification of rat liver "gap junctions" (97).

Cationic detergents

This class of compounds has not been widely applied to membrane studies, although their strong bactericidal action appears to derive from their membrane-disruptive action, and Salton (98) has demonstrated their high membrane affinity. However, Fairbanks and Avruch (99) report that 1% cetyltrimethyl ammonium bromide (CTAB) appears to solubilize human erythrocyte membrane proteins as effectively as SDS, and Heller (100) finds that 1.5% CTAB is superior to urea, guanidine-HCl, digitonin deoxycholate, and SDS in the extraction of rhodopsin from bovine retinal rod membranes.

Chemical Modification

Diverse oligomeric and also supramacromolecular structures can be solubilized and/or monomerized by chemical modification of some or many subunits. Classic examples of this approach are the separation of immunoglobulin chains by reduction of interchain —S—S— linkages and separation of the α- and β-chains of human hemoglobin after complexing their —SH groups with parachloromercuribenzoate. —S—S— reduction is unlikely to prove generally useful in membrane solubilizations, since most membranes lack disulfide linkages. However, —SH modification does solubilize some erythrocyte membrane proteins.*

However, succinylation of membrane proteins appears a strategy with some promise. Here a negative COO^- is substituted for a positive group under mild condition by reaction of succinic anhydride with the ε-amino groups of lysines, as well as other free amino groups of membrane peptides, phosphatidyl ethanolamine, and phosphatidylserine. All of these moieties then bear a negative charge at neutral pH. The resulting electrostatic repulsions appear sufficient per se to separate membrane components or alter protein architecture to a degree allowing solubilization (101, 102).†

* Parachloromercuribenzene sulfonate (pCMBS; 1–5 mM) effects a selective but partial release of the same erythrocyte ghost polypeptides as 6 M guanidine HCl, i.e., SDS-PAGE bands 1, 2, 2.1, 4.1, 4.2, 5, and 6 (J. R. Carter, Jr. *Biochemistry* **12:** 171, 1973; T. L. Steck and J. Yu, *J. Supramolec. Structure* **1:** 220, 1973). The major glycoproteins are not eluted, nor are SDS-PAGE bands 3, 5, and 7. Parachloromercuribenzoate (pCMB) acts similarly. Mersalyl, another organic mercurial, solubilizes about 66% of erythrocyte ghost protein (A. C. Cantrell, *Biochim. Biophys. Acta* **311:** 381, 1973), but apparently lacks some of the specificity of pCMB and pCMBS in that it elutes glycoproteins also.

† Isopeptide bond formation, an important element in the structure of some proteins, has now been observed in the case of some membrane proteins (P. J. Birkbichler, R. M. Dowben, S. Matacic, and A. G. Loewy, *Biochim. Biophys. Acta* **291:** 149, 1973). Data obtained by cyanoethylation indicate that a large proportion of the lysine residues in the proteins of L-cell plasma membranes and endoplasmic reticulum bear covalently linked

(*continued on overleaf*)

Moldow et al. (102) have studied this approach in greatest detail. They define "soluble" as "nonretained" by a 0.45 nm Millipore filter. This material could, however, be pelleted at 7.2×10^7 g-min and, according to negative staining electron microscopy, consists of small vesicles. Recoveries were near 90% for total ghost mass, 75–100% for protein, 85–100% for cholesterol, 70–95% for phospholipid, and 75–100% for carbohydrate. The solubilized membrane protein exhibited optical activity and infrared changes indicative of an increase in "unordered" peptide folding as well as appearance of some β-structuring. ATPase and Rh activity are lost, but ABO antigenicity is retained.

Succinylation alone appears of limited utility; it does not achieve extensive membrane disruption or separation of lipids from proteins and at the same time destroys important membrane functions. Nevertheless, the technique may find further utility in conjunction with other solubilization procedures.

Fractionation of Membrane Proteins

General

No aspect of membrane biology presents more obstacles and frustrations than membrane protein fractionation. Ideally, one would subject these substances to the same procedures that have proved so successful with soluble proteins, namely, electrophoresis, ultracentrifugation, isoelectric precipitation, "salting-in" or "salting-out," molecular sieving, ion-exchange chromatography, etc. Indeed, many investigators have attempted to apply one or more of these procedures to membrane proteins and have met some success in what are probably proteins of low "membrane affinity" (Fig. 2.7). However, the "core" proteins have proved extraordinarily intractable to analysis, and rather few methods show much promise to resolve this dilemma.

We have already addressed the techniques that make up the prerequisities for membrane protein fractionation, namely, (a) membrane

---

substituents on their ε-amino groups. 40% of these lysine residues appear involved in isopeptide linkages forming ε (γ-glutamyl) lysine. The remainder of the covalent linkages remain to be established.

Isopeptide formation can occur within a single peptide segment or by cross linking between peptide chains. Both mechanisms are intrinsically important and both introduce additional complications into the interpretation of SDS-PAGE. Thus cross linking might account for the puzzlingly high proportion of large "peptides" (apparent mol wts in excess of 150,000, according to SDS-PAGE) found in membranes. In addition intramolecular isopeptide bridges could cause anomalous migration. Also, isopeptide formation would markedly alter the hydrophobicity of the involved proteins and should thus be considered in any evaluation of this property. Finally, the presence of lysine isopeptides would hinder succinylation.

isolation, fractionation, and purification and (b) membrane solubilization. We shall now turn to membrane protein fractionation, discussing first the various approaches available and then turning to some selective extraction schemes. It will become evident that at present, membrane protein fractionation can very often be equated with electrophoresis in SDS-laden polyacrylamide gels (SDS-PAGE). Accordingly, we shall examine the strengths and weaknesses of this method in some depth.

For those proteins that can be selectively extracted into true solution in aqueous solvents, the large armamentarium of protein fractionation techniques, listed earlier, can be applied. These techniques have been widely described (103–105).

The situation differs critically in the case of what we call "core" proteins, whose solubilization requires chaotropic agents, detergents, or special solvents. If we remove the solubilization agents, e.g., by dialysis against water, the proteins reaggregate. Fractionations must, thus, generally be carried out in the presence of the solubilizers. Since many of the effective solubilizing agents bear a charge, ion-exchange chromatography, true electrophoresis, "salting-in or -out," and isoelectric precipitation cannot be utilized in their presence. Uncharged dispersing agents, in general, do not effect separation of membranes into their molecular subunits, and thus lack potency in fractionation maneuvers. However, such agents occasionally permit selective fractionation of membrane components. For example, liver microsomal cytochrome b_5 and cytochrome b_5 reductase solubilize in 1.5% Triton X-100 (106). When the mixture is applied to DEAE-cellulose, equilibrated against aqueous buffers, the cytochrome reductase elutes immediately, while cytochrome b_5 adsorbs, but can be later eluted by 0.25 M thiocyanate containing 0.25% deoxycholate.

Gel-permeation chromatography on polyacrylamide and agarose, using SDS-containing buffers, has been applied to membrane fractionations (54, 93), but lacks the effectiveness of SDS-PAGE. The most successful application of this approach has been in certain "selective" extraction procedures to be discussed below.

Ultracentrifugation in the presence of detergents and chaotropic agents has not proved satisfactory in membrane fractionation, but has been utilized for this purpose (107). As far as detergents are concerned, only SDS appears totally effective, this only at levels above its critical micelle concentration. The detergent micelles sediment in a centrifugal field, thus perturbing solubilizing conditions. The same problem applies to chaotropic agents, which, as we have already pointed out, must be utilized at high concentrations. A centrifugal field, thus, produces both density and solubilization gradients of these agents.

Fractionations in organic solvents have not yet proved very useful, presumably because adequate auxiliary techniques are lacking. We have already commented on this for the case of trifluoroacetic acid and molecular sieving on polyacrylamide. The use of controlled-pore glasses offers some hope here. Also, electrophoresis in 80–90% 2-chloroethanol using specilized apparatus (108) with suitable anticonvectants would appear to be a fruitful area of exploration.

Electrophoresis

Efforts to fractionate membrane proteins commonly employ electrophoretic techniques, particularly in the *zonal* mode, which separates components according to their mobility in an electric field. *Isoelectric focusing*, which brings proteins to rest in a pH gradient at their isoelectric point, has also been introduced into the membrane field (109). Electrophoresis of membrane proteins can be further classified into approaches, where separation occurs (a) according to *charge* and (b) according to *size*. Finally, electrophoresis may be either *free* or on some kind of medium preventing *convective and diffusion* artifacts. Today, microporous gels of cross-linked polyacrylamide enjoy particular favor as electrophoresis supports, i.e., in "polyacrylamide gel electrophoresis" or PAGE (110, 111).

PAGE attracts through its sensitivity, speed, simplicity, economy, reliability, as well as considerable versatility and reproducibility. Gels can be rapidly prepared with a wide range of porosities, as well as in a number of shapes and sizes. They lack fixed charges, thus avoiding endoosmotic interference. Chemical resistance is considerable permitting use of solvents containing chaotropic agents, detergents, and ranging in pH from 1 to 10. Purity is generally sufficient and difficulties arise only when highly sensitive techniques, e.g., fluorescence, are employed in quantification procedures. The gels may be stained with a variety of reagents, scanned spectrophotometrically, spectrofluorometrically and photographically, radioautographed, sliced, or comminuted mechanically. In a new technique (112), using *N, N'-diallyl-tartardiamide* mole for mole instead of methylene bis-acrylamide, the gels can be dissolved by treatment with 2% periodic acid for 20 min at 37°C. Use of this approach after electrophoresis allows for more facile analysis of separated components. Periodate may, of course, degrade the carbohydrate moiety of glycoproteins.

PAGE is suitable for the separation of model compounds in the molecular weight range 10^3–2×10^5 daltons and can be applied in submicrogram amounts. However, it does not constitute a panacea any more than other fractionation method. In membrane fractionations, a number of difficulties derive from the additives required to keep membrane components in solution. We shall comment further on this matter. More generally, many proteins adsorb nonspecifically to polyacrylamide, giving poor recoveries (113); many strategies minimize this difficulty. Quantification is always complicated in analytical PAGE, and extraction of reactive, amino-containing gel contaminants interferes with analysis of eluted components by dansylation or with ninhydrin procedures.

Unlike soluble proteins, membrane proteins require special maneuvers to maintain them in a sufficiently dispersed state for electrophoresis. The relatively high ionic strengths required for electrophoresis add to this complication. Accordingly, the electrophoretic separation of membrane proteins requires the presence of solubilizing agents. Some of these permit separations according to the *intrinsic charge* of a protein at a given pH (as in conventional electrophoresis); others provide fractionation approximately according to particle *size*.

SEPARATION BY CHARGE

Here the solubilizing agent either should bear no net charge or should not impose its charge upon the components to be separated. Chaotropic ions (including guanidine-salts) are not useful for this application, because they solubilize only at concentrations that provide a conductivity, which is excessive for electrophoretic separations. Urea is thus generally used in this mode of electrophoresis. A typical procedure for *zonal* electrophoresis is that of Takayama et al. (75), who solubilize lipid-free mitochondrial proteins in phenol:acetic acid:water (2:1:1, w/v/v), containing 2 *M* urea, and electrophorese on 7.5% polyacrylamide, containing 35% aqueous acetic acid, 5 M in urea. Neville (38) has employed a closely related approach, and Tuppy et al. (114) proceed similarly, substituting acetic acid with proprionic acid. Some workers have amplified the Takayama system with nonionic detergents ± reducing agents (115).

The high concentrations of organic acids serve to give all proteins a positive charge, to improve the solubility of membrane proteins, and to block accumulation of cynate due to urea decomposition; cyanate reacts with amino groups and would modify the cationic charge of the membrane proteins.

Merz et al. (109) employ 8 M urea as solubilizing agent in separating erythrocyte ghost proteins in a pH gradient of 3–10 on 2.5% PA, by *electrofocusing*; 0.02 M EDTA and 0.2% mercaptoethanol are used throughout their system, but the anodal buffer was 0.2% H_2SO_4, and the cathodal buffer is 0.4% ethanolamine, degassed to prevent CO_2 bubble formation in the pH gradient at the isoelectric point of bicarbonate.

Zonal electrophoretic separations according to intrinsic charge do not as yet provide high resolutions, presumably because membrane proteins remain aggregated in the electrophoresis media. Although Merz et al. (109) claim high separation efficiency for their *electrofocusing* approach, this contention is not convincingly supported by their data. All in all, we feel that this area requires extensive refinement and correlation with other fractionation procedures.

SEPARATION BY SIZE

This methodology is restricted primarily to electrophoresis in SDS-laden PA gels (SDS-PAGE), although some investigators have extended the approach to cationic detergents (99), and others have applied it to "free" electrophoresis (116).

The attractiveness of SDS include (i) rapid and extensive, if not complete, protein-lipid dissociation; (ii) the fact that many proteins bind SDS in proportion to their molecular weight and can, thus, be classified according to the latter criterion. In addition, bound SDS confers a high negative charge to most proteins allowing rapid separation. Finally, the detergent is bactericidal and denaturing at levels above its critical micelle concentration, thus protecting against proteolysis artifacts.

The assumed common properties of uniform charge density and an assumed regular relation between electrophoretic mobility and molecular weight form the basis for the hypothesis that the electrophoretic mobility of proteins/peptides in SDS-PAGE systems varies exponentially with molecular weight (117, 118), i.e.,

$$\text{mol wt} = K \cdot 10^{-bx} \tag{2.3}$$

where K = a constant, x = migration distance, and b = the slope of the calibration curve. We shall show further on that this relationship is not universal, as generally thought, but before continuing discussion of SDS-PAGE, we note that rather analogous results can be obtained by substitution of SDS with cetyl trimethylammonium bromide (99), although the experimental variance in the latter report is unexpectedly large.

SDS-PAGE electrophoresis occupies such a dominant position in current approaches to membrane protein fractionation that a few comments concerning *practice* are in order.

GELS

Most workers utilize approximately 5% acrylamide, of which 2.5–5% comprises the cross-linking agent N, N′-methylene-bisacrylamide. Such gels are suitable for the fractionation of molecules in the range of 2×10^4–2×10^5 daltons.

About 2–3% acrylamide gels permit fractionation of larger particles but, because of their fragility, require inclusion of agarose for stability (119). Higher acrylamide concentrations are suitable for the separation of smaller molecules. We have already commented on the possible utility of periodate-soluble gels.

We routinely employ gels 3 mm in diameter, formed in chemically clean Pyrex tubes. These allow fractionation of ~5 μg protein and, after a migration of about 35 mm, resolution is equivalent to that obtained with larger gels. However, 6 mm and 10 mm gels are used more commonly. We prefer the 3 mm gels because (a) they can be photographed at 1:1 magnification in a format suitable for our computerized gel-scanning procedure and (b) temperature control is optimal.

Various investigators differ in their preferences for electrophoresis buffers; these are usually employed near neutrality [but see Fairbanks and Avruch (99)]. Phosphate and Tris buffers appear most widely used in continuous systems. Electrofocusing has not been applied to SDS-PAGE so that buffer gradients serve little purpose in this system. Most investigators prefer to work at low ionic strength, which minimizes protein aggregation. For the same reason, many *workers* add EDTA to complex possible contaminating polyvalent metals; such metal ions also form SDS complexes. Sodium salts are generally preferred because of the relatively high affinity of SDS for K^+.

SAMPLE PREPARATION

We have already questioned the generalization that 1.4 g SDS is bound per gram protein at saturation. Certainly, some proteins bind more and others less detergent. As pointed out by Nelson (90), maximal binding occurs at higher ionic strength, but this is not compatible with optimal electrophoretic conditions. Notably, the critical micelle concentration of SDS, ~6 mM, varies very little with ionic strength (120). Most commonly, samples containing 0.5–3 mg protein per milliliter are made to 1–3% SDS at ionic strength <0.1. Lipid-free samples may not dissolve readily, and because of the slow monomerization of many oligomeric proteins (90), samples may need to be maintained in SDS *well beyond* the few minutes or hours commonly employed. Monomerization may be enhanced by boiling, pH extremes, and urea, but this matter has not been explored systematically.

In general, 0.5–5% mercaptoethanol or 5–50 mM dithiothreitol are included in the solubilization medium. This is to rupture any S—S links present (to obtain "valid" molecular-weight estimates) and to prevent adventitious S—S formation. Some investigators use 50–100 mM iodoacetamide to alkylate —SH groups for the latter purpose, but

SDS itself is alleged to prevent —SH oxidation (94). O-phthalaldehyde, used as covalent label (121), also complexes —SH groups and prevents their oxidation.

A low-molecular-weight "tracking dye" is usually included in the sample mixture to mark the electrophoresis front.

For electrophoresis in cetyl trimethylammonium bromide (99), membranes are prepared essentially as for SDS-PAGE.

ELECTROPHORESIS

For SDS-PAGE, proteins are electrophoresed toward the anode, and in CTAB-PAGE toward cathode. We routinely operate at temperatures controlled to 15–25 ± 1°C by means of a constant-temperature circulator. Temperatures less than 15°C are undesirable because of SDS precipitation. Field strengths of 5–12 v/cm are typical. We generally electrophorese each gel until the tracking dye has come to a defined position. This allows optimal comparison of gels. Some workers use visible internal standards for the same purpose.

SLAB GELS AND TWO-DIMENSIONAL PAGE

Analytical and micropreparative PAGE can be advantageously conducted, using 3–5 mm gel slabs contained between two cooling plates (122, 123). A number of such devices can be obtained commercially.

In the one-dimensional mode, up to 10 samples can be run side by side, allowing for more *precise comparisons* than with conventional methods. The samples are usually introduced into pockets formed by spacers during polymerization of the slab.

Our interest in the slab approach derives from the fact that one may use it for two-dimensional SDS-PAGE, i.e., a sample is run first in one direction under one set of conditions and then at 90° under a different set of conditions. This generally necessitates removal of a gel containing the already separated components, reinserting it across the top of a slab, and then proceeding to electrophoresis under the second condition. The approach applies particularly well to cases where one wishes to electrophorese first in a gel of one porosity and then in one of another porosity, thereby extending the molecular-weight range for separation (122). We utilize the procedure without changing PA concentration and use the approach to electrophorese at two different pHs or with and without added urea or reducing agent (Fig. 2.10).

DETECTION, QUANTIFICATION, RECOVERY

Absorptive Staining. Most workers use this technique despite the serious obstacles discussed further on. Since the procedures take a long

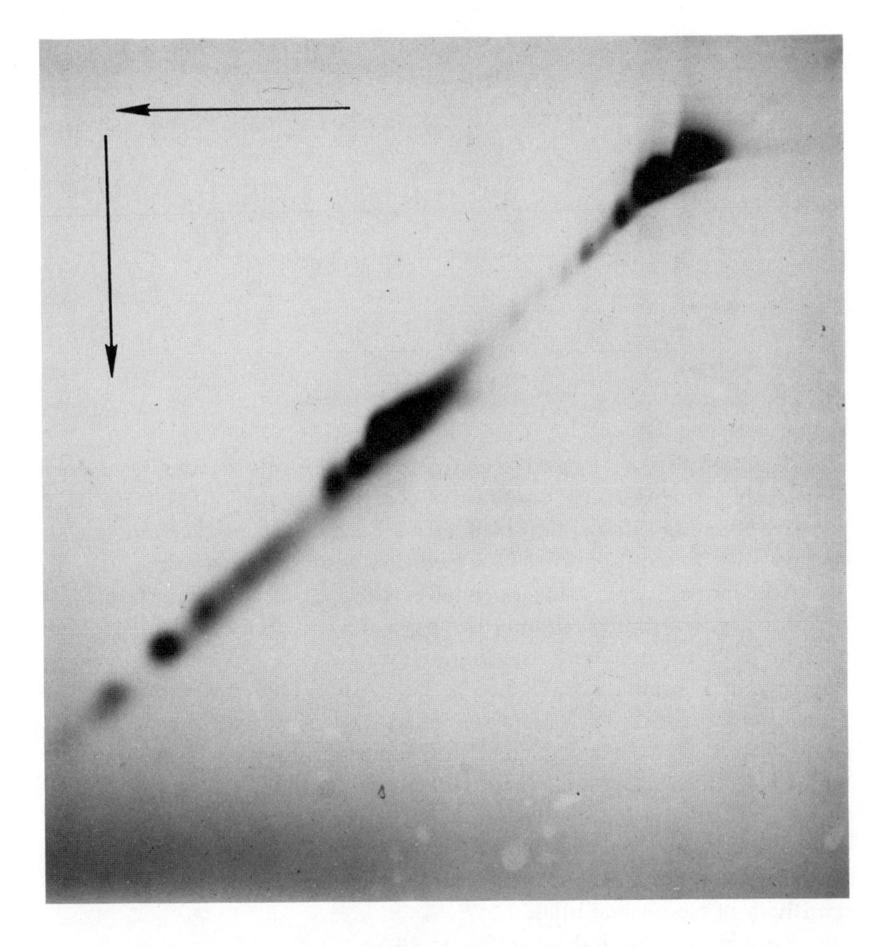

Fig. 2.10 Two-dimensional SDS-PAGE of erythrocyte membrane proteins. The samples are first run in 1% SDS in one direction, and then at 90° in 1% SDS containing 50 M dithiothreitol to reduce any possible S—S linkages. Note that the final pattern forms an exact diagonal, i.e., the mobilities of all components are identical in both directions. If S—S linkages had been present, new components should have arisen during separation in the second dimension. Migration direction toward anode. Horizontal arrow = first direction; vertical arrow = second direction. (Courtesy of H. Knüfermann.)

time, many workers avoid protein diffusion and band broadening by treating the gels with agents such as 20% trichloroacetic acid (124) or sulfosalicylic acid (125).

The most common staining technique utilizes the dye Coomassie brilliant blue (e.g., G250).

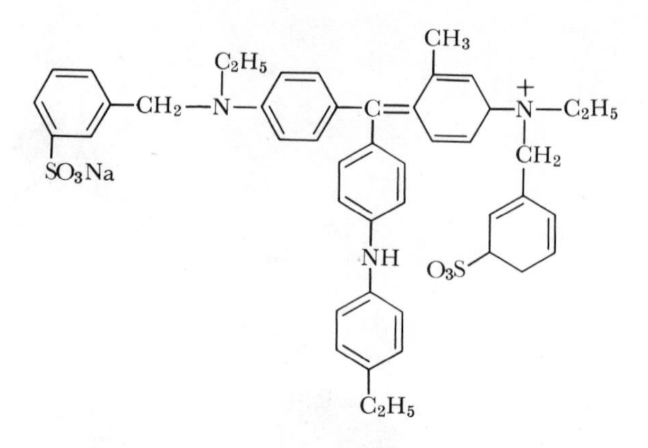

Because of its net negative charge, this dye stains optimally at acid pH, where proteins bear a net positive charge; it provides high sensitivity because of its high absorption coefficient. Removal of excess stain can be carried out by electrophoresis or by elution with 5–10% acetic acid.

Adsorptive stains color optimally when SDS has been removed. Accordingly, Fairbanks et al. (47) have devised a technique for concurrent SDS removal and staining. Here the gels are soaked in 25% isopropanol containing 0.02–0.05% Coomassie blue for about 15 hr. Subsequent washing with isopropanol at 15-hr intervals completes destaining in 3 days. Gels stained with Coomassie blue should be stored in about 15% acetic acid and shielded from strong light.

Other anionic adsorptive protein stains, e.g., Ponceau red (sodium cumeneazo-β-naphthol disulfonate) and Amido Schwartz (2-naphthyl amine 6,8-disulfonate) are occasionally employed (47). They lack the sensitivity of Coomassie blue.

1-anilino-8-naphthalene sulfonate (ANS)

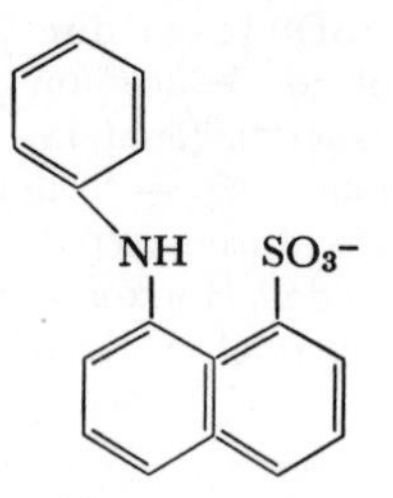

can be used as an absorptive protein stain, of high sensitivity due to its fluorescence characteristics (126). In SDS-PAGE, the detergent should be removed to reduce background (although SDS binds ANS poorly).

Carbohydrate Staining. In general, glycoproteins stain poorly with

Coomassie blue, perhaps because of the negative charges conveyed by sialic acid. These can be localized by the *periodic Schiff reaction* (PAS) (47), in which vicinal glycols of the sugar moiety are first periodate oxidized, allowing them to form colored adducts with Schiff's reagent (1% *p*-rosaniline = triamino-diphenyl-tolylcarbinol).

The PAS procedure is usually less sensitive than protein stains, since there are fewer reactive sites, some sugars do not stain, and others yield evanescent reactions. Glycolipids with suitable carbohydrate moieties also react, but much of the staining at the electrophoretic "front" derives from phosphatide plasmalogens.

A superior method for glycoprotein detection utilizes the dye Alcian blue (127), an undisclosed derivative of copper phthalocyanin. In this procedure, one oxidizes vicinal glycols conventionally (1% periodic acid in 3% acetic acid, 60 min, followed by water washing, a soak in 0.5% metabisulfite for 30 min and another water wash). Staining requires 0.5% Alcian blue in 3% acetic acid for 4 hr; gels are destained with 7% acetic acid. Two major advantages derive from this technique: stability of the reagent, and stability of staining.

Intrinsic Protein Fluorescence. As we shall discuss in some detail, quantitation is a serious problem in SDS-PAGE: Some proteins, e.g., glycoproteins and pepsin, stain very poorly with adsorptive stains; others color intensely. Thus, the intensity of a band after, e.g., Coomassie staining, may not truly reflect its protein content. One way to avoid this dilemma involves scanning gels for their intrinsic fluorescence (excitation ~290 nm). Knüfermann and associates (128) have devised a scanning attachment to the Hitachi-Perkin Elmer MPF 2A spectrofluorometer for this purpose. Clearly, this approach requires SDS with minimal fluorescent contamination. Present data indicate that the approach is feasible, but that adequate resolution requires further refinement. This approach may, of course, give false values with proteins of unusual tryptophan content. Use of mechanical slicing and periodate solubilization has not offered major advantages as yet.

Radioactive Labeling—General. Samples can be labeled by radioiodination or treatment with ^{14}C or ^{3}H acetic anhydride [cf. (54)] prior to electrophoresis. —SH groups can also be tagged by radioactive compounds not dissociated by the —SH reagents in the electrophoresis medium. Radioactivity can then be localized as follows:

(a) periodate solubilization of N, N′-diallyl tartardiamide X-linked gels, elution of the soluble material in dropwise fashion and counting;
(b) mechanical "slicing" or pulverizing the gels serially along their length and counting the segments;
(c) lengthwise slicing of the gel followed by radioautographs.

The first two approaches provide maximal quantification, but yield poor resolution. The third gives high resolution but poor quantification; moreover, it requires considerable time.

Radioactive Labeling—Specific. In Chapter 1, we have discussed the merits and disadvantages of diverse techniques designed to label the external surfaces of animal cells. Despite our reservations about the vectorial nature of these reagents, particularly in cells capable of pinocytosis, they have clearly been established to react with membranes and can serve as labels in the electrophoretic separation of membrane proteins. Indeed, the diazonium salt of [^{35}S]-sulfanilate (129, 130), [^{35}S]-formyl-methionyl sulfone methyl phosphate (131), and ^{125}I, introduced by enzymatic iodination (132) have been used repeatedly in this manner and show a different labeling pattern when employed with intact erythrocytes and membranes isolated therefrom [cf. in (133)].

Recently Arrotti and Garvin (134) showed that two reagents designed to label the membrane exterior ^{3}H trinitrobenzene sulfonate and ^{3}H picryl chloride actually penetrate the erythrocyte membrane (but at different rates), react with its components during transit, and then become depleted by reaction with hemoglobin at the membrane interior. Not unexpectedly, the nonpolar picryl chloride permeated more slowly than trinitrobenzene sulfonate.

More recently, these authors showed by SDS-PAGE that trinitrobenzene sulfonate preferentially labels band 3 of Fairbanks, Steck, and Wallach (47), but not the glycoprotein ("glycophorin") that migrates in this region. In contrast, the erythrocyte membrane components of high apparent molecular weight (bands 1 and 2) label freely with *picryl chloride*, but not trinitrobenzene sulfonate. Finally, low-molecular-weight components label strongly with both reagents, as do lipid components (135).

Since both reagents react equivalently with free —SH and —NH_2 residues, the authors conclude with reason that different proteins within intact membranes exhibit diverse reactivities to permeating reagents. This point, well recognized for soluble proteins, is also brought out by the dansylation studies of Schmidt-Ullrich et al. (136).

In regard to the studies of Arrotti and Garvin (134, 135), we note that a ^{32}P-labeled intermediate of the erythrocyte $Na^+ - K^+$-sensitive ATPase also migrates with approximately the electrophoretic velocity of the component binding [^{3}H]-trinitrobenzene sulfonate (137). Arrotti and Garvin (134, 135) suggest that the band-3 protein(s) constitute all or parts of a hydrophilic transmembrane channel [q.v. (65)].

Covalent Fluorescent Labeling. The time required to visualize separated membrane components constitutes a major annoyance in SDS-PAGE. This problem has been circumvented by the development of techniques where covalent fluorescent adducts are formed prior to electrophoresis; fluorescence techniques are required to provide sufficient sensitivity. These methods add great versatility as well as speed to SDS-PAGE and allow one to follow electrophoretic separation in progress. However, proteins with heme groups, which are not released by SDS, may quench fluorescence.

4-Acetamido, 4'-Isothianostilbene-2,2'-Disulfonate. This diazonium dye, reacting with amino, histidyl, and guanido groups of membrane proteins [q.v. (138)], the first fluorescent dye specifically designed for membrane studies, has found no important application in SDS-PAGE.

Dimethyl Amino Naphthalene Sulfonyl Chloride (DANSCL). This dye reacts predominantly with amino groups and can be combined with membrane proteins in two ways. The first follows the procedure of Talbot and Yphantis (139). Here, proteins in Tris- or phosphate-buffered 3–5% SDS, pH 8.2, are combined with excess 10% DANSCL in acetone, boiled 1–5 min, made 1% in mercaptoethanol, boiled again briefly and unreacted DANSCL, or any DANS-OH formed, removed by gel permeation on Sephadex G-25. The green fluorescence of the dansylated proteins is readily visualized by illuminating at ~365 nm. Dansylation allows ready detection of 8 ng protein in 5 mm gels.

A more selective dansylation procedure, specifically designed for membrane studies, has been developed by Schmidt-Ullrich et al. (136). To avoid the membrane lysis usually generated by the solvents that solubilize the apolar DANSCL, these authors incorporated the dye into cholesterol-lecithin micelles, whence active DANSCL molecules combine with intact erythrocytes (without hemolysis) or isolated membranes. The labeling patterns differ in the two cases, presumably due to a change in membrane protein reactivity and/or accessibility during hemolysis. In the case of dansylated ghosts, all known proteins appear labeled. The dansylated proteins can easily be observed during migration.

O-Phthaladehyde (OPT)

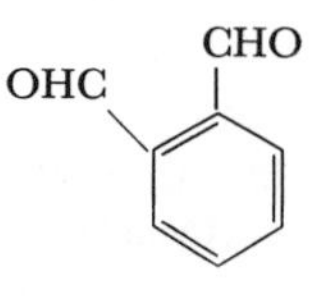

This interesting reagent has been introduced into the fields of protein analysis and SDS-PAGE by Weidekamm et al. (121). OPT lacks fluorescence except in the presence of proteins and glutathione. Our NMR and IR data suggest that fluorescence requires one aldehydic group to form a Schiff base, and the other a thiohemiacetal.

In the protein case, including membrane proteins, intense fluorescence develops after reacting 0.1–50 μg protein in 0.05 M phosphate, pH 8.5 with 36 μmoles mercaptoethanol, and then followed with 0.19 μmoles OPT. This can be done in 1% SDS. Under such circumstances, all proteins form fluorescent adducts, although quantitative differences from one species to the next are quite evident. We have found maximal fluorescent yields with erythrocyte membrane proteins (121). To detect fluorescence, excitation at 336 μm and emission at 431 μm are optimal. Electrophoresis can be initiated without further manipulation immediately after formation of the OPT complexes.

An interesting and important aspect of OPT fluorescence is the extensive transfer of excitation energy from tryptophan (excitation 286 μm) to OPT. At full OPT saturation, nearly all protein fluorescence excited at 286 μm is transferred to OPT.

The general fluorescence reaction of OPT with proteins presumably involves formation of a Schiff base with a protein amino residue, plus a thiohemiacetal with mercaptoethanol. However, OPT yields fluorescent adducts with SH-containing proteins in the absence of mercaptoethanol, in an —SH-specific reaction. In this case, the fluorescent adduct presumably involves formation of a Schiff base with a protein $—NH_2$ and a thiohemiacetal with a nearby protein —SH.

Fluorescent OPT adducts are photosensitive. Hence the electrophoreses should be run in the dark or subdued light, and gels should be properly photographed not more than 30 min after exposure to light.

The fluorescence distribution in PAGE gels can be evaluated by scanning attachments for diverse fluorometers. However, we know of no unit with satisfactory resolution; also, scanning takes considerable time. For dyes with *visible* fluorescence, photographic techniques offer many advantages.

When gels contain fluorescent bands and a low concentration of fluorophore in the bands (typically in the case of PAGE), illumination with exciting radiation produces emission of fluorescent light in direct proportion to the concentration of fluorophore in the gel. One can, therefore, photograph fluorescent gels with suitable film, controlling the focusing so that a constant proportion of the emitted light impinges on the film. The darkening D of the film varies logarithmically as the intensity I of the impinging light and the duration t of exposure. Accordingly,

$$D = j \log I \cdot t \tag{2.4}$$

where j is a constant for the films, depending on its intrinsic characteristics and developing conditions. D can be measured by densitometric scanning (128) or, much more rapidly, by computerized image recognition devices (140). Since D is a logarithmic function of I, which varies linearly with chromophore concentration, the latter must be computed from the antilogarithm of D.

Colored gels can be rapidly "scanned" by photographing them during transillumination under standard conditions and analyzing the negatives by a computerized image recognition device (140).

The concentration c of dye in a band relates to the light absorption A as follows:

$$A = \varepsilon \cdot cl \tag{2.5}$$

where $A = \log I/I_0$, the logarithm of the ratio of light intensity transmitted by the sample to the incident light intensity. $\varepsilon =$ the absorption coefficient, and $l =$ the optical path length in cm. Film darkening D varies logarithmically with I and t, where t is the exposure time, i.e.,

$$D = \log I \cdot t \tag{2.6}$$

Accordingly, incorporating our previous equation

$$D = j \log t + j\varepsilon cl \tag{2.7}$$

where j is the constant describing the film properties. Film darkening is, thus, directly proportional to the stain concentration in a band.

QUANTITATION—GENERAL

Quantitation represents a major problem in SDS-PAGE; indeed, as generally applied, the technique is essentially "quantitative" only in providing relative electrophoretic mobility. The questions of "how much" protein resides in a given electrophoretic component and what proportion of the whole this represents require further resolution.

The problem of quantitation depends upon the technique used to detect the separated proteins, but nearly all practical approaches present difficulties ranging from minor to severe. The only ideal situation we can imagine is one where all peptide components are *equivalently labeled covalently prior* to electrophoresis.

Radioactive Labeling. We doubt that uniform labeling of all membrane components can be achieved by chemical manipulations of membrane isolates, because no inherent law dictates that all of the diverse polypeptides in a membrane should, for example, acetylate, dansylate, or pyridoxilate equivalently. Rifkin et al. (141) address this problem realistically, pointing out that their pyridoxilation-tritiation procedure would label diverse proteins differentially and not necessarily in proportion to their abundance. This is because of (a) inherent variations in the reactivity of diverse amino groups and (b) differences in amino content from protein to protein.

The same considerations clearly apply to the use of ^{35}[S]-diazonium sulfanilate (129), ^{35}[S]-formyl-methionyl-sulfone-phosphate (131), and enzymatic iodination (132). The ideal may, perhaps, be approached when working with membranes derived from microorganisms or animal cells cultured continuously in media containing radioactive precursors that incorporate uniformly. However, as already noted, one cannot yet resolve diverse components as well by radioactivity "scanning" as by optical scanning. Moreover, the distribution of a given peptide after electrophoresis cannot be readily predicted; it may be Gaussian or it may not.

Absorbed Dyes. These are most commonly used in SDS-PAGE and offer the greatest obstacles to quantitation. At best, one can only define a dye-binding curve for a given set of conditions, but this cannot be identified unambiguously with protein/peptide distribution because of the following:

(a) For any one stain, the degree of staining can vary drastically from one protein to the next, e.g., some proteins do not adsorb anionic stains such as Coomassie blue.
(b) The degree of staining of diverse proteins depends sensitively upon the conditions used.
(c) Staining is rarely stable, and fading depends on gel thickness, the protein component involved, and also other variables.
(d) One cannot quantify by measuring merely the peak absorbance of a given band, since a band has a finite width. For this reason, gels are "scanned" photometrically, i.e., to obtain a distribution of absorbance (fluorescence) vs. migration distance and the areas under the various peaks determined by mechanical or electronic integrators, or by cutting out the "peaks" of a recording and weighing them. One can also integrate using the photographic techniques described on p. 107. Several scanning devices commercially available possess adequate resolution for integration. However, with unknown proteins, one cannot assume a Gaussian (or any other) distribution. Integration is, thus, to a large extent empirical.
(e) With conventional round gels, the optical path for absorbance varies from a maximum at the diameter to zero at the periphery, making alignment critical even to obtain reliable dye-absorption curves.
(f) A dominant component closely similar in molecular weight to a minor one will obscure the latter.

Covalently Bound, Nonradioactive Dyes. To date, only radioactive and fluorescent markers have been found to possess the sensitivity required to label proteins prior to PAGE and to monitor their fractionation by this procedure. We here address the special problem involved in fluorescence scanning, noting parenthetically that these also apply to the use of non-covalently absorbed fluorescent dyes such as ANS. We also remind the reader that the Trp fluorescence of peptides containing this amino acid has also been used to monitor PAGE fractionations (see above). However, DANSCL and OPT enjoy three advantages. First, excitation of fluorescence does not require quartz or reflecting optics. Second, the emitted fluorescence is visible. Third, both agents can also be excited via energy transfer from tryptophan.

Pyridoxilation of membrane proteins as described in Chapter 1 (141) and on p. 107 may serve to monitor membrane protein fractionation, but has not yet been utilized for this purpose. The excitation maximum for pyridoxal phosphate lies at ~330 nm, and emission occurs at ~385 nm.

DANSCL and OPT also offer major advantages when compared to adsorptive dyes such as Coomassie blue, namely:

(a) Both methods are speedy.
(b) They are sensitive.
(c) They give direct visualization of separations.
(d) Both methods can be used for protein assays by direct spectrofluorometry.
(e) Proteins vary less among one another in reactivity, e.g., glycoproteins stain. However, one cannot assume that all species react equivalently.
(f) "Staining" conditions are well defined, and the degree of reaction can well be defined by direct fluorometry.
(g) Gel geometry is less critical.

Problems of overlapping bands, bandwidth, and bandshape are as severe here as with Coomassie blue, etc.

RECOVERY

Kapada and Crambach (113) find that many proteins adsorb nonspecifically to PAG and cannot be fully eluted after PAGE; they suggest a number of techniques that largely obviate this problem.

This matter has not been extensively investigated for SDS-PAGE, presumably because so many workers utilize adsorptive staining for protein/peptide detection. However, we have satisfied ourselves that nonspecific adsorption of membrane proteins does not represent a significant complication in SDS-PAGE (142). We labeled the membranes with OPT prior to electrophoresis and photographically monitored the rate of migration of diverse fluorescent components into the gel. We then reversed migration by changing the polarity of the electric field. Within the limits of our technique, resolved bands migrated out of the gel at the same rate they entered.

We also eluted low-molecular-weight components through the bottom of the gel before reversing polarity to elute high-molecular-weight components through the top of the gel. Our recovery of OPT-labeled proteins was 95%, and the eluted gels showed no residual fluorescence.

ARTIFACTS—INCOMPLETE MOLECULAR DISSOCIATION

Most workers in the field implicitly or explicitly accept the following crucial premises concerning SDS-PAGE.

(a) SDS binds to protein polypeptide chains in proportion to their molecular size.

(b) SDS, plus S—S reducers, dissociates all proteins into their constituent polypeptide chains.

(c) SDS completely separates membrane proteins from membrane lipids.

(d) In binding the detergent under S—S reducing conditions, the peptide chains refold to form particles of identical shape, but differing in size according to their molecular weight.

We have addressed these items in our comments on SDS solubilization, and we do not doubt that these generalizations approximate the behavior of dozens of soluble proteins, but cannot accept this as proof for the behavior of membrane proteins. Moreover, Nelson (90) (Table 2.3) clearly shows that some proteins, e.g., papain, do not bind SDS significantly, that oligomeric proteins generally require much longer to split into their subunits than is allowed in preparing for SDS-PAGE, and that some oligomeric proteins, e.g., glucose dehydrogenase, are not monomerized by SDS.

In addition, Wallach and Weidekamm (91) report that when freshly isolated washed human erythrocytes are rapidly transferred into 1–2% SDS, 0.04 M dithiothreitol, their hemoglobin migrates electrophoretically as two principal components, one corresponding to monomer and one to dimer. Importantly, both bands still contained significant proportions of heme. Since this prosthetic group binds predominantly by apolar interactions, these data further indicate that SDS not only does not monomerize fully but also that it does not unfold the peptide chain significantly enough to release the prosthetic group. We find, in contrast, that heme-free globin migrates predominantly as monomer.

The specialized amino acid sequences found in various proteins also leads us to suspect that one cannot freely assume that all polypeptide chains in all proteins will form similarly structured entities. The interaction of poly-L-lysine with SDS vividly illustrates this point: SDS forms strong complexes with poly-L-lysine, which exhibit the typical optical activity and infrared characteristics of the β-conformation (67). As one might expect, these migrate anomalously in SDS-PAGE, because of both their 2° structure and their tendency to aggregate at high SDS levels.

At least one well-studied human erythrocyte membrane protein, the major glycoprotein, must also be suspected of existing as a dimer during SDS-PAGE. By this method, apparent molecular weights of ~90,000 are obtained (e.g., 47, 131, 132). Segrest et al. (89) correct this value to about 55,000 on the basis of their studies, showing poor SDS binding by the protein. However, even this estimate virtually doubles the value of 31,400 established by several workers who have used phenol extraction to isolate this material [cf. review by Winzler (143)]. This latter value is very well established by standard physical techniques as well as chemical analyses and suggests that the SDS-PAGE value is incorrect.

The experimental evidence pointing to incomplete peptide dissociation and/or unfolding may derive from at least two facts established from the study of proteins in general.

(a) In certain proteins, such as the major glycoprotein of human erythrocytes, or mellitin (cf. Chapter 3), apolar amino acid residues are concentrated in one region of the polypeptide chain, and polar, as well as charged, residues lies elsewhere.

(b) In many proteins, e.g., the hemoglobins (144) and murein lipoprotein (31), apolar amino acid residues occur regularly at every third or fourth position over long portions of the amino acid sequence. When such sequences are arrayed α-helically, the resulting helices exhibit a polar and an opposite apolar face.

Wallach and associates (33, 65) have discussed the possible role of such arrays in membrane protein structure. Here, we merely wish to draw attention to these structural specializations that might interfere with molecular dissociation of proteins by SDS. They could also account for the 10–20% of phosphatide that apparently remains protein associated during SDS-PAGE (142) and gel-permeation chromatography in SDS (54).

In this connection, we stress that, although Shapiro et al. (117) originally claimed that SDS variation between 0.1% and 1% does not markedly alter electrophoretic mobility, Fairbanks et al. (47) and others find that components, migrating as single bands in 1% SDS, split into multiples at ≤0.2% SDS. This appears reasonable since the critical micelle concentration of SDS lies near 0.23%; below this level, the detergent is fully monomeric. At higher total SDS levels, the monomer concentration stays at ~0.23%.

ARTIFACTUAL PEPTIDOLYSIS

As emphasized early by Fairbanks et al. (47), SDS-PAGE of biomembranes must always be suspect of peptidolysis artifacts. Indeed,

SDS-PAGE is particularly susceptible to such artifacts, and cleavage of one peptide bond per peptide chain can drastically alter a PAGE pattern. The reasons are as follows.

First, SDS tends to unfold most proteins, making them susceptible to proteolytic attack. This might be of no consequence, but for the fact that peptidases are generally less susceptible to SDS denaturation than other proteins (145).

Second, most biomembranes derive from tissues and/or cells containing peptidases.

Concerning the first problem, peptidolysis can be minimized by transferring the material to be analyzed rapidly to *high* ($\geq 1\%$) SDS levels. If one wishes to promote solubilization by heating, the sample should be brought to 100°C as quickly as feasible to prevent peptidase activation before denaturation begins. Ionic strengths should be maintained as low as feasible, since high salt levels appear to potentiate peptidolysis (47).

Regarding tissue proteases, one would ideally like to avoid all peptidase contamination. This, however, appears to be wishful thinking, *first* because virtually all "tissues," blood included, contain cells rich in lysosomes and also because many biomembranes are believed to contain "intrinsic peptidases."

Lysosomes, which bear many peptidases, abound in many cells, e.g., macrophages and blood granulocytes, which are unduly fragile. This fact is rarely appreciated. Thus, most techniques used to prepare erythrocyte employ conditions that are known to damage leucocytes, releasing their complement of peptidases. The situation is even more unpredictable in more complicated tissues, where virtually all subcellular fractions may include some lysosomal contaminants. Few consider this possibility or test for it, but it remains a crucial aspect of SDS-PAGE.

The matter of intrinsic proteases appears somewhat obfuscated. In our view, all reports claiming that plasma membranes contain such components lack the controls necessary to exclude lysosomal contamination.

In discussing the action of proteases on biomembranes, one should recognize the fact demonstrated by Speth et al. (146) that membranes can maintain their physical identity through noncovalent associations, despite extensive peptide cleavage. Addition of SDS to such membrane preparations separates the diverse peptide fragments, and this is clearly apparent upon subsequent SDS-PAGE.

While artifactual proteolysis can create major confusion, controlled proteolysis can be used to analyze the distribution of proteins perpendicular to the membrane plane (147). Although the cited study unwittingly introduced some ambiguities, it also developed certain fundamental approaches. Moreover, many well-characterized proteases can be obtained bound to large, inert particles and can be inhibited by specific reagents, thus allowing their use as highly specific "probes."

MOLECULAR AGGREGATION—ARTIFACTUAL AND DELIBERATE

Several sources of *artifactual* aggregation may be cited. Denaturants, SDS included, can lead to physical entanglement of polypeptide chains, producing large aggregates which may not enter gel, or can produce smearing and even multiple faint bands. This problem increases with the concentration of the sample applied.

Unless S—S-reducing conditions prevail, nonspecific interpeptide cross-linkages may form. Thus, an apparently unique major polypeptide aggregates irreversibly in concentrated guanidine-HCl (148). This process can be blocked by prior reduction and acylation of the peptide's SH, and Steck (148) suggests that unfolding of this peptide allows intermolecular S—S bridges to form; these do not break even in 1% SDS containing dithiothreitol or mercaptoethanol.

Several workers employ *deliberate* intermolecular cross-linking to evaluate the spatial relationship of diverse membrane proteins. This approach was first introduced by Marfey (149), who reacted intact human erythrocytes with 1,5-difluoro-2,4-dinitrobenzene (DNDFB) under physiologic conditions. He found that this reagent links some membrane proteins to the β-chains of hemoglobin through Cys β-93. α-chains do not react, and no other cross-links were observed, although Lys-Lys, Tyr-Lys bonds, etc., might have been expected.

Knüfermann et al. (95) have extended these studies; they find dramatic differences between intact erythrocytes and erythrocyte ghosts. In the latter case, a concentration of 12 μM of DNDFB sufficed to link all membrane peptides into aggregates, which would not enter a 5.8% PAG. In contrast, treatment of intact cells with this DNDFB level generates 3 new SDS-PAGE bands of very high molecular weight, presumably cross-linked products of lower-molecular-weight peptides, which diminish quantitatively. Simultaneously, a component with apparent mol wt 48,000 disappears, to be replaced by one of about 65,000. This could correspond to a complex with a Hb β-chain.

Steck (150) employs a number of cross-linking procedures, assaying their effect by SDS-PAGE. His data indicate that many components seen upon SDS-PAGE of normal human erythrocyte ghosts form new higher-molecular-weight components upon cross-linking (Table 2.4). Interestingly, the major glycoproteins do not react.

The cross-linking studies performed to date do not yield unambiguous evidence as to the in vivo disposition of various electrophoretically

Table 2.4 Effect of Cross-Linking on Some Major Membrane Proteins[a]

Component	Apparent Mol Wt ($\times 10^{-5}$)	Cross-Linking Reagent	Presumed Product Apparent Mol Wt ($\times 10^{-5}$)
I	2.40	Formaldehyde	
		Glutaraldehyde	3.5–7.0
II	2.15	Oxidation	
II_1	1.95	Oxidation	3.5–7.0
III	0.88	Oxidation	1.65
IV_1	0.78	Gluturaldehyde	Lost
IV_2	0.72	Formaldehyde	2.7
V	0.43	Oxidation	3.5
VI	0.35	All reagents	No effect
VII	0.29	Glutaraldehyde	Lost
Glycoproteins		All reagents	No effect

[a] From Steck *et al.* (151).

defined peptides, particularly since these may be mobile in the membrane plane and since native cross-linking by isopeptide bonds may occur more frequently than anticipated (p. 93). Nevertheless, it appears clear that the β-chains of some Hb molecules lie in close proximity to certain membrane proteins at the interior membrane face.*

Knüfermann et al. (151) have explored the possible use of reactive groups coupled to inert carbohydrate chains (153) as "long span" cross-linking reagents to explore membrane topology (Fig. 2.11).

ARTIFACTS IN MOLECULAR-WEIGHT ESTIMATION

While one generally assumes that SDS-PAGE systems allow reasonable peptide molecular-weight determinations, we feel that much skepti-

* Unlike glutaraldehyde, the bifunctional reagent dimethyl adipimate does cross-link erythrocyte membrane glycoproteins (T. H. Ji, *Biochem. Biophys. Res. Commun.* **53:** 508, 1973). This may be due to a different reactivity of dimethyl adipimate or to the greater separation of its reactive groups (8.6 A). Cross-linking increases with time. This may be due to fluctuations in the distances between the proteins and glycoproteins and/or the rates of the chemical processes involved.

Cross-linking reagents can be used to evaluate the proximity of membrane proteins and certain phospholipids (G. V. Marinetti, R. Baumgarten, D. Sheeley, and S. Gordesky, *Biochem. Biophys. Res. Commun.* **53:** 302, 1973). Dinitrodifluorobenzene and suberimidate link 8.4% and 2.3% of total lipid phosphate, respectively, to membrane protein. This corresponds to 20% and 5.8% of the aminophospholipids. However, essentially all of the phosphatidylethanolamine + phosphatidylserine *react* with the cross-linking reagents. This approach should allow assessment of which membrane proteins are closely associated with aminophosphatides and also what phosphatides are primarily involved.

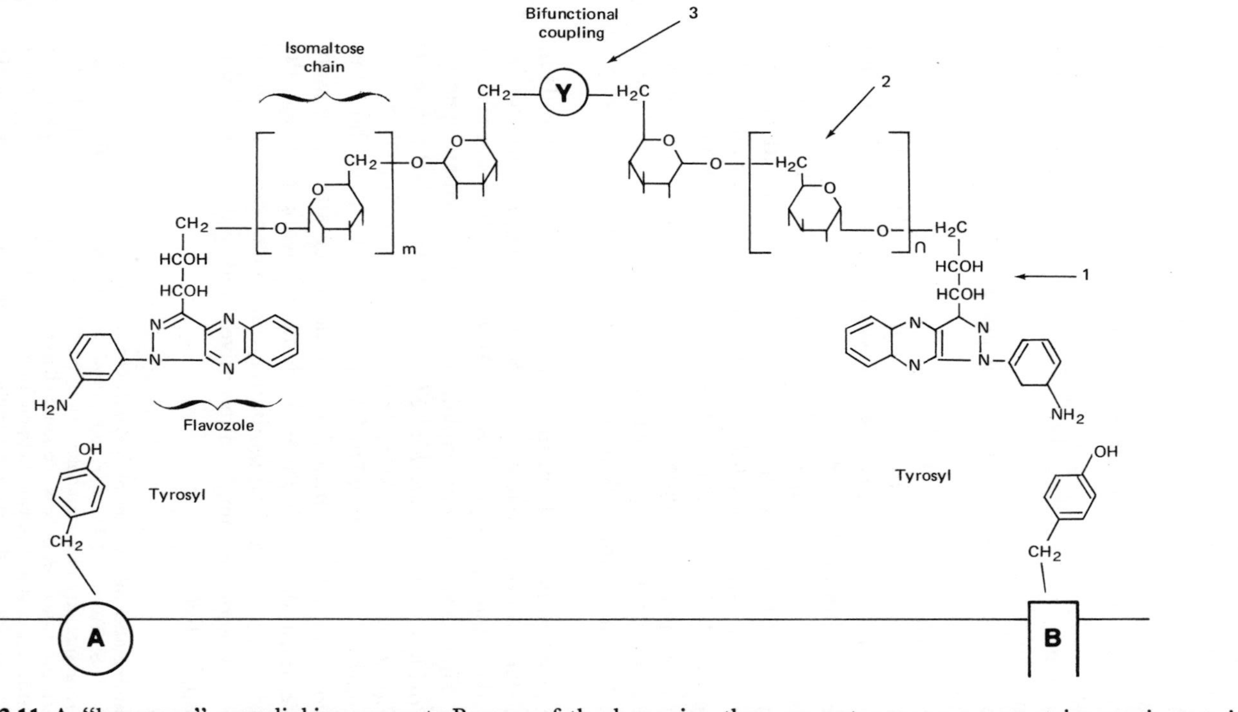

Fig. 2.11 A "long span" cross-linking reagent. Because of the large size, these reagents are nonpermeant in nonpinocytosing cells. Also, they react under physiologic conditions. A and B are membrane proteins with reactive groups (tyrosyls in the illustration) exposed at the external membrane surface. Y symbolizes a bifunctional link between isomaltose chains via w, w^1—CH_2— groups; this can be S—S, —NH_2—CH_2—NH_2—. → represents possible experimental cleavage sites, e.g., (1) vicinal hydroxyl (periodate), (2) isomaltose links (dextranase), (3) S—S reduction.

cism must be instilled into this notion. We have already emphasized some major points, namely:

(a) SDS +S—S reducers do not necessarily reduce all oligomeric proteins into their subunits.

(b) SDS may affect subunit dissociation only over prolonged time periods.

(c) SDS does not necessarily effect complete lipid-protein separation.

(d) SDS does not necessarily unfold all proteins or reduce them to the same shape.

(e) Aggregation may occur in SDS.

All of these effects may introduce error in molecular estimates by SDS-PAGE. But other ambiguities exist, the worst being that we do not know with certainty how SDS works.

Certainly, many proteins migrate in suitable SDS-PAGE systems so that the logarithm of their molecular weight varies inversely with electrophoretic mobility. In such cases free electrophoresis in SDS and SDS-PAGE gives concordant results (116). However, Weber and Osborn (118) find that a low proportion of cross-linking in the gels made the log vs. molecular-weight curve concave, while high levels of cross-linking avoided this, but made the gels unsuitable for high-molecular-weight components.

Another problem derives from the SDS binding of diverse proteins. Segrest et al. (89) point out that the poor SDS binding of glycoproteins accounts for their low mobility in SDS-PAGE, but other alternatives clearly exist. Thus, how can one explain the fact that papain, pepsin, and glucose oxidase, *which do not bind SDS, migrate in SDS-PAGE as if they were SDS complexes properly reflecting protein molecular weight* (90)?*

To summarize, we consider SDS-PAGE a useful, facile method, but one which would be far less attractive if some better alternative were available. We must do much more fundamental work to make the technique unambiguous and quantitative, to understand its basic mechanisms, to provide meaningful alternatives, and to develop suitable preparative applications.

* Possible ambiguities inherent in SDS-PAGE continue to generate concern. Thus, careful analyses (J.-S. Tung and C. A. Knight, *Anal. Biochem.* **48**: 153, 1972) indicate that electrophoretic mobility of macromolecules in SDS-PAGE correlates to molecular weight only when the molecules possess the same charge: mass ratio and hydrodynamic configuration. Although many proteins happen to react with SDS in a comparable manner, others do not, leaving the technique empirical. A notable example of the type of anomaly likely to arise with membranes comes from studies of sarcoplasmic reticulum (W. Hasselbach and A. Migala, FEBS Letters **26**: 20, 1972). Two functionally distinct sarcoplasmic proteins which are separable by chromatography on DEAE cellulose in Triton X-100, exhibit identical mobilities upon SDS-PAGE.

Selective extraction

If our view of membrane organization (Fig. 2.7) bears any semblance of reality, extraction procedures selective according to membrane "affinity" would appear to represent a reasonable approach to membrane fractionation. In designing such a strategy, one must assume that "membrane-associated" proteins and "core proteins" bind to the membrane each through some dominant interaction, which might be selectively disrupted.

This type of approach is exemplified by the work of Rosenberg and Guidotti (101) on the purification of human erythrocyte membrane proteins.*

These workers *first* prepared hemoglobin-free erythrocyte "ghosts," in the process eluting many compounds, e.g., hemoglobin, which may be membrane associated in vivo, but lack high "membrane affinity."

Secondly, they eluted the high-molecular-weight component "spectrin" using 1.0 mM EDTA, 5.0 mM mercaptoethanol (11% of ghost protein).

In the *third* step, all protein extractable with 0.8 M NaCl was eluted (41% of ghost protein).

Fourthly, the residue was lipid extracted, a step eluting 7% of the protein.

In a *fifth* step, the lipid-free residue was dissolved in 3% SDS and subjected to gel-permeation chromatography, yielding the five major components listed in Table 2.5.

SELECTIVE PROTEOLYSIS

Certain membrane protein components, such as antigens, other specific receptors, portions of pumping systems, etc., extend through the membrane "permeability barrier"; such elements may be cleaved off by selective proteolysis, allowing their further characterization and elucidating their functional role. This approach has been used extensively and with considerable success. We do not plan to review it exhaustively, but wish to present some pertinent examples.

* Protein perturbants, such as guanidine-HCl, NaOH, lithium diiodosalicylate (5–20 mM), acid anhydrides, and certain organic mercurials effect the selective partial extraction of the same erythrocyte membrane polypeptides (Ref. 148; T. L. Steck and J. Yu, *J. Supramolec. Structure* **1:** 220, 1973). These polypeptides comprise roughly half the membrane protein and are thought to lie at the cytoplasmic membrane surface. Extraction of human erythrocyte ghosts with Triton X-100 at ionic strength near 0.04 preferentially elutes membrane glycoproteins, apparently as oligomeric complexes. At an ionic strength near 0.008, certain non-glycosylated polypeptides are also released (J. Yu, D. A. Fischman and T. L. Steck, *J. Supramolec. Structure*, **1:** 233, 1973). The elution achieved with Triton X-100 is essentially opposite to that produced with protein perturbants.

Table 2.5 Molecular Sieving of Presumptive "Core" Proteins of Human Erythrocyte Membranes on Sephadex G-200 Using 1% SDS[a]

Component	% of Residue	Range of Apparent Mol Wt
I	20	170,000–130,000
II	12	99,000– 75,000
III	31	38,000– 30,000
IV	17	25,000– 20,000
V	20	15,000– 10,000

[a] After Rosenberg and Guidotti (101).

Proteolytic enzymes have been used to unravel the protein architecture of *external* membrane surfaces and parts of *membrane cores* by evaluating the actions of peptidases on intact cells or purified, sealed membrane vesicles, and *internal* membrane surfaces, by applying the enzyme intracellularly by microperfusion, when possible.

In general, externally exposed plasma membrane proteins appear more resistant when peptidases are added to intact cells than to purified membrane vesicles, but there are major differences between cell types.*

Action of Proteases on Axons. Several workers have examined the effects of various proteases, applied externally, as well as introduced

* The greater protease action on isolated erythrocyte ghosts, compared with the membranes of intact cells, continues to be ascribed to the ability of the proteases to penetrate to the interior faces of the purified membranes. However, there is considerable species variation in the phenomenon, as well as a dependence upon the type of protease (R. B. Triplett and K. L. Carraway, *Biochemistry* **11:** 2897, 1972). The possibility that the membrane proteins become intrinsically more susceptible to digestion during membrane isolation remains to be tested adequately.

The effect of low levels of trypsin (1 μg/ml) on erythrocyte ghosts provides novel information, obscured in experiments employing exhaustive proteolysis (J. Avruch, H. D. Price, D. B. Martin, and J. R. Carter, *Biochim. Biophys. Acta* **291:** 494, 1973). The procedure appears to attack two major membrane proteins/peptides (mol wt 89,000 and 77,500 by SDS-PAGE), with associated exocytotic vesiculation of the membrane but no release of protein into solution. The data suggest that new charge groups appear on the internal surfaces of the vesicles and that the two proteins/peptides attacked are essential for membrane integrity. Interestingly, trypsinization appears to *reduce* the vesicles' permeability to glucose.

Carter et al. (J. R. Carter, Jr., J. Avruch, and D. B. Martin, *Biochim. Biophys. Acta* **291:** 506, 1973) have further examined the effects of trypsinization upon glucose transport by erythrocyte ghost vesicles. Untrypsinized vesicles do not exhibit preferential uptake of D-glucose. However, specific glucose transport begins after treatment with 1 μg trypsin/ml in correlation with the above noted cleavage of two membrane proteins. D-glucose transport persists even after exposure to 100 μg trypsin/ml and shows the following features: Accumulation and release of D-glucose occurs more rapidly than for L-glucose; a number of sugars competitively inhibit D-glucose transport; known inhibitors of glucose uptake, e.g., phlorizin, exert their typical action on the trypsin generated vesicles; countertransport of D-glucose can be demonstrated.

intracellularly via micropipettes, upon the electrical functions of nerve cells axons, to localize functionally significant proteins on the outer or inner plasma membrane surfaces. These experiments have addressed the electrical potential across the axonal membrane at rest, axonal excitability, and the amplitudes as well as forms of action potentials following nerve excitation. Because of size restrictions, these studies have primarily utilized the "giant axons" of various squids and crustaceans, nerve fibers which, though unmyelinated, bear a thin layer of satellite cells closely apposed to their external surfaces. Such "giant axons" exhibit normal potentials for hours under continuous intracellular perfusion, when perfusion fluids of appropriate ionic composition are used.

A number of proteases have been tested, e.g.,

(a) *nonspecific endopeptidases*, such as *Bacillus subtilis* protease (154);
(b) *specific endopeptidases*, such as *trypsin* and *chymotrypsin* splitting peptides involving aromatic amino acids (154);
(c) *exopeptidases*, such as *carboxypeptidases* A and B (154).

The results are consistent, regardless of species: *extraaxonal application of the various peptidases does not change axonal functions*. This observation is most reasonably attributed to a lack of exposed or protease-sensitive protein on the external surface, or to some type of steric shielding, possibly by carbohydrate.

Indeed, certain data show directly that enzymes can permeate through this layer and that the external surface of the axonal plasma membrane per se is not sufficiently susceptible to proteolysis to produce significant functional changes. Thus, Tobias (155) has shown that lobster giant axons exposed to trypsin remain excitable and capable of impulse conduction, although their surface structure is micromorphologically altered by this manipulation and the protease penetrates into the axon interior, as demonstrated by the appearance of protease activity in axoplasm extruded after treatment. These observations imply preexistence or enzymatic production of channels large enough to permit the permeation of as large a molecule as trypsin (molecular weight about 20,000).

There can be no question that proteases may seriously disrupt axonal function *when allowed to act upon its cytoplasmic surface*; however, we do not know the degree of peptidolysis required or the proteins involved in the functional impairment.

Roxas and Luxoro (156) show that the injection of trypsin (1 mg/ml) into squid giant axons causes decreased excitability in 2 min and full functional collapse within 5 min. Further extending these studies, Tasaki and Takenaka (154) find that intracellular perfusion of squid giant axons with trypsin, chymotrypsin, papain, ficin, carboxypeptidases A and B, as well as leucine aminopeptidase, at a concentration of 1 mg/ml, in various media, block conduction across the perfused zone, within 2.5–6 min (trypsin and chymotrypsin) to 30 min (leucine amino peptidase). Alterations of resting and action potentials and, in some cases, repetitive firing generally precede the conduction block.

Takenaka and Yamagashi (157) have further extended their approach, using enzyme concentrations of 0.01–1.0 mg/ml. They confirm previous data on trypsin, finding conduction block after 2–3 min, even at 0.01 mg/ml. Surprisingly, however, mixture of nonspecific endopeptidases such as *B. subtilis* (strain N′) protease, as well as bromelin, papain, and ficin, are considerably less active on axon function than trypsin, despite their greater peptidolytic action.

Thus, short (5 min) perfusions with *Bacillus proteus* protease and bromelin produce *minimal changes* on resting potentials, excitability, or action potentials, whereas longer perfusions leave the resting potential unaltered and produce only small alterations in action potential.

All these enzymes are inert when applied externally. The authors cannot explain the unimpressive action of the "pan-proteases," but further implicate internally exposed membrane proteins in the specific electrical properties of the membrane.

Antigens

Certain membrane constituents extend to the outer surface of the plasma membrane, there exposing specific chemical groups or providing unique patterns that comprise *antigenicity* of the cell. Antigens might be structural entities, enzymes, or parts thereof, hormone receptors, or combinations of these or other substances. In general, their functional role remains to be specified, although the L-antigen of sheep erythrocytes appears closely linked to active cation transport (158, 159).

The action of peptidases on intact cells supports the localization of certain antigens at the external surface of plasma membranes, a fact previously established by their accessibilities to immunoglobulins, with subsequent cell agglutination and/or immune lysis. However, certain proteases additionally allow the release, isolation, and characterization of peptide molecules bearing certain blood-group haptens, histocompatibility specificities, and other antigenic determinants.

For example, trypsinization of human erythrocytes releases sialoglycopeptide fragments of molecular weight about 31,000, which are about 78% carbohydrate and bear the *M*- or *N*-specificity of the parent cells (160–163). These fragments and their parent glycoproteins contain some of their carbohydrate linked to the peptide chain via *O*-glycosidic linkages between *N*-acetylgalactosamine and hydroxyl groups of serine and threonine (162–164). The amino acid composition of these glycopeptides is considerably more hydrophilic than that of the total glycoprotein, and Winzler (143) suggests that the carbohydrate-bearing peptide moiety locks into the membrane via a highly apolar peptide segment.

Interestingly, even prolonged treatment of intact, human erythrocytes with pronase, trypsin, or chymotrypsin destroys nearly all the cells' acetyl cholinesterase activity, but with no alteration of choline or Na^+ transport and very little hemolysis (165). In contrast, the Na^+, K^+-sensitive ATPase of red cell membrane fragments quickly deteriorates upon exposure to trypsin (166); such data tend to localize the cholinesterase to the external membrane surface.

Proteolysis has found extensive application in the study of *histocompatibility antigens*. Polypeptide fragments bearing mouse H-2 or human HLA determinants can be cleaved off the surfaces of suitable cells, or membrane isolates therefrom, by careful *papain* treatment (167–171). However, trypsinization damages the cells and destroys the antigens. In both systems, two types of peptidolysis release antigen-bearing glycopeptides, namely,

(a) *class I fragments*, with a molecular weight of 57,000, containing 85–90% polypeptide and 10–15% carbohydrate, including neutral sugars, sialic acid, and glucosamine;

(b) *class II fragments*, with the same overall composition, in the case of H-2, but a molecular weight of 35,000.

Class I fragments from mouse cells bear more than one H-2 determinant, whereas class II peptides appear to carry only one specificity.

These peptide fragments absorb appropriate allo-antibody and induce allograft immunity in vivo (172). However, the potency of the soluble isolates appears considerably less than that of antigen-bearing membrane fragments (173).

Nathenson and associates report similar amino acid compositions for both class I and class II peptides isolated from human or mouse cells, but tryptic or cyanogen bromide digests of the papain fragments from mouse strains differing at the H-2 locus ($H\text{-}2^b$ vs. $H\text{-}s^d$) yield different peptide maps (174), implicating amino acid sequence in the determination of the histocompatibility specificities; this could be direct or through attached carbohydrate (175). Thus, Sanderson et al. (176) have obtained highly purified HLA-glycopeptides using *particle-bound pronase*, but find that their specific antigenicity *declines but does not disappear*, even after maximal peptidolysis (177); this could be because steric hindrance by carbohydrate prevents complete enzyme action, or because sugars participate in determining histocompatibility specificity.

The above antigens appear to comprise membrane "core" entities, but we also know of loosely bound membrane-associated components that are polypeptide or protein in character and subject to cleavage by extrinsic proteases. For example, in some instances of Burkitt lymphoma (178, 179), the lymphocytes bear immunoglobulin M, with μ- and κ-chain specificity, on the outer surface of the plasma membrane. This resists extraction by gentle procedures, but can be cleaved off using papain digestion (180).

Several distinct chemical groupings in/on membranes combine with their specific ligands only under unusual conditions. They either reside too deeply within the plasma membrane and/or are sterically hindered to be normally inaccessible. They do, however, become more exposed after limited protease action on cell surfaces, normally abound on the surfaces of certain nontumorous cells, as well as during certain stages of mitosis, and become easily accessible upon *neoplastic conversion* of cells by oncogenic viruses and other carcinogenic agents. Moreover, they become more "strongly expressed" after permissive infections with certain viruses, whether potentially oncogenic or not. One of us has recently reviewed this matter in extenso (181).

Epilogue

Endowed with a formidable array of techniques for the study of membrane lipids, the membrane biologist lacks the technology to deal in detail with membrane proteins. This contrasts strikingly with the situation facing those interested in soluble proteins and often leads the membrane specialist to almost desperate maneuvers and unwitting pitfalls. For this reason, we have rather strongly stressed the ambiguities of the few techniques available, hoping also to stimulate the activity necessary to achieve the technological progress so urgently needed here.

References

1. Van Deenen, L. L. M. *Progr. Chem. Fats Other Lipids* **8:** pt. 1, 1965.
2. Kates, M., and Wassef, M. K. *Ann. Rev. Biochem.* **39:** 323, 1970.
3. Lennarz, W. J. *Ann. Rev. Biochem.* **39:** 359, 1970.
4. Florkin, M., and Stotz, E. H., eds. *Comprehensive Biochemistry* **18:** 1970.
5. Johnson, A. R., and Davenport, J. B. In *Biochemistry and Methodology of Lipids.* New York: Wiley, 1971.
6. Johnson, A. R., and Davenport, J. B. In *Methods in Enzymology,* vol. 14, J. M. Loewenstein, ed. New York: Academic Press, 1969.
7. Zwaal, R. F. A., Roelofson, B., Comfurius, P., and van Deenen, L. L. M. *Biochim. Biophys. Acta* **223:** 474, 1971.
8. Woodward, C. B., and Zwaal, R. F. A. *Biochim. Biophys. Acta* **274:** 272, 1972.
9. Fiehn, W., and Hasselbach, W. *Eur. J. Biochem.* **13:** 510, 1970.
10. Hasselbach, W., and Heimberg, K. W. *J. Memb. Biol.* **2:** 341, 1970.
11. Weidekamm, E., Wallach, D. F. H., and Fischer, H. *Biochim. Biophys. Acta* **241:** 770, 1971.

12. Eibl, H., Hill, E. E., and Lands, W. E. M. *Europ. J. Biochem.* **9:** 250, 1969.
13. Ferber, E. In *Biomembranes,* D. Chapman and D. F. H. Wallach, eds. London: Academic Press, 1973 (in press).
14. Roelofson, B., de Gier, J., and van Deenen, L. L. M. *J. Cell Comp. Physiol.* **63:** 233, 1964.
15. Fleischer, S., and Fleischer, B. In *Methods in Enzymology,* R. W. Estabrook and M. E. Pullman, eds. New York: Academic Press, 1967, p. 10.
16. Folch-Pi, J., Lees, M., and Sloane-Stanley, G. H. *J. Biol. Chem.* **226:** 497, 1957.
17. Folch-Pi, J., and Lees, J. M. *J. Biol. Chem.* **191:** 807, 1951.
18. Redman, C. M. *Biochim. Biophys. Acta* **282:** 123, 1972.
19. Mokrash, L. *Prep. Biochem.* **2:** 19, 1972.
20. Zahler, H. P., and Wallach, D. F. H. *Biochim. Biophys. Acta* **135:** 371, 1967.
21. Singer, S. J. *Adv. Prot. Chem.* **17:** 1, 1962.
22. Robinson, P. J., Dunnill, P., and Lilly, M. D. *Biochim. Biophys. Acta* **242:** 659, 1971.
23. Bruckdorfer, K. R., and Green, C. *Biochem. J.* **104:** 270, 1967.
24. Tsukagoschi, N., and Fox, C. F. *Biochemistry* **10:** 3309, 1971.
25. Wirtz, K. W. A., and Zilbersmit, D. B. *J. Biol. Chem.* **243:** 3596, 1968.
26. Taylor, R. B., Duffus, P. H., Raff, M. C., and de Petris, S. *Nature* **233:** 225, 1971.
27. Zahler, W. L., Fleischer, B., and Fleischer, S. *Biochim. Biophys. Acta* **203:** 283, 1970.
28. Wallach, D. F. H. In *The Specificity of Cell Surfaces,* B. D. Davis and L. Warren eds. Englewood Cliffs, N.J.: Prentice-Hall, 1967, p. 129.
29. Steck, T. L., and Wallach, D. F. H. In *Methods in Cancer Research.* New York: Academic Press, 1970, vol. V, p. 93.
30. Wallach, D. F. H., and Gordon, A. In *Regulatory Functions of Biological Membranes,* J. Järnefelt, ed. Amsterdam: Elsevier, 1968, p. 87.
31. Braun, V., and Busch, V. *Europ. J. Biochem.* **28:** 51, 1972.
32. Danielli, J., and Davson, H. M. *J. Cell Comp. Physiol.* **5:** 495, 1935.
33. Wallach, D. F. H., and Zahler, P. H. *Biochim. Biophys. Acta* **150:** 186, 1968.
34. Haydon, D. A., and Taylor, J. *J. Theoret. Biol.* **4:** 281, 1963.
35. Davies, J. T., and Rideal, E. K. In *Interfacial Phenomena.* New York: Academic Press, 1961, ch. 2, p. 56.
36. Maddy, A. H., and Kelly, P. G. *Biochim. Biophys. Acta.* **241:** 290, 1971.
37. Neville, D. M., Jr. *Biochim. Biophys. Acta* **133:** 168, 1967.
38. Ibid. **154:** 540, 1968.
39. Ibid. *Biochem. Biophys. Res. Commun.* **34:** 60, 1969.
40. Mazia, D., and Ruby, A. *Proc. Natl. Acad. Sci.* (U.S.) **61:** 1005, 1968.
41. Hamaguchi, H., and Cleve, H. *Biochim. Biophys. Acta* **233:** 320, 1971.
42. Harris, J. R. *Biochim. Biophys. Acta* **188:** 31, 1969.
43. Harris, J. R. *J. Mol. Biol.* **46:** 329, 1969.
44. Reynolds, J. A., and Trayer, H. *J. Biol. Chem.* **246:** 7337, 1971.
45. Reynolds, J. A. *Ann. N.Y. Acad. Sci.* **195:** 75, 1972.
46. Marchesi, V. T., and Steers, E., Jr. *Science* **159:** 203, 1968.
47. Fairbanks, G., Steck, T. L., and Wallach, D. F. H. *Biochemistry* **10:** 2606, 1971.

48. Racker, E., ed. In *Membranes of Mitochondria and Chloroplasts*. New York: Van Nostrand, 1970.
49. Kirk, J. T. *Ann. Rev. Biochem.* **40:** 161, 1971.
50. Dawson, R. M. In *Biological Membranes: Physical Fact and Function*, D. Chapman, ed. New York: Academic Press, 1968, p. 203.
51. Mitchell, C. D., and Hanahan, D. J. *Biochemistry* **5:** 51, 1966.
52. Tanford, C. *Adv. Prot. Chem.* **23:** 121, 1968.
53. Ibid. **24:** 1, 1970.
54. Triplett, R. B., Summers, J., Ellis, D. E., and Carraway, V. L. *Biochim. Biophys. Acta* **266:** 484, 1972.
55. Gwynne, J. T., and Tanford, C. *J. Biol. Chem.* **245:** 3269, 1970.
56. Marchesi, V. T., and Andrews, E. P. *Science* **174:** 1247, 1971.
57. Hatefi, Y., and Hanstein, W. G. *Proc. Natl. Acad. Sci.* (U.S.) **62:** 1129, 1969.
58. Shooter, E. M., and Einstein, E. R. *Ann. Rev. Biochem.* **40:** 635, 1971.
59. Maddy, A. H. *Biochim. Biophys. Acta* **117:** 193, 1966.
60. Rega, A. F., Weed, R. J., Reed, C. F., Berg, G. G., and Rothstein, A. *Biochim. Biophys. Acta* **147:** 297, 1967.
61. Hamaguchi, H., and Cleve, H. *Biochem. Biophys. Res. Commun.* **47:** 459, 1972.
62. Hamaguchi, H., and Cleve, H. *Biochim. Biophys. Acta* **278:** 271, 1972.
63. Fletcher, M. A., and Woolfolk, B. J. *J. Immunol.* **107:** 842, 1971.
64. Fletcher, M. A., and Woolfolk, B. J. *Biochim. Biophys. Acta* **278:** 163, 1972.
65. Wallach, D. F. H., and Zahler, P. H. *Proc. Natl. Acad. Sci.* (U.S.) **56:** 1552, 1966.
66. Lenard, J., and Singer, S. J. *Proc. Natl. Acad. Sci.* (U.S.) **56:** 1828, 1966.
67. Sarkar, P. K., and Doty, P. *Proc. Natl. Acad. Sci.* (U.S.) **55:** 981, 1966.
68. Graham, J., and Wallach, D. F. H. (unpublished).
69. Weidekamm, E., and Wallach, D. F. H. (to be published).
70. Kramer, R., Schlatterer, Ch., and Zahler, H. P. *Biochim. Biophys. Acta* **282:** 146, 1972.
71. Schnaitman, C. A. *Proc. Natl. Acad. Sci.* (U.S.) **63:** 412, 1969.
72. Zwaal, R. F. A., and van Deenen, L. L. M. *Biochim. Biophys. Acta* **150:** 323, 1968.
73. Winzler, R. J. In *Red Cell Membrane: Structure and Function*, G. A. Jamieson and T. Z. Greenwalt, eds. Philadelphia: J. B. Lippincott, 1969, p. 157.
74. Demus, H., and Mehl, E. *Biochim. Biophys. Acta* **203:** 291, 1970.
75. Takayama, K., MacLennan, D. H., Tzagaloff, A., and Stoner, C. D. *Arch. Biochem. Biophys.* **114:** 223, 1966.
76. Blumenfeld, O. O., Gallop, P. M., Howe, C., and Lee, L. T. *Biochim. Biophys. Acta* **211:** 109, 1970.
77. Wallach, D. F. H., and Zahler, P. H. (unpublished).
78. Drey, W. J., Papermaster, D. S., and Kühn, H. *Ann. N.Y. Acad. Sci.* **195:** 61, 1972.
79. Kundig, W., and Roseman, S. *J. Biol. Chem.* **246:** 1407, 1971.
80. Salton, M. R. J., and Schmitt, M. D. *Biochem. Biophys. Commun.* **27:** 529, 1967.
81. Philipott, J. *Biochim. Biophys. Acta* **225:** 201, 1971.
82. Philipott, J. *Bull. Soc. Chim. Biol.* **50:** 1481, 1968.

83. Robinson, D. R., and Jencks, W. P. *J. Amer. Chem. Soc.* **87:** 2470, 1965.
84. Rosai, J., Tillack, T. W., and Marchesi, V. T. *Fed. Proc.* **30:** 453, 1971.
85. Wallach, D. F. H. *J. Genl. Physiol.* **54:** 35, 1969.
86. Dunnick, J. D., Marinetti, G. V., and Greenland, P. *Biochim. Biophys. Acta* **266:** 684, 1972.
87. Reynolds, J. A., and Tanford, C. *Proc. Natl. Acad. Sci.* (U.S.) **66:** 1002, 1970.
88. Dunker, A. K., and Rueckert, R. R. *J. Biol. Chem.* **244:** 5074, 1969.
89. Segrest, J. P., Jackson, R. L., Andrews, E. P., and Marchesi, V. T. *Biochem. Biophys. Res. Commun.* **44:** 390, 1971.
90. Nelson, C. A. *J. Biol. Chem.* **246:** 3895, 1971.
91. Wallach, D. F. H., and Weidekamm, E. in *Erythrocytes, Thrombocytes, Leukocytes*, E. Gerlach, K. Moser, E. Deutsch, W. Wilmanns, eds. G. Thieme: Stuttgart, Germany, 1973, p. 2.
92. Katzman, R. L. *Biochim. Biophys. Acta* **266:** 269, 1972.
93. Reynolds, J. A., and Tanford, C. *J. Biol. Chem.* **245:** 5161, 1970.
94. Fish, W. W., Reynolds, J. A., and Tanford, C. *J. Biol. Chem.* **245:** 5166, 1970.
95. Knüfermann, H. (personal communication).
96. Tremblay, G. Y., Daniels, M. J., and Schaechter, M. *J. Mol. Biol.* **40:** 65, 1969.
97. Goodenough, D. A., and Stoeckenius, W. *J. Cell Biol.* **54:** 646, 1972.
98. Salton, M. R. J. *J. General Physiol.* **52:** 227S, 1968.
99. Fairbanks, G., and Avruch, J. *J. Supramolecular Structure* **1:** 66, 1972.
100. Heller, J. *Biochemistry* **7:** 2906, 1968.
101. Rosenberg, S. A., and Guidotti, G. *J. Biol. Chem.* **243:** 1985, 1968.
102. Moldow, C. F., Zucker-Franklin, D., Gordon, A., Hospelhom, V., and Silber, R. *Biochim. Biophys. Acta* **255:** 133, 1972.
103. S. Kolowick and N. O. Kaplan, eds. *Methods in Enzymology*. New York: Academic Press, vol, 2, 1967.
104. Ibid. vol. 11, 1967.
105. Ibid. vol. 22, 1971.
106. Spatz, L., and Strittmatter, P. *Proc. Natl. Acad. Sci.* (U.S.) **68:** 1042, 1971.
107. Dunham, P. B., and Hoffman, J. F. *Proc. Natl. Acad. Sci.* (U.S.) **66:** 936, 1970.
108. Wallach, D. F. H., and Garvin, J. E. *J. Amer. Chem. Soc.* **80:** 2157, 1958.
109. Merz, D. C., Good, R. A., and Litman, G. W. *Biochem. Biophys. Res. Commun.* **49:** 84, 1972.
110. Wipple, H. E., ed. *Ann. N.Y. Acad. Sci.* **121:** 305, 1964.
111. Crambach, A., and Rodbard, D. *Science* **172:** 440, 1971.
112. Anker, H. S. *FEBS Letters* **7:** 293, 1970.
113. Kapadia, G., and Crambach, A. *Anal. Biochem.* **48:** 90, 1972.
114. Tuppy, H., Swetly, P., and Wolff, I. *Eur. J. Biochem.* **5:** 339, 1968.
115. Panet, R., and Selinger, Z. *Eur. J. Biochem.* **14:** 440, 1970.
116. Banker, G. A., and Cotman, C. W. *J. Biol. Chem.* **247:** 5856, 1972.
117. Shapiro, A. L., Vinuela, E., and Maizel, J. V. *Biochem. Biophys. Res. Commun.* **28:** 815, 1967.
118. Weber, K., and Osborn, M. J. *J. Biol. Chem.* **244:** 4406, 1969.
119. Peacock, A. C., and Dingman, C. W. *Biochemistry* **7:** 668, 1968.

120. Harkins, W. D. *The Physical Chemistry of Surface Films*. New York: Reinhold, 1952, p. 305.
121. Weidekamm, E., Wallach, D. F. H., and Flückiger, R. *Anal. Biochem.* **54:** 102, 1973.
122. Raymond, S. *Ann. N.Y. Acad. Sci.* **121:** 350, 1964.
123. Ritchie, R. F., Harter, J. G., and Bayles, T. B. *J. Lab. Clin. Med.* **68:** 842, 1966.
124. Lenard, J. *Biochemistry* **9:** 1129, 1970.
125. Maizel, J. V., Jr. In *Fundamental Techniques in Virology*, K. Habel and N. P. Salzman, eds. New York: Academic Press, 1969, p. 334.
126. Hartman, B. K., and Udenfriend, S. *Anal. Biochem.* **30:** 391, 1969.
127. Wardi, A., and Michos, G. A. *Anal. Biochem.* **49:** 607, 1972.
128. Knüfermann, H., Wallach, D. F. H., and Schmidt-Ullrich, R. (to be published).
129. Berg, H. C. *Biochim. Biophys. Acta* **183:** 65, 1969.
130. Bender, W. W., Garan, H., and Berg H. C., *J. Mol. Biol.* **58:** 783, 1971.
131. Bretscher, M. S. *J. Mol. Biol.* **58:** 775, 1971.
132. Phillips, D. R., and Morrison, R. *Biochemistry* **10:** 1766, 1971.
133. Wallach, D. F. H. *Biochim. Biophys. Acta* **265:** 61, 1972.
134. Arrotti, J. J., and Garvin, J. E. *Biochim. Biophys. Acta* **255:** 79, 1972.
135. Arrotti, J. J., and Garvin, J. E. *Biochem. Biophys. Res. Commun.* **49:** 205, 1972.
136. Schmidt-Ullrich, R., Knüfermann, H., Wallach, D. F. H., and Fischer, H. *Biochim. Biophys. Acta* **307:** 353, 1973.
137. Avruch, J., and Fairbanks, G. *Proc. Natl. Acad. Sci.* (U.S.) **69:** 1216, 1972
138. Maddy, A. H. *Biochim. Biophys. Acta* **88:** 390, 1964.
139. Talbot, D., and Yphantis, D. A. *Anal. Biochem.* **44:** 216, 1971.
140. Weidekamm, E., Wallach, D. F. H., Neurath, P., Flückiger, R., and Hendricks, J. *Anal. Biochem.* (in press).
141. Rifkin, D. B., Compans, R. W., and Reich, E. *J. Biol. Chem.* **247:** 6432, 1972.
142. Weidekamm, E., Wallach, D. F. H., and Flückiger, R. (to be published).
143. Winzler, R. J. *Internat. Rev. of Cytol.* **29:** 77, 1970.
144. Perutz, J. F. *J. Mol. Biol.* **13:** 646, 1965.
145. Stark, G. R. In *Methods in Enzymology*, ch. II, 590, 1967.
146. Speth, V., Wallach, D. F. H., Weidekamm, E., and Knüfermann, H. *Biochim. Biophys. Acta* **255:** 386, 1972.
147. Steck, T. L., Fairbanks, G., and Wallach, D. F. H. *Biochemistry* **2617:** 1971.
148. Steck, T. L. *Biochim. Biophys. Acta* **255:** 553, 1972.
149. Marfey, P. S. In *The Red Cell Membrane: Structure and Function*, G. A. Jamieson and T. J. Greenwalt, eds. Philadelphia: J. B. Lippincott, 1969, p. 309.
150. Steck, T. L. *J. Mol. Biol.* **66:** 295–305, 1972.
151. Knüfermann, H., Himmelspach, K., Schmidt-Ullrich, R., and Wallach, D. F. H, in *Twenty-first Colloquium on Protides in the Biological Fluids*, H. Peelers, ed. (in press).
152. Knüfermann, H. K., Himmelspach, K., and Wallach, D. F. H. Quoted in D. F. H. Wallach, *The Plasma Membrane*. New York: Springer-Verlag, 1972, p. 60.
153. Himmelspach, K., Westphal, O., and Teichman, B. *Eur. J. Immunol.* **1:** 106, 1971.

154. Tasaki, I. A., and Takenaka, T. *Proc. Natl. Acad. Sci.* (U.S.) **52:** 84, 1964.
155. Tobias, J. M. *J. Genl. Phys.* **43:** 575, 1960.
156. Roxas, E., and Luxoro, M. *Nature,* **199:** 78, 1963.
157. Takenaka, T., and Yamagashi, S. *J. Genl. Physiol.* **53:** 81, 1969.
158. Lauf, P. K., and Tosteson, D. C. *J. Membr. Biol.* **1:** 177, 1969.
159. Lauf, P. K. *J. Membr. Biol.* **3:** 1, 1970.
160. Maekela, O., Miettinen, T., and Pesola, R. *Vox Sang.* **5:** 492, 1960.
161. Ohkuma, S., and Ikemoto, S. *Nature* **212:** 198, 1966.
162. Winzler, R. J., Harris, E. D., Pekas, J. D., Johnson, C. A., and Weber, P. *Biochemistry* **6:** 2196, 1966.
163. Thomas, D. B., and Winzler, R. J. *J. Biol. Chem.* **244:** 5943, 1969.
164. Kathan, R. H., and Adamy, A. J. *J. Biol. Chem.* **242:** 1716, 1967.
165. Martin, C. K. *Biochim. Biophys. Acta* **203:** 182, 1970.
166. Marchesi, V. T., and Palade, G. E. *J. Cell Biol.* **35:** 385, 1967.
167. Nathenson, S. G., Shimada, A., Yamane, K., Muramatsu, T., Cullen, S., Mann, D. L., Fahey, J. L., and Graff, R. *Fed. Proc.* **29:** 2096, 1970.
168. Shimada, A., and Nathenson, S. G. *Biochemistry* **8:** 4048, 1969.
169. Yamane, K., and Nathenson, S. G. *Biochemistry* **9:** 1336, 1970.
170. Mann, D. L., and Nathenson, S. G. *Proc. Natl. Acad. Sci.* (U.S.) **64:** 1380, 1969.
171. Mann, D. L., Rogentine, G. N. J., Fahey, L., and Nathenson, S. G. *J. Immunol.* **103:** 282, 1969.
172. Graff, R. J., and Nathenson, S. G. *Transplantation Proceedings* **3:** 249, 1971.
173. Simmons, T., and Manson, L. A. *Transplantation Proceedings* **3:** 253, 1971.
174. Shimada, A., Yamane, K., and Nathenson, S. G. *Proc. Natl. Acad. Sci.* (U.S.) **65:** 691, 1970.
175. Muramatsu, T., and Nathenson, S. G. *Biochemistry* **9:** 4875, 1970.
176. Sanderson, A. R., Cresswell, P., and Welsh, K. I. *Transplantation Proceedings* **3:** 220, 1971.
177. Cresswell, P., and Sanderson, A. R. *Biochem. J.* **117:** 43, 1970.
178. Klein, E., Klein, G., Nadakarni, J. S., Nadakarni, J., Wigzell, H., and Clifford, P. *Cancer Res.* **28:** 1300, 1968.
179. Hammond, E. *Exptl. Cell. Res.* **59:** 359, 1970.
180. Nadakarni, J. S., Svehag, S. E., Nadakarni, J., and Klein, G. *Immunology* **20:** 667, 1971.
181. Wallach, D. F. H. In *Biological Membranes,* D. Chapman and D. F. H. Wallach, eds. New York: Academic Press, 1973, p. 253.

CHAPTER 3

INTRODUCTION TO MEMBRANE SPECTROSCOPY AND SPECTROSCOPIC PROBES

General

Molecules exist in a series of discrete, discontinuous, quantized energy states, which differ in the energies of orbital electrons, in the electrostatic fields of atomic nuclei, the energies of vibration of these nuclei relative to each other, and the rotational energies of the molecules as a whole. Moreover, further discrete energy levels are generated upon exposure of the molecules to external electric or magnetic fields, due to interaction of such fields with the molecules, displacing orbital electrons to higher energy levels, altering the spins of electrons and nuclei, or the rotation of the molecule as a whole.

In representative small molecules, the various electronic levels are separated by energies in the order of 10^{-11} ergs; vibrational levels differ by 10^{-12}–10^{-13} ergs; the separation of rotational levels is in the order of 10^{-16} ergs, and the separation between energy levels *induced* by external magnetic or electric fields is smaller still.

When a molecule is exposed to electromagnetic radiation with an energy matching that required for the transition from one energy level to another, this energy *may* be absorbed, inducing the transition, or it *may not*. Whether it does so, depends upon the *transition probability*, which can be computed for simple systems by quantum mechanics. Many transitions are *forbidden*; i.e., they cannot occur at all. In general, one can predict which transitions are allowed under given conditions, using relatively simple mathematical *selection rules*.

In terms of transfer of energy to a molecule, electromagnetic radiation behaves as if it were composed of discrete energy "packages," *quanta* or *photons*. The energy E of each quantum relates to the *wave characteristics* of the radiation as follows:

$$E = h\nu = hc/\lambda = hc\omega \tag{3.1}$$

where E = energy in ergs; ν = frequency in sec^{-1}; λ = wavelength in cm (10^8 A); ω = wave number in cm^{-1}; $c = 2.9979 \times 10^{10}$ cm/sec, the velocity of propagation of electromagnetic radiation in vacuo; and h = Planck's constant (6.624×10^{-27} ergs-sec).

From the above, one can see that *ultraviolet radiation* ($\lambda \sim 2000$–3000 A) provides photons with sufficient energy to induce *electronic transitions*; however, some energy levels are so far apart that only quanta in the experimentally inaccessible frequency range between the ultraviolet and X-ray could effect the transition. Similarly, in some molecules, energy levels are sufficiently close together so that the electrons involved can be excited by visible light (e.g., ~4000–5000 A). Infrared radiation [λ = 1–20 μm (microns)] consists of photons that can induce typical vibrational transitions. Radio-frequency radiation (λ = 1 mm–1000 m) provides photons that can promote rotational transitions and transitions between energy levels in imposed electric and magnetic fields. Diverse molecules can, thus, exhibit characteristic energy absorptions in *all* of these spectral regions, the precise frequencies at which these occur depending upon the separation of the various intramolecular energy levels.

Absorption spectra present a function of radiant energy along the ordinate vs. frequency or wavelength along the abscissa. One commonly plots absorbance (A) or transmittance (%T) against wavelength (nm–m) or frequency in cm^{-1} (wave number). If I_0 is the radiant intensity incident on the sample and I the transmitted intensity, according to the Beer-Lambert-Bouguer law,

$$I = I_0 \cdot e^{-alc} \quad \text{or} \quad \frac{I}{I_0} = e^{-alc} \tag{3.2}$$

where a = absorptivity at a given frequency in units of one's choosing; l = path length of the sample in cm; and c = concentration of the sample. Further I/I_0 = the transmittance T, conveniently expressed as % transmittance (%T). Thus, $T = e^{-alc}$, and $\log_e T = \log_e e^{-alc} = alc$. Also, $\log_{10} 1/T = alc$, and $\log_{10} 1/T$ is defined as the absorbance A, the $\log_{10}$ of the reciprocal of transmittance.

Vibrational energy transitions can also be induced by visible or UV light through the *Raman effect*. When transparent molecules are exposed to a beam of light, most of the photons pass through. However, a small proportion of the incident light is *scattered* at the same frequency as the incident radiation. A still smaller fraction of photons interact with the molecules, giving up enough of their energy to induce a vibrational transition. Such scattered photons exhibit frequencies generally less than that of the incident light. The frequency difference corresponds to the energy-level separation involved in the transition [Eq. (3.1)]; in some cases, energy may also be added to the scattered radiation. This phenomenon is currently assuming major importance in *laser Raman spectroscopy*, which, apart from its intrinsic significance, forms an important supplement to *infrared* spectroscopy, since the two approaches are governed by different selection rules.

Exact treatments of molecular dynamics by quantum mechanics has only been achieved for very simple substances. The use of spectroscopy in the study of macromolecules or supramacromolecules remains mostly empirical and assumes that the energy levels of various bonds, etc., depend on the array and environment of only few atomic nuclei, so that structural features of localized regions of large molecules may be inferred by comparing the spectral properties of the macromolecule to those of appropriate small reference molecules.

Not only are spectroscopic analyses of large molecules and molecular arrays empirical, but they are also *averaging techniques*. If this is true for isolated macromolecules, it is particularly so for membranes, which often exhibit a striking, environmentally dependent topologic heterogeneity (Fig. 3.1) and comprise supramolecular arrays, including numerous, diverse lipids and peptides (Fig. 3.2). The full power of spectroscopic tech-

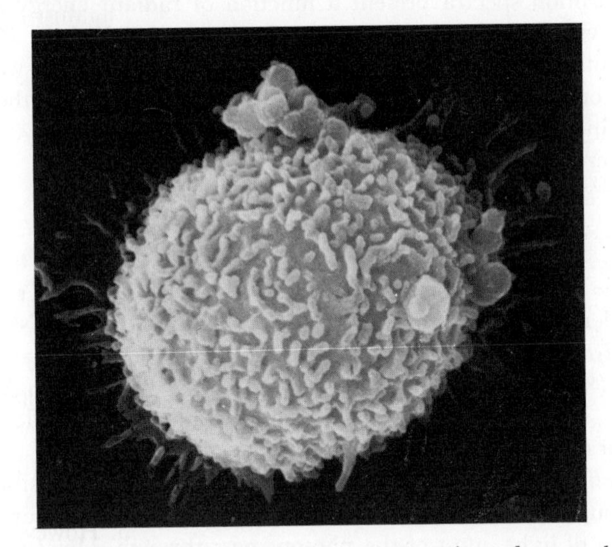

Fig. 3.1 *Scanning electron micrograph* of a mouse peritoneal macrophage in the process of attaching to a glass surface. The cell, fixed at 25°C with 2% gluturaldehyde before preparation for scanning electron microscopy, shows a stunning surface morphology. Much of the cell surface is elaborated by branching, undulating microvilli, ~ 100 nm in diameter, enormously increasing surface area of the cell and serving as contact points with other cells. Upon contact with glass, the cell spreads and flattens out, as we begin to see here. Above all, the photograph demonstrates the striking surface heterogeneity of certain cells. Magnification 10,000. (Courtesy of Dr. P. S. Lin.)

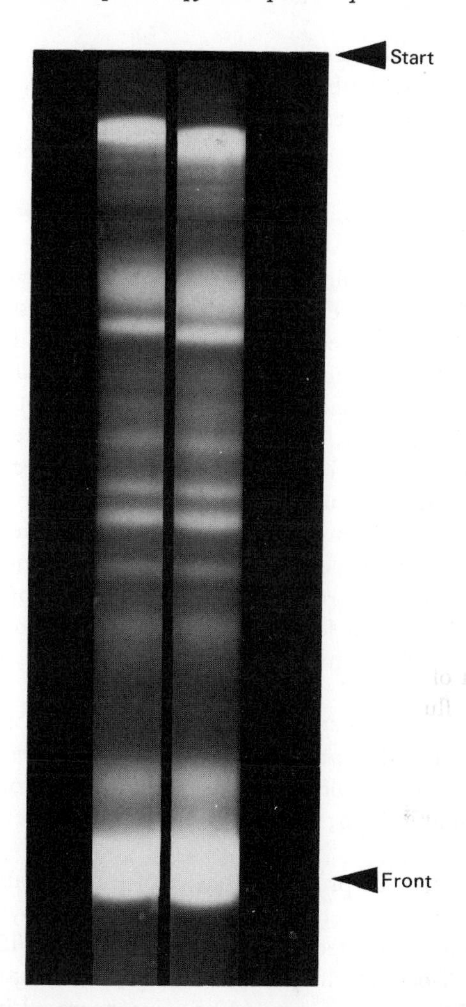

Fig. 3.2 *Electrophoretic separation* of the protein/peptide components of human erythrocyte membranes in 5.7% polyacrylamide gels using a buffer which is 1% in sodium dodecyl sulfate (SDS). Under the conditions employed most, but not all, proteins separate into their peptide subunits, which migrate as rod-shaped SDS complexes roughly according to the molecular weights of the detergent-peptide complexes. Proteins/peptides are visualized as fluorescent complexes with *o*-phthalaldehyde and 2-mercaptoethanol. The left gel was loaded with 3-μg protein, the right with 4 μg. The cathode is at the top. The intense band at the front contains little protein. While demonstrating the diversity of proteins/peptides in even a relatively simple membrane, these electrophoretic patterns must be interpreted cautiously—not all polymeric proteins dissociate into their minimal peptide entities in SDS, and proteins/peptides with unusual intrinsic charge properties (e.g., glycopeptides), unusual SDS binding and entities that remain globular in the presence of detergent, all exhibit anomalous migration in this system (E. Weidekamm and D. F. H. Wallach, unpublished).

niques can, thus, be realized only through their application in conjunction with each other and with biophysical, biochemical, and biological approaches, such as discussed in Chapters 1 and 2.

Spectroscopic Probes

Aims

Certain small molecules reflect the polarity, viscosity, and other features of their immediate environment in their spectroscopic properties. This realization has long been utilized in determining, for example, the critical micelle concentrations of amphiphilic molecules (1). A large armamentarium of such "molecular probes" exists—optically absorbing, optically active, fluorescent, paramagnetic—which have been bound to biological macromolecules or inserted into macromolecular systems, in efforts to determine molecular architecture and the relations between structure and function. This approach has yielded considerable insight into the dynamic structure of some proteins; indeed, in certain cases, it has been possible to insert probes into active centers, or other well-defined sites, and to correlate their location with X-ray crystallography. The use of probes in the study of membranes is less advanced, but under very active investigation.

The optimal use of a "molecular probe" requires that it be stable in its membrane location and that it *not disturb those membrane properties under study*. The probe should reflect the characteristics of its binding site in an interpretable fashion; ideally, it should be restricted to a unique, defined site.

We suspect that all probes fall short of these ideals in most cases. This is brought out in a very important, basic study by Lesslauer et al. (2). They examined the effect of the fluorescent probes 12, 9-anthranoyl-stearate (AS), *N*-octadecylnaphthyl-2- amino-6-sulfonate (ONS), and 1-anilino-8-naphthalene sulfonate (ANS), incorporated into vesicles and oriented multilayers of dipalmitoyl lecithin, upon the X-ray diffraction of these phosphatide arrays. Their data show that all of the probes perturb the native bilayer structure of the phosphatides.

AS appears to lie with the carboxyl residue in the plane of the polar ends of the phosphatides, while the anthracene ring lies in the apolar core, where it perturbs the hydrocarbon chain packing. The probe also introduces an apolar moiety into the relatively polar glycerol region. Because of its length, ONS apparently fits rather poorly, with the sulfonate in the plane of the choline residues and the fluorophore also in the polar region. ANS appears to lie as in lecithin-cardiolipin dispersions (3), with the sulfonate packed among the polar head groups, but the bulky fluorophore penetrating only slightly into the hydrocarbon chain region. As consequence, the latter is quite anomalously structured.

The studies of Cadenhead and Katti (4) show this problem to be equally serious for spin labels. They compared pure monomolecular films of nitroxide derivatives of 5-α-androstan-3-one and 5-α-cholestan-3-one, as well as films of these substances mixed with myristic acid, with established data on cholesterol and cholesterol-myristic acid films, arguing with reason that their monolayers are analogous to ½ of the type of lipid bilayers postulated for membranes. Not unexpectedly, the pure nitroxide films are considerably more expanded than those of cholesterol under similar conditions. When introduced into myristic acid films, they also tend to expand these, unlike the condensing effect produced by cholesterol under similar conditions.

The crucial question of whether spin-labeled lipids are suitable analogues for natural molecules has been further investigated by Tinoco et al. (5). For this they compared the pressure-area curves of two widely used spin labels, 3-nitroxide cholestane and 12-nitroxide stearate with the natural lipids, all in monolayers at air-water interfaces. The data indicate that the spin labels are not ideal analogues for, and may indeed be rather unrepresentative of, the natural molecules.

Thus, 3-nitroxide cholestane occupies a greater area/molecule than cholesterol at the same pressure and appears less well anchored at the interface. Also, film expansion occurs when cholesterol and the spin

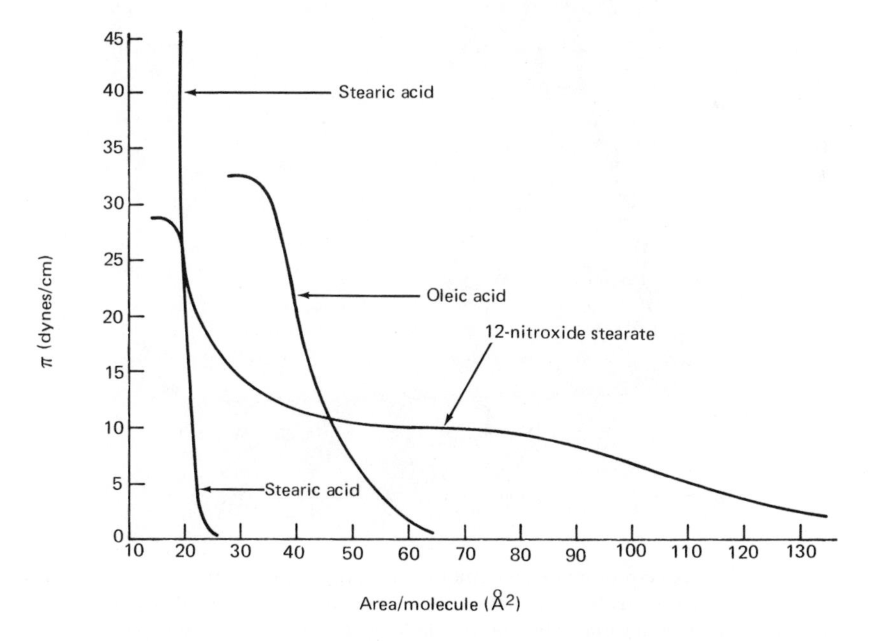

Fig. 3.3 Pressure-area curves of oleic acid, stearic acid, and 12-nitroxide stearic acid at the air-water interface. [From Tinoco et al. (5). Courtesy of Elsevier Publishing Co.]

label are spread as mixed films, indicating greater affinity between unlike molecules than like. Finally, the interaction between the spin label and lecithin differs markedly from that of cholesterol. It also appears that this spin label would report preferentially from cholesterol-containing domains.

The nitroxide stearate yielded a pressure-area curve much expanded over that of the natural lipid (Fig. 3.3). Most likely, both the nitroxide and the carboxyl residues localize at the air-water interface with the hydrocarbon extended horizontally over the interface for considerable distances (Fig. 3.4).

Clearly, the studies of Lesslauer et al. (2, 3), Cadenhead and Katti (4), and Tinoco et al. (5) represent essential efforts, basic to the meaningful utilization of fluorescent and spin-label probes. Importantly, they clearly show (a) *that the probe molecules differ intrinsically from the molecules they are supposed to represent;* (b) *that probes can act as molecular "tumors" perturbing the structure they are intended to characterize.*

These studies illustrate the uncertainty factor implicit in the entire "probe approach." Thus, most fluorescent and spin-label probes could

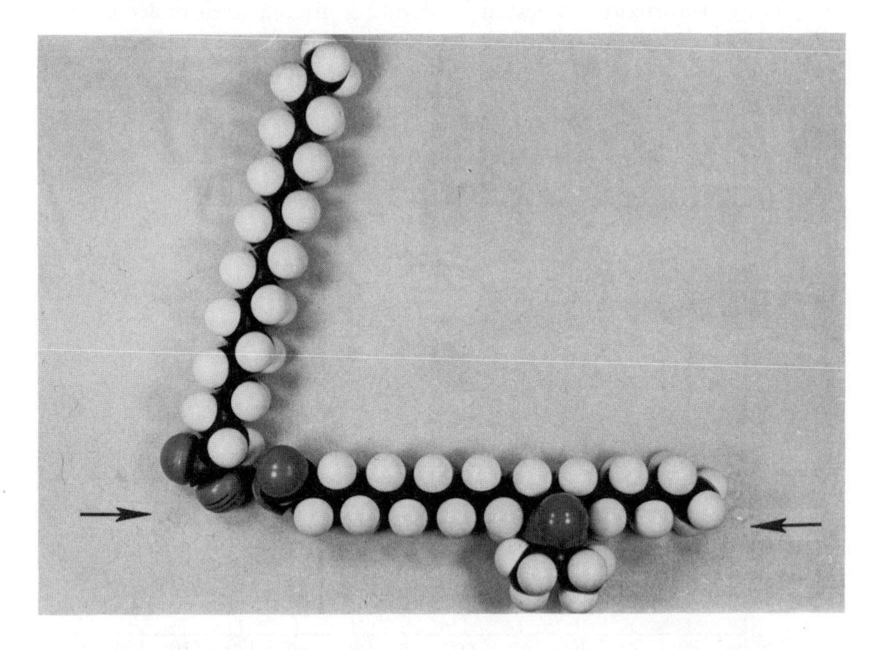

Fig. 3.4 Space-filling models demonstrating the preferred positions of stearate and 12-nitroxide stearate in monolayers at air-water interfaces upon partial film compression. The interface lies perpendicular to the plane of the page at the position of the arrows. The stearate has its polar group anchored in water and the fatty acid chain in air. The nitroxide derivative most likely anchors to water at two sites, namely, the carboxyl group and the polar spin-label group.

Fig. 3.5 (a) Space-filling models of some stearic acid "probe" derivatives. The native molecule is shown on the extreme right. *Fluorostearic* acid, left, an NMR probe, differs little sterically and should serve as a useful stearic acid analogue. In *anthranoyl stearate,* second from left, an enormous residue is attached to the fatty acid chain. It must be suspected as perturbant and moreover is unlikely to fit into well-ordered hydrocarbon regions. *12-nitroxide stearate,* second from the right, also appears like a "molecular tumor." Here the perturbing effect of the reporter group's polarity must be considered also.

artifactually produce a "fluidizing effect" in lipid and/or lipid-protein regions. Moreover, they may only be able to penetrate into rather amorphous regions, giving no information about other domains. A comparison of space-filling models of diverse probes with their natural analogues [Fig. 3.5 (a, b, and c)] emphasizes this point. We therefore urge considerably greater caution in the interpretation of results produced by diverse probes than hitherto employed and view the many ingenious probe experiments summarized below with a certain reserve. Indeed, we suggest that *unless proved otherwise,* a probe cannot be considered representative of *all* membrane domains and must be suspect of acting as a perturbant of the region it is monitoring. A cogent biological example of the latter is the immobilization of spermatozoa by membrane-bound naphthalene sulfonates, the most commonly used fluorescent probes (6), as well as the coenzyme competitive action of ANS on yeast glyceraldehyde-3-phosphate dehydrogenase (7).

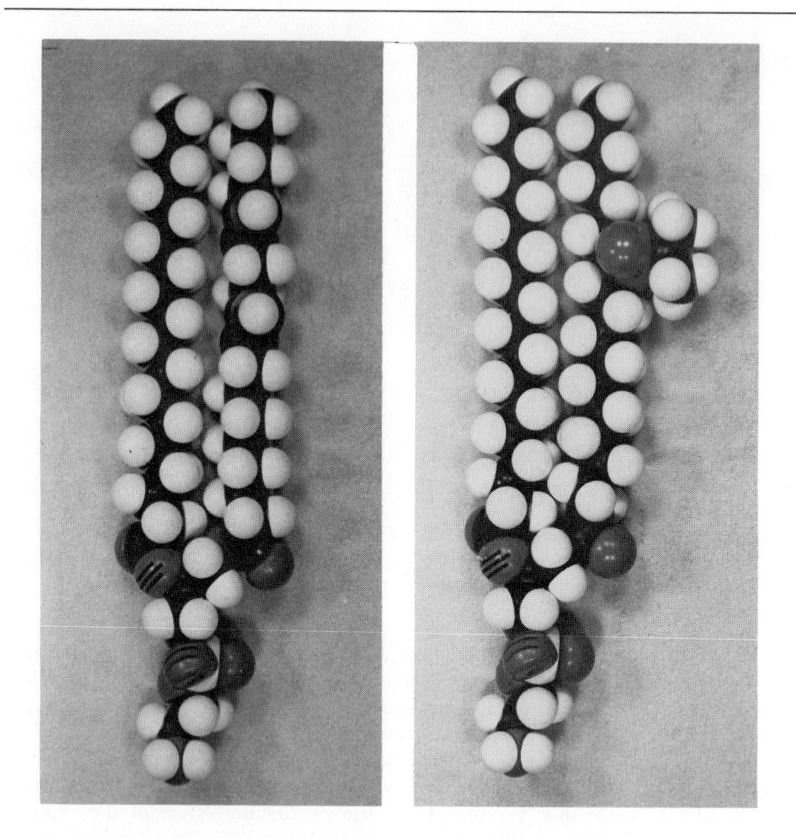

Fig. 3.5 (b) Space-filling models of some of the phosphatidylethanolamine "probe" derivatives. The *native* unsaturated molecule is shown on the extreme *left*. Next, on the right, is the molecule with the *nitroxide* residue on one of the *hydrocarbon* chains. This would interfere with packing of normal chains about it and may interfere with ready insertion into a well-ordered lipid region. To the *right center* lies a phosphatidylethanolamine with the nitroxide attached to the *polar* head group. We find it difficult to see how such a molecule could adequately

Fluorescent Probes

Fluorescent probes yield excitation and emission spectra at very low concentrations and provide information about the polarity and viscosity in the vicinity of the probe, the flexibility of the binding site, and the proximity of the probe to other fluorophores (8).

Optical Absorption

Optically absorbing, nonfluorescent, nonparamagnetic substances can also serve as probes (1), but relatively high probe concentrations are

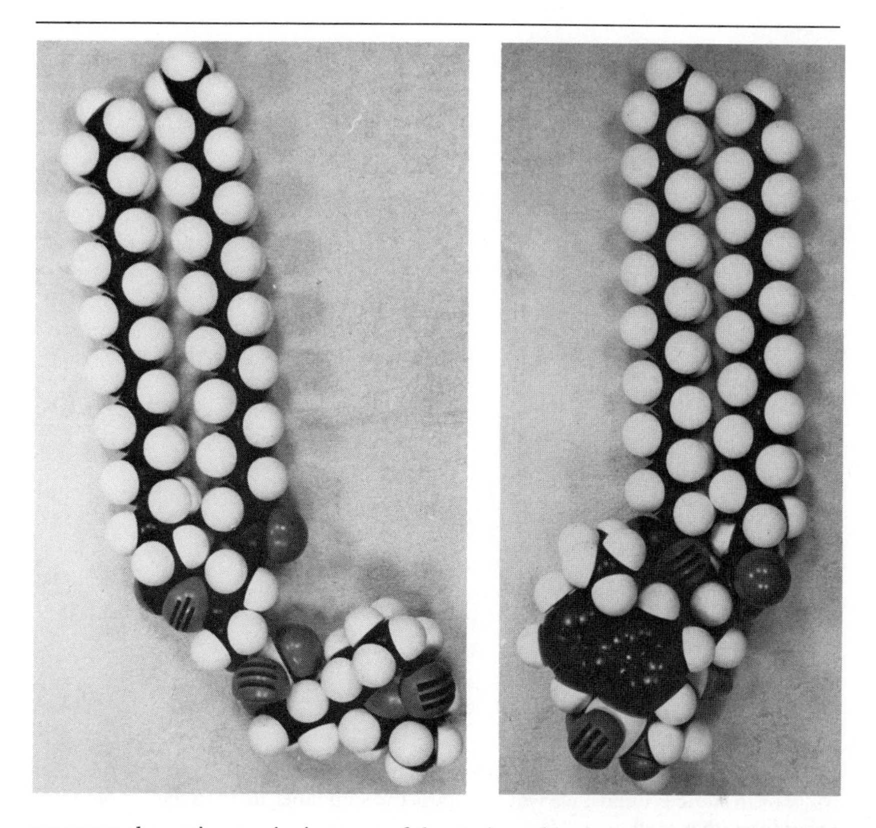

represent the native species in terms of the steric and ionic interactions of the head group, particularly since the —NH_2 group is eliminated, yielding a molecule with net — charge. To the *extreme right* lies DANS—phosphatidylethanolamine. The massive, apolar DANS group clearly changes the steric properties of the polar region dramatically, and, eliminating the native —NH_2 group, creates a species with a net — charge. Current data suggest that the DANS residue folds over and lies in the "glycerol" region when DANS phosphatidylethanolamine becomes incorporated into a lipid bilayer.

required experimentally. They also lack the versatility of fluorescent probes, since the absorption spectra reflect only the conditions of the ground state.

Spin Labeling

The spinning charge of an electron induces a magnetic field. When electrons are paired, as in most chemical bonds, their spins are opposite, and their magnetic moments cancel. However, molecules such as nitroxides ($\supset N^+—O^-$) contain linkages with solitary electrons. Such

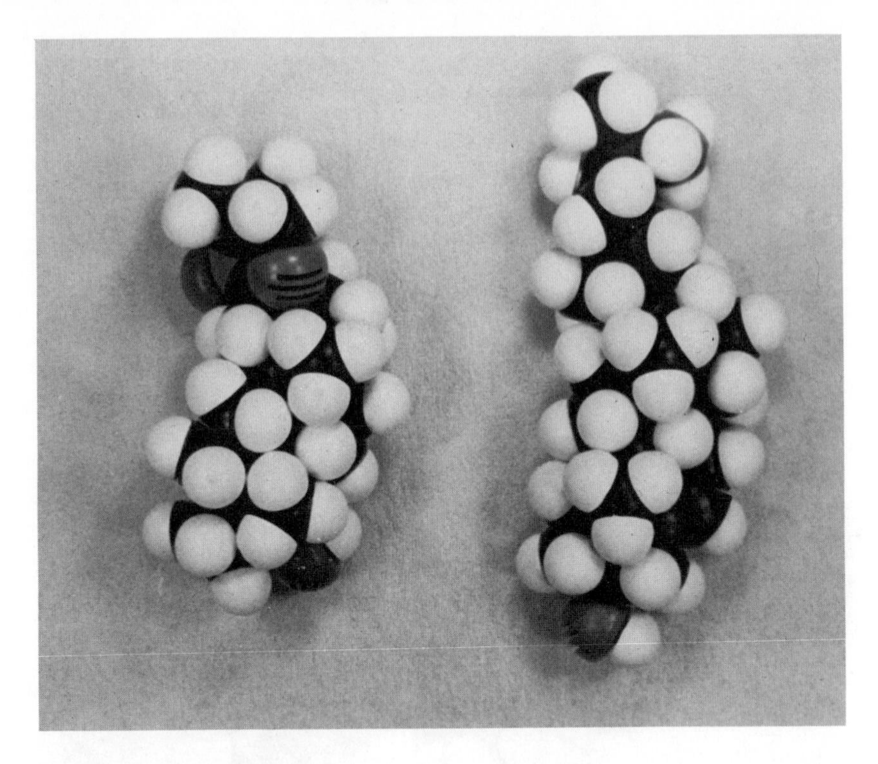

Fig. 3.5 (c) Space-filling models of a steroid nitroxide (left) and cholesterol (right). On steric grounds, we would expect the spin label to represent cholesterol behavior rather well.

paramagnetic substances are the subject of electron spin resonance (ESR) spectroscopy (electron paramagnetic resonance spectroscopy). When placed in a magnetic field, the unpaired electrons orient themselves in the direction of the field. This occurs because an electron has an angular momentum, *its spin*, with a quantum value of $+\frac{1}{2}$ or $-\frac{1}{2}$, i.e., the spin can have either of two opposite senses. The spin gives the electron a magnetic moment. In an applied magnetic field, the electron has lowest energy when this moment is aligned parallel (rather than antiparallel) to the applied field vector. When exposed to electromagnetic radiation of appropriate (resonant) frequency, energy is absorbed, and the electron can flip to align its magnetic moment antiparallel to the applied field.

The magnetic properties of an unpaired electron depend upon the electrical and magnetic fields of nearby electrons and nuclei, its mobility and its relation to external magnetic and electrical fields. Accordingly, McConnell and associates (9, 10) developed the important technique of "spin labeling," in which sterically protected—hence relatively stable—

nitroxide free radicals are covalently coupled to appropriate carrier molecules to give information about the number of probes in a given region, the polarity of the binding sites, the molecular motion of the probe (i.e., the effective viscosity of the medium), the orientation of the probes in an external magnetic field, and the interaction among various spin-labeled molecules themselves and between spin labels and nearby fluorophores*. Spin labeling, in general, has been recently reviewed (11–14).

Proton Paramagnetic Probes

This recent and promising technique utilizes the distinctive PMR signals of certain small, membrane-active molecules to provide information about mechanisms of membrane perturbation.

References

1. Corrin, H. L., and Harkins, W. D. *J. Am. Chem. Soc.* **69:** 679, 1947.
2. Lesslauer, W. L., Cain, J. E., and Blasie, J. K. *Proc. Natl. Acad. Sci.* (U.S.) **69:** 1499, 1972.
3. Lesslauer, W. L., Cain, J. E., and Blasie, J. K. *Biochim. Biophys. Acta* **421:** 547, 1971.
4. Cadenhead, D. A. C., and Katti, S. S. *Biochim. Biophys. Acta* **241:** 709, 1971.
5. Tinoco, J., Ghosh, D., and Keith, A. D. *Biochim. Biophys. Acta* **274:** 279, 1972.
6. Edelman, G. M., and Milette, C. F. *Proc. Natl. Acad. Sci.* (U.S.) **68:** 2436, 1971.
7. Nakradova, N. K. N., Asryant, R. A., and Ivanov, M. V. *Biochim. Biophys. Acta* **263:** 622, 1972.
8. Brand, L., and Withhold, B. In *Methods in Enzymology*, vol, XI, C. H. W. Hirz, ed. New York: Academic Press, 1967, p. 776.
9. Hamilton, C. L., and McConnell, H. M. In *Structural Chemistry and Molecular Biology*, A. Rich and N. Davidson, eds. New York: W. H. Freeman, 1968, p. 115.
10. Griffith, O. H., and Waggoner, A. S. *Acc. Chem. Res.* **20:** 17, 1969.
11. McConnell, H. M., and McFarland, B. G. *Quart. Rev. Biophys.* **3:** 91, 1970.
12. Smith, I. C. P. In *Biological Applications of Electron Spin Resonance Spectroscopy*, J. R. Balton, D. Borg, and H. Schwartz, eds. New York: Wiley, 1971.
13. Griffith, O. H., and Waggoner, A. S. *Accounts Chem. Res.* **2:** 17, 1969.
14. Jost, P., and Griffith, O. H. In *Methods of Pharmacology*, C. C. Chignell, ed. New York: Appleton, 1971, vol. 2, ch. 7.

*A spin-labeled, fluorescent derivative of 5– dimethylaminonaphthalene sulfonamide has been synthesized as a fluorescent probe for membranes (R. A. Long and J. C. Hsia, *Can. J. Biochem.* 51, 876, 1973). The compound is fairly water soluble, but partitions into the polar regions of phosphatide multibilayers. Incorporation of cholesterol into the bilayers does not alter the fluorescence or electron paramagnetic resonance of the compound, but decreases its solubility in the multibilayers. The interaction between the nitroxide and the fluorophore are not investigated.

CHAPTER 4

INFRARED AND LASER RAMAN SPECTROSCOPY

Introduction to Infrared Spectroscopy

Absorption of Infrared (IR) Radiation

The interatomic distances within molecules fluctuate cyclically about average values through one or more simultaneous vibrational motions—stretching, deformation, torsion, etc. When such motions change the dipole moment of a given bond, its electric field oscillates at the same frequency as the bond vibration (10^{10}–10^{11} Hz, corresponding to wavelengths of 2–20 μ). Moreover, upon exposure of the bond to electromagnetic radiation of the same frequency as the dipole oscillation, it absorbs some of the radiant energy in a quantal fashion, promoting it to a higher energy state. The energies involved in IR spectroscopy are of the order of tenths of electron volts and, unlike ultraviolet and visible light, IR radiation effects only vibrational and rotational transitions within molecules. Indeed, at normal temperatures, IR absorption usually arises principally from vibrational rather than rotational transitions. IR spectroscopy deals with these processes at ranging levels of sophistication. The vibrational frequencies of polyatomic molecules derive from (a) the masses of the linked atoms, (b) their geometric interrelationships, and (c) the forces due to deviation from equilibrium bond configurations. In simple molecules up to 20 atoms, these factors can be approximated by classical mechanics (1–3), but interpretation of IR spectra of larger molecules is complicated by the enormous number of different, fundamental vibrations ($3n - 6$ for a molecule with n atoms). The utility of IR spectroscopy of complex molecules depends upon (a) use of symmetry considerations where possible, (b) group frequency approaches when symmetry is low, and (c) use of small model compounds. Several publications (4–6) give detailed listings of the characteristic absorption frequencies corresponding to the group vibrations of organic molecules. Some frequencies of general interest are listed in Table 4.1.

Table 4.1 Wavenumber values for some important IR-active groups

Wavenumber (cm^{-1})[a]	Functional Group
	O—H stretching
3636–3571	Free O—H
3546–3448	Bonded O—H···O of single-bridged dimer
3390–3077	Bonded H···O—H···OH of double-bridged polymer
	C—H stretching
3105–3077	$R_2{=}CH_2$
3049–3012	$R_2{=}CHR$
2941–2899	R—CH_3
2924–2857	R_2—CH_2
2899–2874	R_3—CH
2857–2703	R—C(=O)—H
2703	C—H and bonded O—H···C combination band
	N—H stretching
3509 and 3390	Free N—H primary amide
3333 and 3175	Bonded N—H···primary amide
3448–3390	Free N—H secondary amide
3333–3279	Bonded N—H···secondary amide, single bridge (trans)
3175–3144	Bonded N—H···secondary amide, single bridge (cis)
3509–3311	N—H primary amine
3509–3311	N—H secondary amine
3390–3205	N—H imines
3125–3030	$\overset{+}{N}H_3$ amino acids
2150	$\overset{+}{N}H$
	C—D stretching
2155	C—D
2066	C—D
	P—H, P—OH stretchings
2469–2353	P—H
2703–2564	P—OH (broad absorption)

[a] Wavenumber (cm^{-1}) = 1/(wavelength in cm).

Table 4.1 (*cont.*)

Wavenumber (cm^{-1})	Functional Group
	<u>C=O and C=C stretching vibrations</u>
	<u>Aldehydes</u>
1739–1724	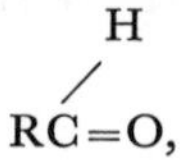 H / RC=O, saturated
1715–1695	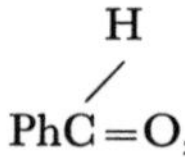 H / PhC=O, aryl
1709–1681	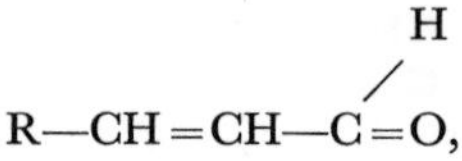 H / R—CH=CH—C=O, α, β-unsaturated
	<u>Ketones</u>
1724–1709	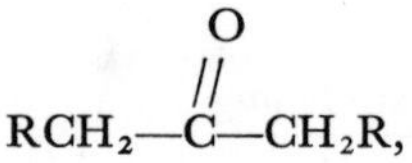 O ‖ RCH_2—C—CH_2R, saturated
1667–1653	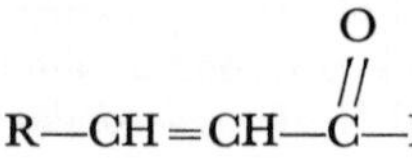 O ‖ R—CH=CH—C—R, α, β-unsaturated
1745	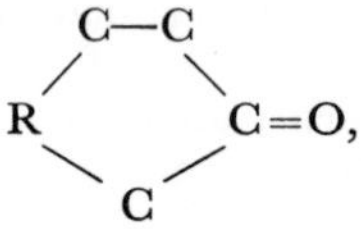 C—C / \\ R C=O, 5-membered, saturated ring \\ / C
1721	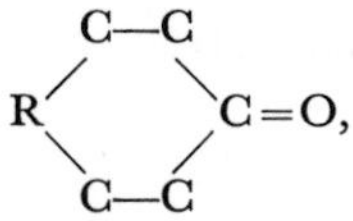 C—C / \\ R C=O, 6-membered, saturated ring or \\ / C—C
	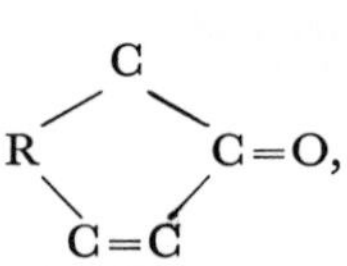 C / \\ R C=O, 5-membered, α, β-unsaturated ring \\ / C=C
1681	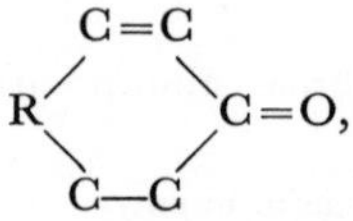 C=C / \\ R C=O, 6- (or7-)membered, α, β-unsaturated ring \\ / C—C

Table 4.1 (*cont.*)

Wavenumber (cm^{-1})	Functional Group
	<u>Acids</u>
1761	$R{-}\overset{\overset{O}{/\!/}}{C}{-}OH$, saturated monomer
1724–1071	$R{-}\overset{\overset{O \cdots HO}{/\!/}}{C}{-}OH,\ O{=}C{-}R$, saturated dimer
1695–1689	$R{-}C{=}C{-}\overset{\overset{OH}{\vert}}{C}{=}O$, α, β-unsaturated
1695–1681	$Ph{-}\overset{\overset{OH}{\vert}}{C}{=}O$, aryl
1667–1653	Chelated hydroxy acids, some dicarboxylic acids
	<u>Esters</u>
1770	$H_2C{=}C{-}\overset{\overset{OCH_3}{\vert}}{C}{=}O$, vinyl ester
1739	$R{-}\overset{\overset{OCH_3}{\vert}}{C}{=}O$, saturated
1724–1718	$R{-}C{=}C{-}\overset{\overset{OCH_3}{\vert}}{C}{=}O$, α, β-unsaturated, or $R{-}COOPh$, aryl
	<u>C—H deformations (saturated)</u>
1492–1449	$-CH_2-$
1471–1429	$-C-CH_3$, asymmetrical deformation
1399–1389	$-C-(CH_3)_3$
1389–1379	$-C-(CH_3)_2$
1379–1370	$C-CH_3$ symmetrical deformation
1342–1333	$-C-H-$

Table 4.1 (*cont.*)

Wavenumber (cm^{-1})	Functional Group
	C—O stretching and C—OH bending
	Alcohols
1205–1124	Tertiary open chain saturated
1124–1087	Secondary open chain saturated
1087–1053	Primary open chain saturated
1205–1124	Highly symmetrically branched secondary
1124–1087	α-unsaturated or cyclic tertiary
1099–1087	Secondary with branching on one α-carbon
	Acids
1290–1282	C—O
1190–1183	C—O
	Esters
1266–1250	C—O
1190–1177	C—O
	Anhydrides
1299–1205	Cyclic
1177–1053	Open chain
	Phosphorus
1300–1230	P=O (hydrogen bonding may shift to lower frequencies; usually two bands)
1050	P—O—CH_3 aliphatic
	Skeletal vibrations
893	Epoxy-oxirane ring derived from internal R—C=C—R (trans only)
833	Epoxy-oxirane ring derived from internal R—C=C—R (cis only)
833	Hydroperoxide
772	Ethyl
769	CH_2 rocking on long carbon chain
741	*n*-propyl
725	Hydroperoxide

Table 4.2 lists the most important IR absorption bands of H_2O and D_2O.

Table 4.2 Infrared Vibrational Spectra of H_2O and D_2O*

Vibration	H_2O	D_2O
O—X stretching (ν_s)	Very broad band with two main maxima and a shoulder (sh) at 25°C	
	$\nu = 3920$ sh	$\nu = 2900$ sh
	$\varepsilon = 0.83$	$\varepsilon = 0.60$
	$\nu = 3490$	$\nu = 2540$
	$\varepsilon = 62.7$	$\varepsilon = 59.8$
	$\nu = 3280$	$\nu = 2450$
	$\varepsilon = 54.5$	$\varepsilon = 55.2$
Association (ν_A)	$\nu = 2125$	$\nu = 1555$
	$\varepsilon = 3.23$	$\varepsilon = 1.74$
X—O—X′ bending (ν_2)	$\nu = 1645$	$\nu = 1215$
	$\varepsilon = 20.8$	$\varepsilon = 16.1$
Libration (ν_L)	Broad band between 300 cm^{-1} and 900 cm^{-1}	
Hindered translation (ν_T)	Prominent shoulder on ν_L band at ~ 190 cm^{-1}; 30°C	

*ν = frequency of band maximum in cm^{-1}.
ε = extinction coefficient $\times 10^{-3}$ cm^2 mol^{-1} at absorption maximum.

Vibrations are classified according to type. *Stretching* vibrations involve lengthening (shortening) of interatomic distances; *deformation* (*bending*) vibrations are those where bond angles change. Often a given atomic group exhibits more than one kind of *stretching* or *bending vibration* as typified by the CH_2 group (Fig. 4.1). In the *scissoring* vibration, the two C—H bonds bend toward each other in the HCH plane. A torsional oscillation about a C—C bond is a *rocking* vibration. *Wagging* implies that both CH bonds bend out of the HCH plane in the same direction. In *twisting*, they bend out of plane in opposite direction.

Infrared absorption bands reflect the separation between vibrational energy levels. If interatomic vibrations were truly harmonic (i.e., if the restoring force were truly proportional to the square of the displacement from an equilibrium position), the vibrational levels would have energy values given by $E = h\nu_0(n+\frac{1}{2})$, where ν_0 is the fundamental vibration frequency, and n may be 0, 1, 2, 3, etc. Since selection rules allow transitions only between adjacent levels, absorption frequencies ν, representing vibrational transitions, would correspond exactly to the fundamental

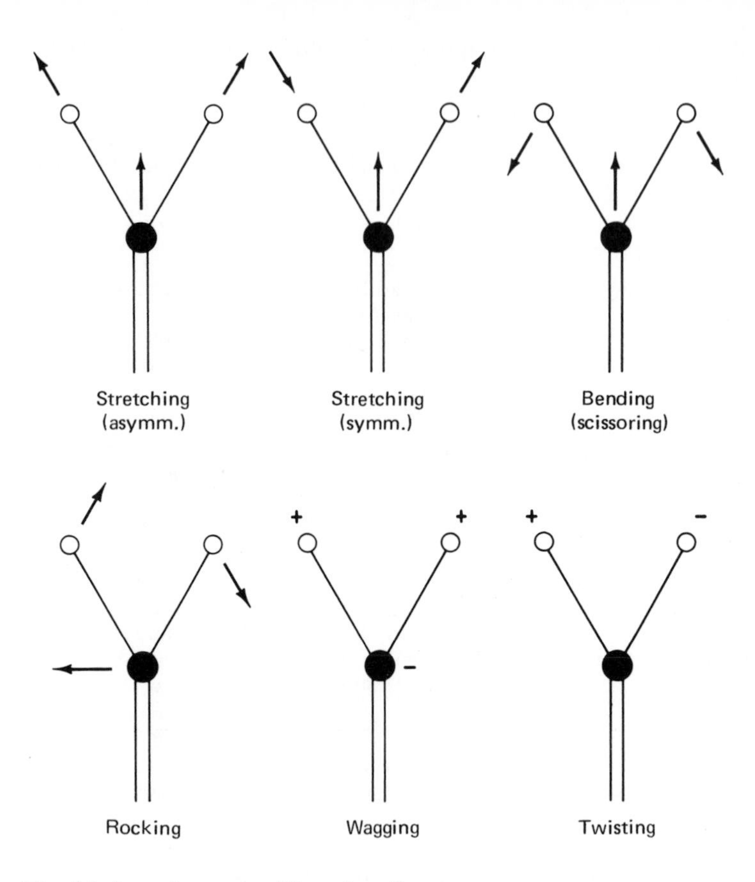

Fig. 4.1 Some important IR-active vibrations.

frequencies ν_0. However, actual molecular vibrations are not truly harmonic, and this relationship is therefore only approximate. Nonetheless, one absorption band should be observed for each fundamental mode of vibration, and ν should be close to ν_0. An additional selection rule for infrared absorption is that only vibrations altering the molecular dipole produce absorption. The selection rule that only a single quantum of vibrational energy can be absorbed at one instant is not absolute for complex molecules. In such, weak absorption bands can be observed corresponding to a change of 2 or more in n, or reflecting simultaneous transitions in two or more fundamental vibrations. Such absorption bands are referred to as *overtones* and *combination* bands, respectively.

The exact IR frequencies observed in complex molecules depend upon the atoms to which a vibrating group is bound in *covalent linkage*. Thus, the —C=O stretching frequency lies at 1750 cm^{-1} for aliphatic, undissociated carboxylic acids, below 1600 cm^{-1} for the corresponding carboxylates, and near 1710 cm^{-1} for carboxyl esters.

Physical state and nature of solvent

Vibrational frequencies depend on whether a given compound is in the pure state (gas, liquid, or solid) or in solution. When in solution, it depends on the nature of the solvent. Thus, the —C=O stretching vibration of amides in dilute solution occurs near 1690 cm^{-1}, but in the solid state the corresponding absorption lies near 1650 cm^{-1}. These differences are due to intermolecular interactions, between absorbing molecules in the solid state and between these molecules and the solvent in the solution case.

Hydrogen bonding

Specific intra- or intermolecular interactions can produce prominent IR changes; of these, hydrogen bonding is the best and most pertinent example and will be discussed further below. Hydrogen bonding is strongest in the solid state and may shift involved —O—H and/or —N—H stretching vibrations by up to 2000 cm^{-1}. The correlation is excellent between such frequency shifts and the corresponding shortening of O—H ··· O, —N—H ··· O, etc., bonds, as determined by X-ray crystallography.

Infrared Dichroism

Light consists of sinusoidally propagated quanta, whose electrical and magnetic vectors are oriented perpendicularly to each other and the direction of propagation. However, the orientation of these vectors from one photon to the next is random. By various processes of photoselection (Fig. 4.2), the light can be divided into components whose electrical vectors are all parallel. Such light is plane or linearly polarized and is absorbed preferentially by chromophores whose transition dipoles are parallel to the polarization plane.

Many polymers, including polypeptides, exhibit *infrared dichroism*. Then the absorption band of a given vibration splits into two components of different frequency, preferentially absorbing light polarized perpendicular and parallel, respectively, to the molecular axis. The selection rules involved have been thoroughly treated according to group theory (7) for ordered polymers, helices in particular. Unfortunately, the fact that globular molecules cannot be oriented properly with regard to the polarization plane of the incident light limits the utility of infrared dichroism.

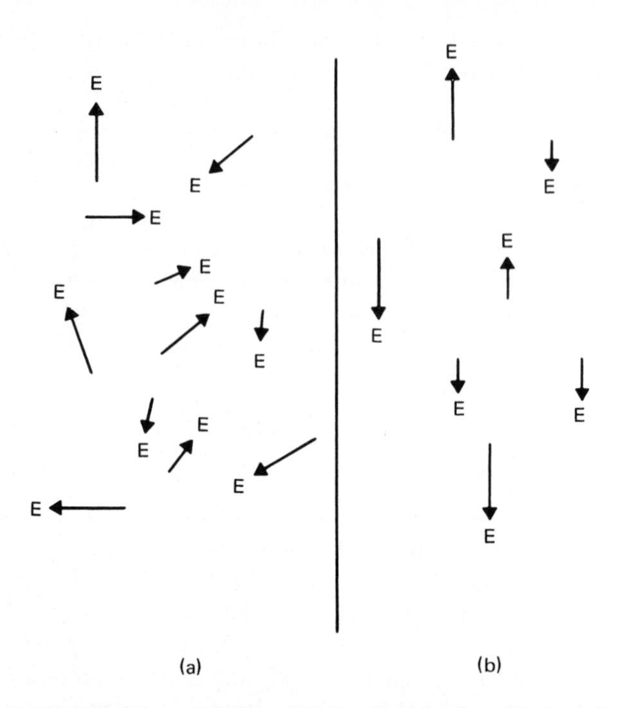

Fig. 4.2 Polarization of infrared light. Randomly polarized light (a) is converted into linearly polarized light, (b) in which the electric vectors **E** of the radiation are all parallel, and perpendicular to the propagation direction of the light (perpendicular to the page-plane). Silver chloride is favored in the manufacture of IR polarizers.

Helices

In an infinite, single-stranded α-helix, coupling along the backbone and through H-bonds splits the intensity of the normal vibration into vectors parallel and perpendicular to the helical axis. In oriented, strongly coupled samples, these two components define the absorption of light polarized at right angles to and parallel to the screw axis of the polymer, and the ratio of the absorbances characterizes the dichroism of the band:

$$\frac{I_{\parallel}}{I_{\perp}} \approx \cot^2 \delta \tag{4.1}$$

where $I_{\parallel}$ and $I_{\perp}$ are the intensity decrements for parallel and perpendicularly polarized light, and δ is the angle between the transition moment of the vibration and the screw axis of the helix.

If δ is the phase angle between vicinal oscillators, the frequency ν of a mode involving these coupled oscillators is given by

$$\nu(\delta) = \nu_0 + \sum_i D_i \cos(i\delta) \tag{4.2}$$

where $D_i = (B_i - A_i)\nu_0$; B_i and A_i are the potential and kinetic energy interaction factors between i neighbors; and ν_0 is the normal frequency of an unperturbed oscillator.

In the case of an infinite helical chain, δ may have any value, but only two yield infrared absorption: $\delta = 0$ and $\delta = \chi$, where χ is angle between successive groups relative to helix axis ($2\pi/3.6$ in the case of α-helices). Since H-bonds here form between every third residue, D_1 and D_3 are dominant, but D_2 is also significant. Two frequencies can be observed:

$$\begin{aligned} \nu(0) &= \nu(\parallel) = \nu_0 + D_1 + D_2 + D_3 \\ \nu(\chi) &= \nu(\perp) = \nu_0 + D_1 \cos\chi + D_2 \cos(2\chi) + D_3 (\cos 3\chi) \end{aligned} \tag{4.3}$$

In this equation $\nu(\parallel)$ is polarized parallel to and $\nu(\perp)$ perpendicular to the helix axis.

Parallel β-conformation

Here two adjacent peptide linkages in a chain can move either in phase ($\delta = 0$) or out of phase ($\delta = \pi$). When groups in adjacent chains, coupled by H-bonds, move in phase, the vibration is infrared active. Two modes result:

$$\begin{aligned} \nu(0,0) &= \nu_0 + D_1 + D'_1 \qquad (\parallel) \\ \nu(\pi,0) &= \nu_0 - D_1 + D'_1 \qquad (\perp) \end{aligned} \tag{4.4}$$

Here D_1 and D'_1 represent the interaction of groups in adjacent chains, and the polarization is relative to the backbone axis.

Antiparallel β-conformation

Inherently, the spectrum arises from four peptide linkages, involving portions of two arms of an antiparallel loop or two antiparallel chains. The resulting frequencies are:

$$\begin{aligned} \nu(0,\pi) &= \nu_0 + D_1 - D'_1 \qquad (\parallel) \\ \nu(\pi,0) &= \nu_0 - D_1 + D'_1 \qquad (\perp;\ \text{in plane}) \\ \nu(\pi,\pi) &= \nu_0 - D_1 - D'_1 \qquad (\perp;\ \text{out of plane}) \\ \nu(0,0) &= \text{inactive} \end{aligned} \tag{4.5}$$

Unordered state

Here the vibrational interactions between peptide groups average to zero, and only ν_0 is observed.

Nonpeptide linkages

As noted, the absorption intensity of an infrared band illuminated with polarized light depends upon the orientation of the dipole moment of the chromophore relative to the incident beam. Thus, in the case of the C=O stretch vibration of esters, the dipole lies along the C=O axis. Hence the orientation of this group can be determined by infrared dichroism. We shall return to this important fact in our discussion of poly-β-benzyl-L-aspartate.

Despite extensive literature on the theoretical and technical aspects of IR spectroscopy in general (6, 8, 9), its potential in the analysis of membrane structure has been insufficiently appreciated until recently, possibly because quantitative intensity measurements remain experimentally difficult, and most studies on biopolymers have been concerned with frequencies. Development of adequate intensity measurements is a critical prospect in IR technique.

Technical

The experimental aspects of IR spectroscopy are well described (e.g., 10). However, a major obstacle to the measurement of IR spectra of biologically important molecules is the strong absorbance due to water over much of the IR spectrum illustrated in Table 4.2. An effective water thickness of more than 0.05 mm exceeds the sensitivity of usual instruments. Only when concentrated solutions can be prepared is it possible to employ water as a solvent; useful spectra of 40% sodium citrate, for example, have been obtained. Such concentrations are unfeasible for most biological polymers, and even at high sample concentrations, the region between 3500 and 1700 cm^{-1} eludes investigation because water absorbs almost completely in this area. The most commonly used techniques to avoid the problem of aqueous interference are (a) substitution of deuterium oxide (D_2O) for water, since D_2O absorbs far less in regions of the IR spectrum that are of interest to the biochemist (Table 4.2), (b) evaporation to dryness of an aqueous layer on a disk of silver chloride (or other IR transparent material), (c) formation of a potassium bromide pellet or oil suspension with a dried sample.

In addition, cooling the sample to $\sim -180°$ drastically sharpens water absorption bands and allows discrimination of otherwise obscured components. The technique of multiple frustrated internal reflection may also be advantageous in some cases (11). Finally, it has proved possible to obtain some protein IR spectra by refined differential spectroscopy.

Polypeptides and Proteins

Synthetic Polypeptides

Assignments

IR analysis has been highly useful in the conformational analysis of polypeptides and proteins, largely because of the rigorous analysis of suitable model compounds such as *N*-methylacetamide (13). This, in its resemblance to the structural repeat of polypeptide chains, provided the clues to the planarity of the CONH group and its trans configuration (13). Complete assignments of all bands between 4000 and 60 cm^{-1} and 4000 and 300 cm^{-1}, respectively, have been made, although some uncertainties remain. A summary of the characteristic IR bands of the peptide bond is given in Table 4.3 (14).

Table 4.3 IR Bands Associated with the Peptide Linkage

Designation	Approximate Frequency (cm^{-1})	Origin[a]	Symmetry
A	3300	NH(s), 2x amide II in Fermi resonance	in plane
B	3100		in plane
I	1650	C=O (s) 80%, C—N (s) 10%, N—H (b) 10%	in plane
II	1560	C—N (s) 40%, N—H (b) 60%	in plane
III	1300	C=O (s) 10%, C—N (s) 30%, N—H (b) 30%, O=C—N (b), 10%, other 20%	in plane
IV	625	O=C—N (b), 40% other 60%	in plane
V	725	N—H (b)	out of plane
VI	600	C=O (b)	out of plane
VII	200	C—N (twist)	out of plane

[a] From Miyazawa (14).
Note: s = stretch; b = bending.

Because of their similarity to the absorption bands of secondary amides, the major polypeptide bands are termed "amide" bands (amide A, B, and amide I–VII). The observed amide absorptions of polypeptides arise principally from (a) *in plane* C=O, C—N, and C—H stretching, OCN bending, and CNH bending, as well as (b) *out of plane* C—N twisting and, in polypeptides, C=O and N—H bending. Suzuki et al. were also able to detect and verify the positions of these seven amide bands in polyglycines I and II [see Table 4.4 (13)].

Table 4.4 Amide Bands of Polyglycine I and II[a]

Amide Band	Polyglycine I Frequency (cm^{-1})	Polyglycine II Frequency (cm^{-1})
I	1685 and 1630	1641
II	1524	1558
III	1297	1309
IV	589	740 (IV and V)
V	710	
VI	628 and 614	701 and 573
VII	217	365

[a] From Suzuki et al. (13).

Conformational dependence of the amide I and II bands in synthetic polypeptides

The absorption bands of polypeptides not only represent contributions from several intrinsic vibrations, but are also sensitive to the folding of the peptide backbone and to the hydrogen bonding with other peptide linkages, i.e., to conformation or 2° structure. Because the peptide backbone of proteins and polypeptides often includes regular arrays which can be described by symmetry space groups, one can frequently distinguish between various conformations. Both synthetic polypeptides and proteins can exist in one particular conformation or a mixture of conformations, depending upon solvents, ionic environment, pH, temperature, segment length, etc. Thus, poly-L-lysine is in an unordered state at low ionic strength near neutral pH, but at pH >11.0 when coulombic repulsion between side chains is small, the α-helical form predominates. However, when heated at 55° at pH 11.3, this polymer assumes the antiparallel β-pleated sheet structure (15). Most synthetic polypeptides cannot be shifted from one conformation to the other simply by environment manipulation, but poly-L-aspartate and poly-L-glutamate are α-helical at acid pH, and unordered at neutral pH and low ionic strength. One can also synthesize polymers such as poly-γ-benzyl-L-glutamate in either α-helical or β-conformation by choice of appropriate conditions of chain initiation (16).

Polypeptides in α-helical and/or unordered conformations can usually be readily distinguished from β-structures by IR spectroscopy. Thus, the amide I band of α-helical poly-benzyl-L-glutamate lies at 1656 cm^{-1} and the amide II band at 1550 cm^{-1}; in the β-structured polymer, the two bands occur at 1630 and 1525 cm^{-1}, respectively (17). Analogously, α-helical poly-L-lysine has an amide I band at 1652 cm^{-1} and the β-form at 1636 cm^{-1} (18). The amide I bands of poly-methyl-L-glutamate in the α-helical conformation correspond to those of poly-benzyl-L-glutamate (19).

While the amide I frequencies of the α-helical or unordered forms of a given polypeptide can be diagnostic under given conditions (e.g., solid films of poly-L-lysine in the α-helical and unordered conformations have amide I frequencies at 1652 and 1657 cm^{-1}, respectively), the proximity of the amide I bands in the *α-helical* and *unordered* conformations, small positional variations from one peptide to the other, and band broadening due to side chains prevent use of this absorption band to distinguish between these two structures, except in cases where infrared dichroism is applicable; thus, the amide I frequencies for solid films of α-helical poly-L-glutamic acid and poly-γ-benzyl-L-glutamate lie at 1652 cm^{-1} and 1656 cm^{-1}, respectively. Also, different workers, using different preparations, locate the amide I band of poly-γ-benzyl-L-glutamate in chloroform between 1656 cm^{-1} and 1651 cm^{-1}.

The amide II band is not very useful in a conformational diagnosis because α-helical and β-structures both give bands near 1510 and 1540 cm^{-1} (20). Also, in proteins, the COO^- group, due to the common aspartate and glutamate residues, interferes through its absorption between 1550 and 1600 cm^{-1}. This is very noticeable in poly-L-glutamate (pH 7.0), where the COO^- absorption is so great that even the amide I band is broadened considerably. The amide II region has yielded conformational information only through measurements of the dichroism of oriented films of polypeptides and proteins. In principle, this technique can discriminate between α-helical, β- and unordered conformations in both the amide I and amide II regions (18, 21), but it is not generally useful for globular proteins, where the axes of secondary structures cannot be oriented uniquely relative to the polarization axis of the IR beam.

The amide I band in the conformational analysis of isolated proteins

IR analyses of protein conformations have generally utilized the amide I band, and studies of proteins whose structures are known through X-ray crystallography have demonstrated the validity of this approach (19). (See Table 4.5.) For reasons already discussed, proteins known to contain both α-helical and unordered peptide sequences do not show significant fine structure in the amide I region.

The presence in proteins of β-structures together with α-helical and/or unordered conformations can readily be distinguished. For example,

Table 4.5 Amide I Frequencies of Various Proteins[a]

Protein	Condition	Frequency (cm^{-1})
Myoglobin	pD 7.4	1650
Bovine serum albumin	Isoionic	1652
	Nujol suspension	1652
	pD 2.0	1648
Carboxypeptidase A	2.4 M LiCl	1650, 1637 shoulder
Lysozyme	Unspecified (21)	1650, 1632 shoulder
Insulin	pD 2.4	1654
	Nujol suspension	1654
	pD 12.5	1644
	pD 2.4 film, heated	1633, 1658
	Acidic CH_3OD	1654
Glucose oxidase		1643, 1648, 1654
Deoxyribonuclease		1637, 1643, 1650
Bovine carbonic anhydrase	pD 7.4, 12.0	1637
	Nujol suspension	1637
	pD 1.8	1646
α-chymotrypsin	pD 4.2	1637, 1685 shoulder
Chymotrypsinogen	pD 8.7	1637
Ribonuclease	pD 4.8	1640, 1685 shoulder
Soybean trypsin inhibitor	pD 7.5 to 10.5	1641
γ-globulin	Oriented film (unpolarized light)	1650, 1635 shoulder
	pD 1.6	1637–1643 (broad)
	pD 10.0 to 12.0	1637
Rennin	pD 1.8, 10.8	1639
β-lactoglobulin B, tryptic core		1615, 1643
α_s-casein C	pD 9–11	1643
	pH 9 (in H_2O)	1656
Phosvitin	pD 3.4; 6.6	1650

[a] From Timasheff et al. (19).

lysozyme, which has about 10% of its peptide linkages in the antiparallel β-conformation (22), exhibits a 1650 cm^{-1} peak with a 1632 cm^{-1} shoulder; carboxypeptidase A, (23) with 15% β-structure, has a 1650 cm^{-1} peak and a 1637 cm^{-1} shoulder. Native β-lactoglobulin A, containing predominantly antiparallel β-structure, has amide I bands at 1632 and 1675 cm^{-1}. When the protein is denatured by raising pH, the 1632 cm^{-1} peak diminishes, and a band around 1643 cm^{-1} becomes the dominant feature in the amide I region, as an unordered structure is formed (19). (The above amide I frequencies were obtained using D_2O solutions. Possible complications due to this sampling procedure will be discussed later.)

Other regions of the IR spectrum that reflect protein conformation

The amide V band, which is readily accessible to many IR spectrophotometers, is of great potential utility in conformational analysis and deserves greater attention, since it appears to distinguish between α-helical and unordered conformations. Thus, the amide V bands of poly-γ-methyl-L-glutamate in the extended β-, unordered, and α-helical conformations exhibit three bands in the amide V region—at 690, 650, and 600 cm^{-1} (24). Consistent with these observations is the fact that lysozyme, a mixture of α-, β-, and unordered segments, yields three amide V bands—at 690, 650, and 600 cm^{-1} (25).

The far infrared has not been readily accessible until recently, but may yield much structural information about proteins, since polypeptides have several conformationally dependent bands in this region. Thus, α-helical poly-L-alanine shows four bands at 190, 167, 120, and 90 cm^{-1}, while the β-form has two, at 442 and 247 cm^{-1} (26). The bands typical of the helical polymer are thought to represent hydrogen bond stretching and bending and the torsional vibrations about the C_α—C and C_α—N axes. The 247 cm^{-1} band may arise from the CO—NH torsional vibration.

Polypeptides at Air-Water Interfaces

At this juncture, we shall introduce the often-neglected work of Malcolm (27–31), which is highly pertinent to the membrane field and relies heavily on infrared techniques. The experiments on poly-β-benzyl-L-aspartate are particularly germane.

This compound, unlike most synthetic polypeptides composed of L-amino acids, normally forms left- rather than right-handed helices, presumably stabilized by entropic effects, other solvent influences and/or electrostatic interactions between the peptide linkages and side chains. The structure of poly-β-benzyl-L-aspartate in water has been unstudied because of its insolubility in aqueous buffers. Accordingly, Malcolm circumvented the problem by spreading the polymer at air-water interfaces, evaluating the force-area curves, as well as surface potentials of such interfacial layers, and analyzing films extracted from air-water interfaces by infrared dichroism and electron diffraction. Using these techniques, Malcolm had previously shown that collapsed films or multilayers of high-molecular-weight polypeptides form arrays of right-handed α-helices. It appears that, with increasing surface pressures, helical molecules at the interface condense into micellar arrays with areas of $\sim$20.5 A/residue. At surface pressures greater than 10 dynes/cm, the monolayer transforms into a bilayer; this produces a plateau in the

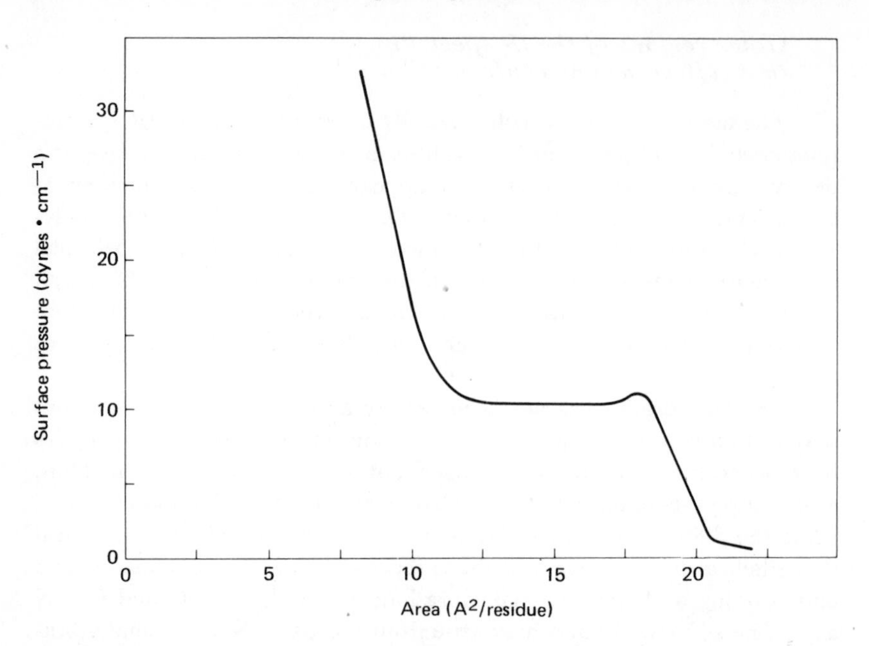

Fig. 4.3 Pressure-area curve of poly-β-benzyl-L-aspartate spread at an air-water interface. A break occurs at 20.5 A^2/residue typical of α-helix. A second inflection at $\sim$ 10 A/residue indicates bilayer formation. See discussion in text. [From Malcolm (30).]

pressure-area curve. The height of this step depends on the hydrophobicity of the side chains and the free energy of the polypeptide-gas interface.

Poly-β-benzyl-L-aspartate of molecular weight 130,000 and 250,000 *also* forms a right-handed α-helix when spread at an air-water interface. Its force-area curve (Fig. 4.3) shows a break giving an area of 20.5 A^2/residue, typical of α-helix. The second break at $\sim\frac{1}{2}$ the area/residue indicates bilayer formation.

When the collapsed mono- or multilayers are oriented uniquely and inspected by infrared dichroism, they yield a pattern characteristic of *right-handed* α-helix: The frequencies of the ester C=O (1740 cm^{-1}), amide I band (1658 cm^{-1}), and amide II maximum (1552 cm^{-1}) do not fit the established values for left-handed α-helical poly-L-aspartate, but are characteristic of right-handed helices. The parallel polarization of the C—O band at 1168 cm^{-1} confirms these conclusions, since the left-handed form exhibits perpendicular dichroism at this frequency. The infrared data are fully supported by electron diffraction studies.

Importantly, the interface yields right-handed α-helices, whether the polymer is spread as *left-handed helix*—from 1:99 dichloroacetic acid: chloroform—or as *unordered* coil—from 10:90 dichloroacetic acid:chloroform (31). However, addition of *isopropanol* to the subphase (1%) abolishes the plateau in the force-area curve, and the collapsed films exhibit the IR characteristics of *left-handed* helices. It is the interfacial relationship, therefore, that produces the change of helical sense (31).

The shift from left- to right-handed helix does not require unwinding of the backbone (which would require high energy) since, if the H-bonds along short helical segments are opened, refolding of that segment in the opposite sense can occur by a simple rotation about the bonds to the α-carbons.

The most appealing interpretation of these studies is that the right-handed helix forms rapidly upon interfacial spreading and that the plateau in the force-area curve represents progressive condensation into a bilayer. The stabilization of the right-handed helix can be reasonably attributed to an interference by the aqueous, polar substrate with the forces normally stabilizing the left-handed helix. The effect of isopropanol is attributed to altered entropic interactions between polymer and water subphase.

These experiments have a general, fundamental relevance to protein structure; depending on amino acid sequence and composition, entropic factors, and microenvironment, some α-helical segments in an unknown protein may actually have a left-handed sense. This matter is particularly pertinent to membrane proteins, where there is good reason to suspect unusual amino acid compositions and/or sequences, as well as a specialized microenvironment due to membrane lipids.

Artificial Lipid-Protein Systems

Fromherz et al. (32) have used infrared spectroscopy to examine solid multilayers of hemoglobin adsorbed unto arachidic acid and methyl-stearate (33). Because of the low absorption of the films, frustrated multiple internal reflection (11) was used to achieve significant signals. The films were deposited on germanium plates.

The films show strong amide I absorption at 1650 cm^{-1}, but the amide II band lies at 1535 cm^{-1} for arachidic acid layers and 1525 cm^{-1} for methyl stearate films. The difference is presumably due to the carboxylate absorption (1540 cm^{-1}) in the former case. The hemoglobin-methyl stearate film exhibits a prominent new band at 1620 cm^{-1}, lacking in native hemoglobin, the pure ester, and the arachidic acid films.

As noted before, IR spectroscopy in the amide I region cannot readily distinguish between the α-helical conformation and unordered structures. Thus, the protein adsorbed unto arachidate might either remain in its normal, predominantly helical state or be uncoiled to some degree. However, the most reasonable explanation for the 1620 cm^{-1}

band in the methyl stearate films would suggest some restructuring into a β-conformation. This is compatible with the rapid penetration of methyl stearate films by hemoglobin upon reduction of the surface pressure on films after the protein had been adsorbed at high surface pressure.

Interestingly, the C=O stretching band of pure arachidate films is much stronger than that of films with adsorbed hemoglobin. This suggests that when the protein binds, the acid dissociates more completely, indicating an ionic lipid-protein interaction. Moreover, in pure arachidate multilayers, lattice forces split the C—H scissoring band into two sharp components at 1458 cm^{-1} and 1467 cm^{-1}, but this effect is perturbed by the protein, resulting in a single band at 1462 cm^{-1}.

The results show that a soluble globular protein can bind to both polar and nonpolar films but by different forces. Polar (ionic) interactions apparently do not perturb the protein, whereas nonpolar interactions induce an important conformational change. These studies are clearly of major importance and indicate new avenues for the study of lipid-protein interactions in biomembranes. They are also directly pertinent to the metabolically linked generation of β-structure in erythrocyte and mitochondrial membranes.

IR Analysis of the Protein Components of Cellular Membranes

The IR spectra of cellular membranes can be correlated with those of synthetic polypeptides and proteins with known conformation to yield information concerning the secondary structure of membrane proteins and their possible conformational transitions. Some data about the effect of lipid upon protein conformation in cellular membranes have also become available.

Plasma Membranes

Erythrocyte membranes

The amide I regions of hemoglobin-free erythrocyte ghosts, air-dried or lyophilized as films on AgCl disks, indicate that such preparations contain no or little β-conformation (34–36). However, since hydrophobic interactions account at least partly for the secondary, and higher structures of membrane proteins, drying for spectroscopy, even though rapid, could induce conformational transitions. Nevertheless, the amide I spectra of erythrocyte membranes suspended in D_2O closely resemble those obtained with dried films (36, 37). Transitions between helical and unordered structures could not be detected in any event.

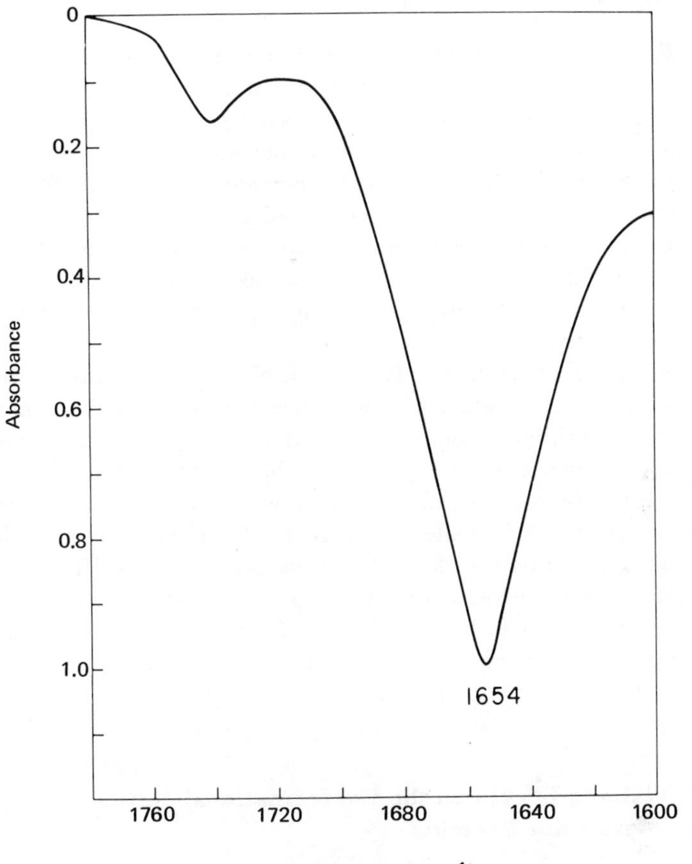

Fig. 4.4 Amide I IR spectrum of erythrocyte ghosts, lyophilized from dispersion in 7.0 mM phosphate, pH 7.2. The spectrum shows no evidence for β-structure.

The amide I region of erythrocyte membranes dried from 7 mM phosphate buffer, pH 7.4, is typified by Fig. 4.4. Maximal absorption occurs at 1652 cm^{-1}, indicative of protein in the α-helical and/or unordered conformation, and no band at 1630–1640 cm^{-1} is observed. From such spectra, together with measurements of optical activity, it has been concluded that the proteins of erythrocyte membranes are mixtures of α-helical and unordered conformations. However, quantitation of the exact proportions of these structures has not been achieved as yet; moreover, the conformations of erythrocyte membrane proteins depend upon their physiological state, as discussed later.

Other plasma membranes

IR studies on purified plasma membranes and endoplasmic reticulum fragments of Ehrlich ascites carcinoma dried at neutral pH (38, 39) yield no evidence for the presence of β-structured peptide, but the spectra of membranes dried from acid solvents indicate some transition to the β-structure (38, 39). Myelin, which is considered the model for the unit membrane concept, with its protein located at the surface of lipid bilayers in an extended β-conformation, also demonstrates an amide I band near 1655 cm^{-1} (40). No unusual absorption occurs near 1630 cm^{-1}, indicating that the assumption of β-structured peptide is not justified in this membrane system.

On the other hand, the IR spectra of dried *Micrococcus lysodekticus* membranes (41) indicate a predominance of α-helical and/or unordered peptide, but exhibit a shoulder near 1635 cm^{-1}, which is interpreted to indicate presence of some peptide in the β-conformation. [Green and Salton (41) obtained identical spectra with membranes pelleted and incorporated into KBr pellets as with the "silver-chloride-disk" technique.] The membranes of *Acholeplasma laidlawii* also show IR evidence of β-structured peptide (42), and their optical activity is best interpreted in these terms. Finally, according to IR criteria, isolated adipocyte plasma membranes contain an appreciable proportion of peptide in the antiparallel β-conformation (43).

Metabolic Effects on the Conformational State of Membrane Proteins

Erythrocyte ghosts

Membrane isolates must be recognized as *membrane derivatives* whose proteins (and lipids) cannot be assumed to exist precisely as in vivo. This is true also for erythrocyte ghosts, although these are derived by a very gentle procedure. These membranes are isolated from chilled, hence metabolically quiescent, cells and deprived during purification of various metabolites necessary for the function of known membrane-bound enzymes. It is perhaps not surprising, therefore, that the hydrolysis of ATP by erythrocyte ghost ATPases, a process during which certain membrane proteins are transiently phosphorylated, induces substantial conformational changes.

Most IR analyses of cellular membranes have been performed on samples dried at ambient temperatures or suspended in D_2O; these methods are not ideal for kinetic measurements. Therefore, Graham and Wallach (44) freeze their membrane reaction mixtures very rapidly in

7 mM phosphate, pH 7.4, upon silver chloride disks, by plunging the latter into liquid nitrogen, and do their IR analyses on samples lyophilized after this freezing step. Although these studies cannot detect possible changes in helicity during freezing and lyophilization, Graham and Wallach (37, 44) have shown that this approach is not destructive to a number of important and sensitive membrane functions. Thus, if lyophilized erythrocyte ghosts are rehydrated by first equilibrating them for several hours at 100% relative humidity at 10°C followed by resuspension in buffer, the membranes are recovered as biconcave disks, which have lost none of their unspecific or ion-specific ATPase (Table 4.6). (Similar rehydration of lyophilized rat liver mitochondria leads to 90% recovery of DNP-stimulated ATPase, indicating no significant impairment of the coupling between electron transport and phosphorylation.) Use of low concentrations of membranes (<3 mg protein/ml) is essential for recovery of these functions. Although incomplete, these data suggest either that dehydration, properly performed, does not induce a change in secondary structure of the membrane proteins in question, or that whatever changes are induced are reversible or do not affect the tested functions.

Addition of ATP and Mg^{2+} (0.5 to 1.0 mM) to erythrocyte membranes induces a shift to the β-conformation, away from an α-helical or unordered structure (Fig. 4.5). This transition is enhanced by Na^+ and K^+ (25 mM and 5 mM, respectively), which stimulate ATP hydrolysis and indeed appears to be a consequence of ATP utilization. It is blocked by EDTA at levels that inhibit ATP hydrolysis.

Table 4.6 Recovery of ion-specific ATPase in erythrocyte ghosts after lyophilization[a]

Ghost Preparation	μmoles ATP hydrolyzed/mg protein/hr		
	Na^+	K^+	$K^+ + Na^+$
Control ghosts	0.374 0.410	0.515 0.562	0.144 0.152
Lyophilized ghosts rehydrated prior to resuspension	0.650 0.657	0.862 0.830	0.211 0.183
Lyophilized ghosts not rehydrated prior to suspension	0.610 0.661	0.770 0.746	0.160 0.087

[a] Ghosts (1.0 mg protein/ml) were shell-frozen in liquid nitrogen and lyophilized. Water was added to the lyophilizate to the original volume without exposure to 100% humidity (10°C for 2 hr).

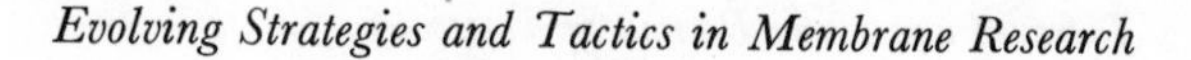

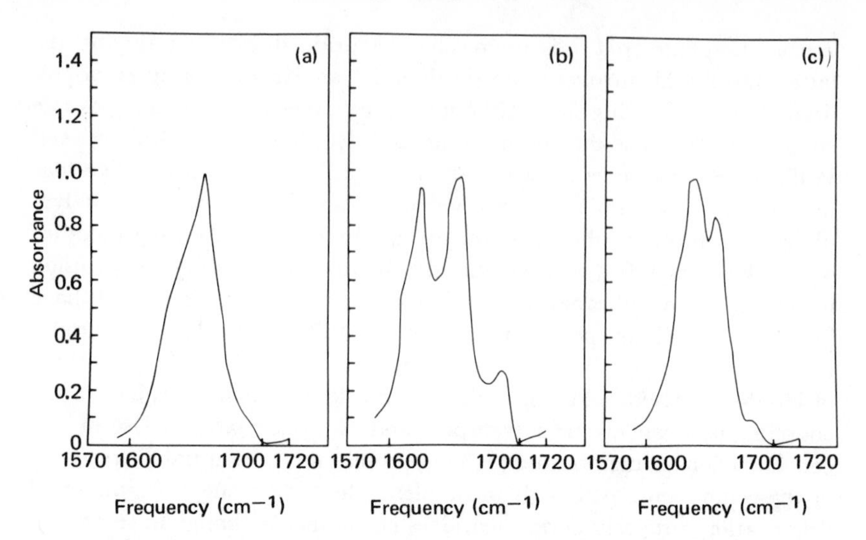

Fig. 4.5 Change in amide I spectrum of erythrocyte ghosts in D_2O dispersion upon addition of certain reagents. See text for details. (a) Control; the peak is at 1654 cm^{-1}; a small shoulder near 1640 cm^{-1} may indicate presence of some parallel β-structure; (b) after ATP hydrolysis in the presence of Na^+ but not K^+. The splitting signifies conversion of a substantial number of peptide linkages to the β-structure; (c) spectrum obtained after ATP hydrolysis in the presence of $Na^+ + K^+$. The spectrum is now dominated by the 1640 cm^{-1} band, indicating even more extensive β-structuring. (From Graham and Wallach (37) by permission of Elsevier Publishing Co.)

The conformational transition is prominent within 2 min after ATP addition when the reaction is carried out at room temperature and stopped rapidly with liquid nitrogen prior to lyophilization for IR examination. The transition is also observed when the metabolic reaction and IR measurements are done in D_2O (37).*

The spectral changes observed in the amide I region are considerable, but might arise without a large rearrangement of the peptide backbone (45). One may, thus, be dealing with structural specializations such as exist in several soluble globular proteins, e.g., carboxypeptidase A, where there is extensive antiparallel folding, but less than maximal β-structured H-bonding (23), and α-chymotrypsin (46), which also contains a region of peptide with a looped, antiparallel tertiary structure

* Infrared spectroscopy has recently been shown to reveal differences between erythrocyte ghosts loaded with glucose and sorbose (G. Zimmer, L. Lacko, and H. Günther, *J. Memb. Biol.* **9:** 305, 1972). Glucose shifts the P—O—C band (1050 to 1060 cm^{-1}) to higher wavenumbers, while reducing the intensity of the P=O, C—O—C region (1220 to 1240 cm^{-1}). Sorbose, in contrast, intensifies absorption at 1220 to 1240 cm^{-}).

but few β-hydrogen bonds. In such cases, a small change in the orientation of the peptide groups by rotations about certain bonds in the backbone would favor more extensive antiparallel-β H-bonding, and this would be reflected in the IR spectra. The feasibility of such alterations emerges from the X-ray analyses of carboxypeptidase A in the presence and absence of substrate (23, 46). Thus, when glycyl-tyrosine is bound to this enzyme as substrate, skeletal rotations about the C_β—C_γ of tyrosine-248 move this residue 12 A, apposing it to an NH_2 group; concurrently 4 hydrogen bonds in a particular region of the protein are broken.

A number of other proteins appear to undergo conformational changes as a normal biologic event. Thus, optical activity measurements (47–49), as well as IR analyses (49, 50) on pure yeast UDP-galactose 4-epimerase, indicate that this enzyme, which contains highly bound NAD, when reduced by the appropriate sugar substrate (or by sodium borohydride), increases its content of antiparallel β-structure, probably at the expense of α-helix. The native, oxidized enzyme exhibits CD minima at 208–210 and 223 nm, i.e., a "helical pattern," whereas the reduced form shows a single minimum at 217 nm, the CD minimum for most β-structures. Concurrent IR measurements on dried films and in D_2O, revealing a shift in absorption maxima from 1652 cm^{-1} to 1630 cm^{-1}, are also compatible with an $\alpha \rightarrow \beta$ transition. The conversions of trypsinogen and chymotrypsinogen to trypsin and chymotrypsin, respectively, also appear to involve small changes in the relative proportions of α-helix and β-structures (51).

Mitochondrial membranes

The proteins of conventionally isolated rat liver mitochondrial membranes contain a significant proportion of peptide in the antiparallel β-conformation (52), the exact proportion depending upon metabolic state (44). The IR spectra from these organelles as isolated exhibit a peak around 1650 cm^{-1} with a pronounced shoulder near 1635–1640 cm^{-1} and a smaller shoulder near 1690 cm^{-1}. The latter two spectral attributes are due principally to the proteins of the inner mitochondrial membrane (53). Elution of soluble, "matrical" mitochondrial proteins by osmotic lysis indicates that the β-structured peptide is membrane associated. Moreover, mitochondrial membranes suspended in D_2O exhibit the same amide I features as film spectra, eliminating the possibilities that the β-structure in these membranes arises during dehydration.

The demonstration of some protein in the β-conformation does not localize the peptide at the membrane surface; it also fits some aspects of the optical activity of mitochondrial membranes, which suggest a hydrophobically stabilized β-conformation (53). In any event, presence in membranes of ordered secondary structures other than the α-helix should not be unexpected, since only a few synthetic polypeptides and fibrous proteins exist in a single conformation. Indeed, as more and more globular proteins are analyzed by X-ray crystallography, it becomes clear that many are rather complex conformational mixtures, containing, among other regular patterns, considerable β-structure. Thus, lysozyme contains helical, unordered, and β-structured segments (22); β-lactoglobulin is a combination of unordered and β-structured peptide (19); carboxypeptidase A is a mixture of helical, unordered, and β-structured regions (23), as is α-chymotrypsin (46). IR measurements also indicate that β-structures are present in deoxyribonuclease, ribonuclease, and carbonic anhydrase (19), as well as in trypsin, chymotrypsin, trypsinogen, and chymotrypsinogen (51). Moreover, in most globular proteins, even the ordered segments may deviate considerably from the ideals seen in synthetic models. For example, not all helical segments of myoglobin and hemoglobin are truly α-helical (54). Also, as already noted, much of the peptide backbone of the carboxypeptidase A is arrayed as antiparallel loops, but few of the H-bonds typical of ideal antiparallel β-structure are completed (23).

The inner membranes of rat liver mitochondria exhibit large differences in the amide I region of the IR spectra, depending upon the metabolic state of the mitochondria at the moment of freezing (44).

Induction of electron transport by the addition of succinate (0.1 mM) in the presence of oxygen shifts the protein conformation toward the antiparallel β-structure, as indicated by the increase in absorption around 1635 cm^{-1} and 1685 cm^{-1}. This can be prevented or reversed by inhibiting electron transport with CN^- or antimycin-A or by inducing phosphorylation (addition of 250 μM ADP). The shift toward the antiparallel β-structure becomes extreme when electron transport is uncoupled from oxidative phosphorylation by addition of 100 μM dinitrophenol. The mechanisms involved in these spectral shifts are thought to resemble those discussed above.

Measurements in D_2O

In order to verify data obtained by film techniques, Graham and Wallach (37) have studied the IR spectra of erythrocytes and mitochondria suspended in D_2O. This technique also demonstrates the presence of some antiparallel β-pleated sheet structure in the protein of mitochondrial membranes. However, erythrocyte ghosts that give no sign of β-conformation by the film method, except during ATP hydrolysis, in D_2O, exhibit a small shoulder near 1635 cm^{-1}, suggestive of some

β-structure. Studies on erythrocyte membranes in deuterated water indicate the unique problems of using this method. If an erythrocyte ghost suspension is transferred as quickly as possible to D_2O by two washings in D_2O, 7 mM phosphate, pD 7.4, it yields a normal amide I spectrum. Over a period of 2 hr, however, the apparent proportion of β-structured peptide increases, according to the rising absorbance near 1640 cm^{-1}, which gradually assumes prominence of the amide I region.

That such a transition should take place in D_2O is not unexpected. *First,* some of the observed spectral changes may be due to deuteration per se, because the amide I frequencies of synthetic polypeptides can shift -2 to -10 cm^{-1} in D_2O compared to the situation in solid films. For example, the amide I band of "unordered" poly-L-glutamate lies at 1657 cm^{-1} in solid films and at 1643 cm^{-1} in D_2O. Such extreme differences are not the rule, and the amide I bands of β-lactoglobulin (extensive antiparallel β-conformation) and myoglobin (primarily α-helical) have almost identical *positions* in D_2O and dry films (20), although in the former the shape of the band and the intensity of the 1685 cm^{-1} shoulder differ in the two situations.

Second, it is established that protein structure in general (55, 56) and β-structures in particular (57, 58) are stabilized through "hydrophobic interactions" and these, as well as H-bonding, vary depending on whether H_2O or D_2O is the solvent (56). Thus, the monomers of tobacco mosaic virus protein polymerize more rapidly in D_2O than in H_2O (59). Further, deuterated ovalbumin denatures less readily in 5 M urea than normal ovalbumin (60), and Vakar et al. (61) show that the viscomechanical properties of wheat gluten change significantly upon deuteration, possibly due to the "greater strength of deuterium bonds." However, one cannot readily predict whether in a given case a deuterium bond is "stronger" or "weaker" than a hydrogen bond. Thus, deuteration increases the length of a hydrogen bond to a greater extent if the original bond was short and strong (62), but in some cases, there is some bond contraction upon deuteration. In poly-γ-benzyl-L-aspartate or poly-β-benzyl-L-glutamate, which are more helical in D_2O than H_2O (63), the N—D—O linkage is 0.025–0.029 A longer than the N—H-·-O linkage (64). Although the energy of the bond is thus decreased, its greater length may change the relative positions of individual atoms so that other effects, which contribute to the maintenance of secondary structure, increase the stability of the molecule as a whole. While deuterium bonds are often more stable than H-bonds (56) and deuteration favors α-helix rather than the unordered coil in poly-γ-benzyl-L-glutamate and RNAse (65), it is still difficult to predict the effect of deuteration upon the strength of hydrogen bonding.

An important effect in the substitution of D_2O for H_2O is upon the hydrophobic interactions that stabilize most proteins (66). It is now recognized that insertion of apolar residues into water increases solvent structuring and, inversely, that there is a gain of entropy when this structuring collapses upon removal of these residues (55). At any given temperature, D_2O is more structured than H_2O (56). Therefore, hydrophobically stabilized interactions would be expected to be stronger in D_2O. Kahil and Lauffer suggest that this might be the reason for the increased polymerization rate of tobacco mosaic virus protein in D_2O and argue that, if hydrogen is involved in the "entropic unions" between apolar groups, deuterated proteins would have different stabilities from their protonated variants (59).

In a pertinent study of the relative effects of D_2O and H_2O on model compounds, Kreschek et al. (66) found a decrease of free energy upon transfer of short-chain hydrocarbons and amino acids from H_2O to D_2O. Unlike the simpler hydrocarbons, however, the free energy of transfer from H_2O to D_2O for the *nonpolar side chains* of amino acids was positive. The data indicate that for simple hydrocarbons, hydrophobic bonding is stronger in H_2O than D_2O, but that the reverse holds when a polar group is attached to the hydrocarbon residue, as in the apolar amino acids. The results also indicate that the architecture of a protein may change when it is transferred from H_2O to D_2O because of its differing entropic interactions with these two solvents.

Returning to erythrocyte membranes, the apparent increase in β-structure is both time and temperature dependent. Freshly prepared ghosts in D_2O at room temperature show only a slight shoulder near 1640 cm^{-1}, but this becomes prominent after an hour. Raising the temperature to 35–40°C markedly accelerates the spectral shift. The conditions that favor the transition in erythrocyte membranes resemble those that promote the $\alpha \rightarrow \beta$ transition in poly-L-lysine, where the β-conformation is very likely stabilized by hydrophobic bonding. The conversion in this synthetic polypeptide occurs readily at 40°C, but even after 1 hr at room temperature, there is a noticeable transition to the antiparallel β-form. On the other hand, Taborsky (67) demonstrates that several freezing and thawing cycles cause a significant transition from an unordered to a β-conformation in the protein phosvitin, and when erythrocyte membranes are treated similarly, the IR spectrum changes, with an increase in absorption near 1640 cm^{-1}, suggesting β-structuring and resembling the D_2O effect.

Importantly, the morphology of erythrocyte ghosts changes as they are transferred from water to D_2O (37). In 7 mM phosphate buffer, pH 7.4 (H_2O), erythrocyte ghosts retain the biconcave shape of the red cells from which they were derived. In 7 mM phosphate buffer, pD 7.4 ($D_2O = 90\%$), the ghosts are spherical and swollen. Endovesiculation, which is prominent when erythrocyte ghosts are suspended in aqueous 0.7 mM phosphate, pH 7.4 at 0°C, is largely eliminated by substitution of D_2O, although some membrane invaginations still occur in this medium.

Lipid-Protein Interactions

It has often been suggested that the structure of membrane proteins depends upon their interactions with membrane lipids (68–70). This topic has also been explored by IR spectroscopy.

Assignments of Lipid Absorption Bands

Not all IR absorption bands of membrane lipids are fully characterized. This is partly due to uncertainty as to the origins and precise frequencies of certain bonds and because of lipid polymorphism. Since the environments of IR-active bands' absorbing moieties influence their absolute frequencies and absorbance, variations in physical state will lead to uncertainties. Phospholipids rarely exist in a single structural state, and their sensitivity to small temperature fluctuations, of a magnitude likely to occur during the scanning of an IR spectrum, introduces complex experimental variables. In spite of these obstacles, a number of infrared bands have been identified, and some of the IR properties of the phosphate groups are well established.

Table 4.7 (71) presents the frequencies of some major vibrations

Table 4.7 IR Bands Due to Fatty Acid Chains of Lipids[a]

Frequency (cm^{-1})	Assignment
1460	CH_2 stretch and bend
1380–1180	CH_2 wag
1300–1170	CH_2 twist
1150–870	C—C stretch and CH rock
~ 3000	C=C—H stretch
1580–1650	C=C stretch
690–980	C=C—H out-of-plane bend
970	CH bend (ethylenic)
2926	CH_2 asymmetric stretch (out of plane)
2853	CH_2 symmetric stretch (in plane)
720	CH_2 rock

[a] From Chapman (71).

within the fatty acid chains of phospholipids. The CH_2-twist frequencies is ill-defined, and it has been suggested (72) that it lies at 1200 cm^{-1} rather than between 1300 and 1170 cm^{-1} as usually thought. The CH_2 bending yields a doublet near 1640 cm^{-1} in condensed lipids, due to interactions between fatty acid chains. The C—C stretching and CH rocking frequencies occur between 1150–870 cm^{-1}, but have not been precisely localized (73). The 720 cm^{-1} CH rocking band reflects the interactions between adjacent hydrocarbon chains. This vibration arises only when at least 4 successive carbon atoms in a hydrocarbon chain are arranged in a planar all-trans-configuration (5). When interactions between chains are strong, as in solids, this vibration yields a doublet separated by as much as 6 cm^{-1} (6). In the liquid state, the 700–1350 cm^{-1} region of long-chain hydrocarbons becomes less distinctive.

The frequency of the phosphate P=O stretching vibration lies between 1350 and 1250 cm^{-1}, but various phosphatides differ. For example, the P=O stretch of phosphatidylethanolamine occurs at 1227 cm^{-1}, and that of phosphatidyl-serine at 1220–1180 cm^{-1}. The frequency displacement may be related to H-bonding, which can shift the P=O stretching absorption to lower frequencies.

The P—O—C bond usually gives rise to strong absorption at 1050–980 cm^{-1} (71). The major lipid contributions in the amide I and II regions are due to the CONH of sphingomyelin, absorbing between 1640 and 1540 cm^{-1}, and the ionized carboxyl group of phosphatidyl-serine, absorbing near 1600 cm^{-1}.

The vibrations associated with the basic groups of phospholipids have received little attention. In dimyristoyl cephalin a weak band, due to NH_3^+, lies at 2100 cm^{-1}, and in some lipoproteins a lipid-associated N—H stretching band has been located at 3330 cm^{-1}.

The most prominent IR absorptions due to cholesterol are the OH stretching band, lying at 3600 cm^{-1} in the free species, and at 3330 cm^{-1} in cholesterol esters (71).

Membranes

Aqueous dispersions of membrane lipids (36) and myelin (40) exhibit a prominent band at 720 cm^{-1}, due to CH_2 rocking. In contrast, this band is not detectable in films of plasma membranes of human erythrocytes (34) and Ehrlich ascites tumors (38, 39), indicating that in these membranes, the planar, all-trans-configuration of hydrocarbon chains is not favored, presumably because of apolar associations between membrane lipids and proteins that make other configurations energetically more probable. Possible contributions of the amide V band in the 700–720 cm^{-1} region of membrane spectra have not been investigated.

In their study of lipid configurations in the membranes of *Micrococcus lysodekticus*, Green and Salton (41) examined not only the $(CH_2)_n$ rocking mode, but also the intensities of the asymmetric and symmetric methylene stretching vibrations (at 2930 cm^{-1} and 2855 cm^{-1}, respectively). They found the asymmetric:symmetric ratios of whole membranes, their lipid extracts, and the extracted residue to be 0.3, 0.1, and 1.4 at liquid nitrogen temperature, but only the lipid extract had significant absorption at 720 cm^{-1}. Green and Salton interpret these results as evidence for close-range, nonpolar interactions between apolar protein side chains and phospholipid hydrocarbon chains.

The IR spectra of erythrocyte membranes display a shoulder at 1711 cm^{-1}, thought by some (34) to be due to lipid carbonyl. However, it is lacking in lipid extracts and may represent some facet of the lipid-protein interaction.

The phosphate ester (P=O) stretching frequency in red cell ghosts (35) and Ehrlich ascites carcinoma membranes (39) lies at 1225 cm^{-1}, as does that of an aqueous dispersion of the membrane lipids (34). This frequency is low for P=O stretching (71), presumably due to H-bonding. This suggests that in these membranes, as in aqueous micellar systems, the phosphatide head groups lie in water, rather than being ionically linked to membrane proteins, as envisaged in many models. Moreover, if the latter were true, the configuration of the phosphate ester group would be seriously distorted. The phosphorus of the P—O^- bond would become more electropositive, changing the P=O dipole moment and its stretching frequency; this is not observed. Finally, when the phosphate is not linked ionically, resonances can occur (see Fig. 4.6). This is consistent with the IR results. All in all, the fact that erythrocyte membranes and their lipid extracts have identical P=O stretching frequencies weighs against prominent involvement of the phosphate groups in lipid-protein interactions in this case. The P=O and P—O—C stretching frequencies deserve closer scrutiny in terms of protein-lipid interactions. This is particularly so for mitochondria with their high content of the polyanion cardiolipin.

The position of the P=O stretching frequency depends not only

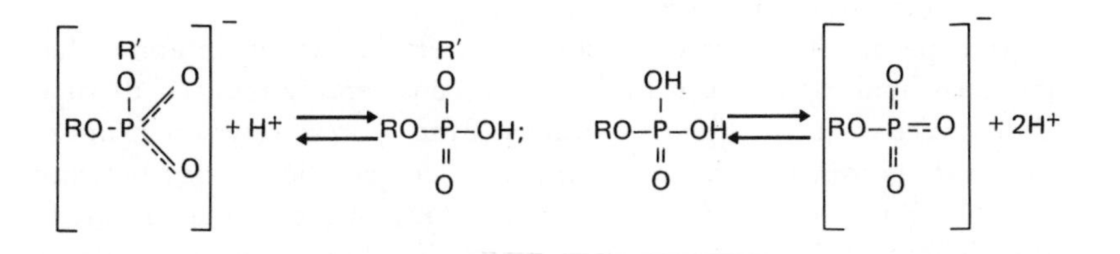

Fig. 4.6 Organic phosphate resonance states.

upon the polarity of its *environment*, but also upon its location in an electric field. May, Kamble, and Acosta (74) have investigated this matter in a novel manner. They examined the effect of direct current fields on the intensities of the phosphate and $(CH_2)_n$ vibrations in cephalin and lecithin films. When natural phosphatides were used, the intensities of the phosphate and $(CH_2)_n$ bands increased with the applied electric field, suggesting alignment of the phosphate-base dipoles and secondary rearrangement of the *hydrocarbon* chains. Their important results suggest that such fatty acid realignment might occur during the passage of nerve impulses and that it might induce structural alterations in axone membrane proteins.

The frequent studies of artificial complexes between mitochondrial phospholipids and cytochrome c (32) suggest that the interaction involved is predominantly ionic. IR studies of membrane phospholipids, especially, should add to our knowledge of lipid-protein interactions in membranes and possible modulations thereof during metabolic activity.

Measurement of Lipid Extraction by IR Spectroscopy

Lipid extraction from membranes and membrane films can be monitored from the intensity of the C=O stretching band near 1730 cm^{-1}, which, at neutral pH, arises almost solely from the fatty ester linkages in membrane phospholipids; its disappearance can, thus, be correlated qualitatively with the removal of ester-containing lipids. The extraction of phospholipids also diminishes the phosphate absorption bands (P=O stretching, 1250 cm^{-1}; P—O—C stretching, 1070 cm^{-1}), but these regions often exhibit residual absorption due to nonlipid phosphate and some nonphosphate compounds (75, 76).

When erythrocytes are extracted in bulk, using 2:1 chloroform: methanol and the residue is cast as a film on a silver chloride disk, the spectrum shows a 1652 cm^{-1} peak and a 1628 cm^{-1} shoulder, suggesting some transition to a β-structure upon lipid removal. In contrast, when erythrocyte or Ehrlich ascites tumor plasma membranes are *first* dried upon AgCl disks and then lipid extracted, the 1628 cm^{-1} peak does not emerge (34, 37, 38). The apparent mechanical conformational stabilization is of interest and requires further study.

It is pertinent in this connection that erythrocyte membranes that have been lipid extracted in bulk show a considerably reduced thermal stability. Yet when the lipid is extracted from a previously dried film of erythrocyte ghosts, this shows no amide I changes when cycled between −110°C and +110°C (34). Chapman (45) suggests that residual, tightly bound lipid might effect this stabilization, but provide no evidence for the presence of such lipid; stabilization by mechanical effects or by protein-protein interactions appears more likely. It should be noted that changes in the proportions of β-structures are the only conformational change detectable by this method and that such structural transitions can occur in previously dried films (39). Thus, immersion of lipid-extracted films of tumor plasma membranes in buffers of pH <3 for 20 min produces a shoulder at 1630 cm^{-1}; similar effects were observed with phospholipid-containing films (39).

Extensive extraction of the membranes of *Micrococcus lysodekticus* with a variety of solvents does not abolish all of the phosphate-ester absorption, a fact that Green and Salton (41) attribute to residual phospholipid. These authors also found that lipid extraction effected slight broadening of the amide I band at 1630 cm^{-1}. The greatest shift in the amide I frequency toward 1635 cm^{-1} occurs in a deoxycholate (DOC)-soluble membrane fraction, containing less than 1% lipid, but the DOC-insoluble fraction shows little change compared to the native membrane. Green and Salton suggest that lipids remaining associated with the insoluble material stabilize the protein, but differential binding of DOC might also produce differences in the IR spectra.

Excitable Membranes

The possible existence of β-structured peptides in excitable membranes has been recently studied, using IR spectroscopy by Papakostides et al. (77). The authors used a membrane isolate from the brains of 5–8-week-old rats, enriched in "nonmyelinated axons" by differential centrifugation (78), and dialysed aliquots of their preparations into (a) 100 mM KCl, (b) 100 mM NaCl, and (c) 50 mM $CaCl_2$. For IR spectroscopy, 600 μg membrane protein was dried on germanium disks at 15°C and 58% relative humidity.

The particles equilibrated with Na^+ or Ca^{++} exhibit a rather sharp amide I band at 1652 cm^{-1}, and no inflection near 1630 cm^{-1}. This indicates that the peptides under observation are in the α-helical and/or unordered structure under these conditions.

In contrast, particles equilibrated with 100 mM KCl showed two amide I absorption maxima at 1648 and 1630 cm^{-1} plus a shoulder at 1695 cm^{-1}, suggesting that a substantial portion of the peptide goes into the antiparallel β-conformation in the presence of K^+. Unfortunately, the experiments, though tantalizingly suggestive, appear inadequately controlled, and one cannot be certain that the alterations observed reflect the behavior of axonal membrane proteins specifically. It would also seem essential to extend the ionic manipulations considerably before one could relate these interesting findings to axonal function.

Summary of IR Results

IR absorption spectroscopy has yielded much information regarding the structure and interactions of lipid and protein components in cellular membranes, although the technique has not been widely explored. Nevertheless, several interesting model systems have been developed. Also, IR spectroscopy has demonstrated the important fact that not all membranes are alike in the conformation of their proteins. The presence of some β-structured peptide appears to be much more widespread than is usually believed. Rat liver mitochondria, rat adipocyte plasma membranes, *Micrococcus lysodekticus* membranes, and *Acholeplasma membranes* certainly contain a significant percentage of their peptide bonds in a β-conformation (definitely antiparallel in the case of mitochondria). Highly significant is the suggestion that in erythrocyte membranes, some peptides assume the antiparallel β-structure during ATP hydrolysis and that the proportion of β-structure in mitochondrial membranes also depends on their metabolic state. It also appears that the conformational stability of some membrane proteins depends upon hydrophobically bound lipid. IR spectroscopy is potentially capable of yielding many more useful results; in particular, conformationally dependent amide vibrations other than the amide I band, the phosphate stretching regions, and various skeletal vibrations deserve closer study.

Raman Spectroscopy

The Raman Effect

Laser Raman spectroscopy has only recently been utilized in the conformational analysis of polypeptides and proteins (e.g., 79–89) and is in its infancy as applied to biomembranes. This is largely due to the fact that suitable instruments have become available only in the past few years. Nevertheless, the potential of the technique for membrane studies is enormous. Like infrared absorption, Raman scattering arises from the interatomic vibrations and rotations of molecules. When a transparent molecule is irradiated with monochromatic light, most of the light passes through, but a small fraction will be scattered. The light-scattering spectrum of the molecule will contain a predominant component at the frequency of the incident light, the Rayleigh peak, but a number of other scattering peaks of lower intensity occur at higher and lower frequencies than the incident beam; these are Raman-scattering peaks. The frequency shift (Raman shift) between the Rayleigh peak and a particular Raman line corresponds to one of the vibrational or rotational frequencies of a particular bond within the molecule that gives rise to that Raman-scattering band.

The shapes and frequency shifts of Raman bands do not depend on the frequency of the exciting radiation, but the intensity of the scattered light does. Raman and other scattering processes, unlike fluorescence, do not involve absorption of radiation. Scattering occurs in $\sim 10^{-12}$ sec, whereas fluorescence emission occurs $\sim 10^{-8}$ sec after excitation. Raman activity appears only if a change in its polarizability occurs during a vibrational motion. Polarizability is an expression of the ability of an applied oscillating electromagnetic field (the incident light) to induce a dipole moment in that bond. A rigorous description of the phenomenon is found in treatises by Woodward (90) and Tobin (91), and we shall only summarize the fundamentals here.

When an oscillating electromagnetic field is applied to a rotating atom with a spherically symmetrical electron cloud, the magnitude of the induced dipole between the electron cloud and the positive nucleus will pass through a maximum, fall to zero, and reverse cyclically with the frequency of the incident radiation, causing scattering of light of the same frequency (Rayleigh scattering). However, if the rotating atom (molecule) possesses an asymmetrical electron cloud, this is not constantly polarizable with respect to the incident radiation. The frequency with which the polarizability changes, i.e., the frequency of rotation, generates a set of Raman-scattering peaks, whose frequencies correspond to the Rayleigh frequency $\pm$ the *rotational frequency*. Also, if the polarizability of a bond changes during a *vibration*, the light scattered from the bond will include the incident light frequency and this frequency $\pm$ the *vibrational frequency*. When the frequencies of the scattered light are higher than that of the incident radiation, the scattering bands are referred to as *anti-Stokes* bands and when they are *lower*, *Stokes* bands (Fig. 4.7). Anti-Stokes bands are invariably weaker than Stokes bands.

As noted, Raman scattering derives from a change of molecular polarizability α during a vibrational motion. Thus, when α varies with Q_c, the normal coordinate of a vibration with frequency ν_c,

$$\alpha = \alpha^0 + \left(\frac{\vartheta\alpha}{\vartheta Q_c}\right)_0 \cdot Q_c \tag{4.6}$$

where α^0 = the polarizability when the nuclei are in their equilibrium position, and $(\vartheta\alpha/\vartheta Q_c)$ is the rate of change of α^0 with Q_c for infinitesimal nuclear displacements.

When a single molecular species is irradiated with light at I_0 w-cm^{-2}, the power scattered at a single Raman frequency ΔI is σI_0 w, where σ

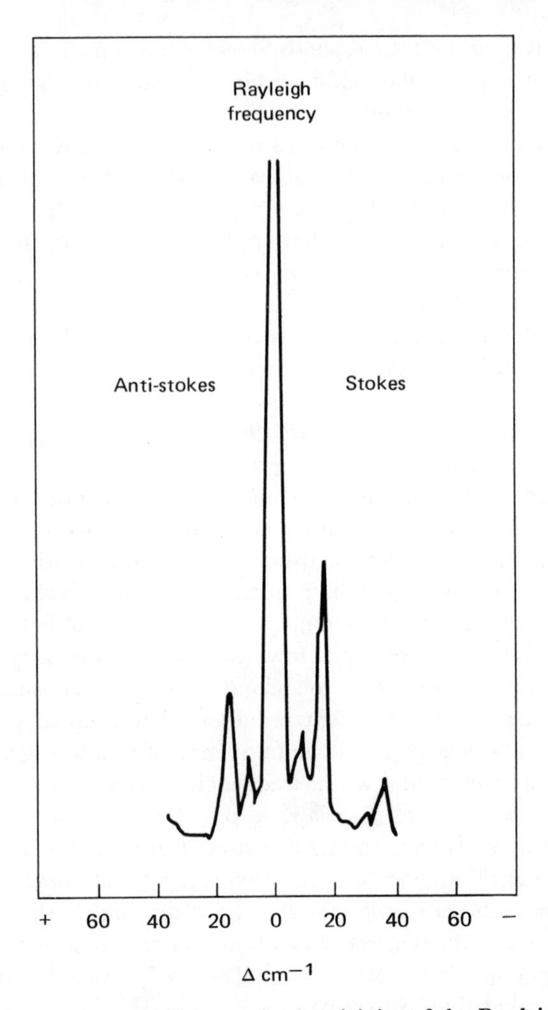

Fig. 4.7 Laser Raman spectrum in the vicinity of the Rayleigh frequency, showing both Stokes and anti-Stokes bands.

has the dimensions of cm². For N molecules, the total scattered power in watts is

$$\overline{\Delta I} = N\sigma I_0 \,\mathrm{w} \tag{4.7}$$

If ρ = the number of molecules/cm³, A = cross-sectional area illuminated in cm², Δx the length of the illuminated volume in cm,

$$N = \rho A\,(\Delta x) \tag{4.8}$$

and

$$\overline{\Delta I} = \rho\sigma\,(\Delta x)\,AI_0 = k\overline{I}\Delta x \tag{4.9}$$

where k = the loss factor in cm^{-1}. Equation (4.9) is equivalent to the Beer-Lambert-Bouguet law for the loss of energy from the exciting beam due to Raman scattering into a given Raman band.

Raman cross sections are not easily determined, and few data are available. The value of σ is $\sim 10^{-28}$ cm^2. Systems with large, polarizable atoms or biologically important groups such as *I*—, —C=C— are likely to be strong scatterers. Ionic bonds tend to scatter poorly.

Raman-scattered radiation is invariably at least partially polarized, irrespective of the polarization of the incident light. In some cases, this can aid in identifying the vibrational mode giving rise to Raman scattering. The polarization of a Raman line is described by p, the depolarization factor, where

$$p = \frac{I_{\perp}}{I_{\parallel}} \tag{4.10}$$

and $I_{\perp}$ and $I_{\parallel}$ are the intensities perpendicular and parallel to the exciting radiation. When the exciting radiation is nonpolarized, $p = \frac{6}{7}$ for all Raman bands arising from *nontotally* symmetrical vibrations. For *totally* symmetrical vibrations, $p < \frac{6}{7}$ and can approach 0. In certain cases, polarization is *inverted*, and p approaches ∞ (see below).

In general, symmetrical vibrations give rise to Raman scattering, while antisymmetrical vibrations usually yield strong infrared absorption. If a molecule has a center of symmetry, no fundamental frequency can appear in *both* the Raman and infrared spectra.

The Raman effect is generally weak; approximately 1% of the incident radiation is scattered and, of this, 99% is Rayleigh scattering. In the 1930s Edsall (92) obtained Raman-scattering data from amino acids and dipeptides using high-intensity Hg-arc light sources. However, the study of large, polymeric molecules became feasible only with the development of stable lasers as monochromatic sources of high intensity, plus recent sophistications in optics and electronics.

An important aspect of Raman spectroscopy is *resonance enhancement* (93–95). This occurs in regions of electronic absorption and is manifested by an often dramatic gain in scattering efficiency ($\sim 10^6$), more than sufficient to compensate for losses in exciting and scattering intensities due to absorption of incident and scattered radiation.

Technical

The basic laser Raman spectrometer consists of (a) a stable gas laser source, (b) an optical system to focus the laser beam on or into the sample, as well as to collect and image the scattered radiation on the slit of (c) a high-quality double monochromator, and (d) a cooled, low-noise photomultiplier to count scattered photons; this is tied to a photon counter and ultimately a display, i.e., a plot of scattering intensity vs. frequency shift (Fig. 4.8). Unless computer facilities are available, a multichannel analyzer, or signal averager, is a desirable option for high-sensitivity measurements, and a necessity for differential spectroscopy. However, the relatively low cost of small computers offers great promise in this technique. Thus, commercial systems, which can perform the following, are now available.

(a) Data can be smoothed, e.g., by least squares analyses.

(b) The systems can combine previously stored or logged spectra by subtraction or ratioing. Solvent background can, thus, be automatically eliminated.

(c) Data may be signal averaged over multiple scans.

(d) Derivative spectra can be obtained, minimizing interference by broad fluorescence bands.

(e) Data may be integrated over a desired spectral region.

(f) Data may be expanded or shrunk in the X axis.

(g) Data may be expanded, shrunk, or shifted in the Y axis.

(h) The systems can automatically select and print out intensity and spectral position of peaks or other significant features as required.

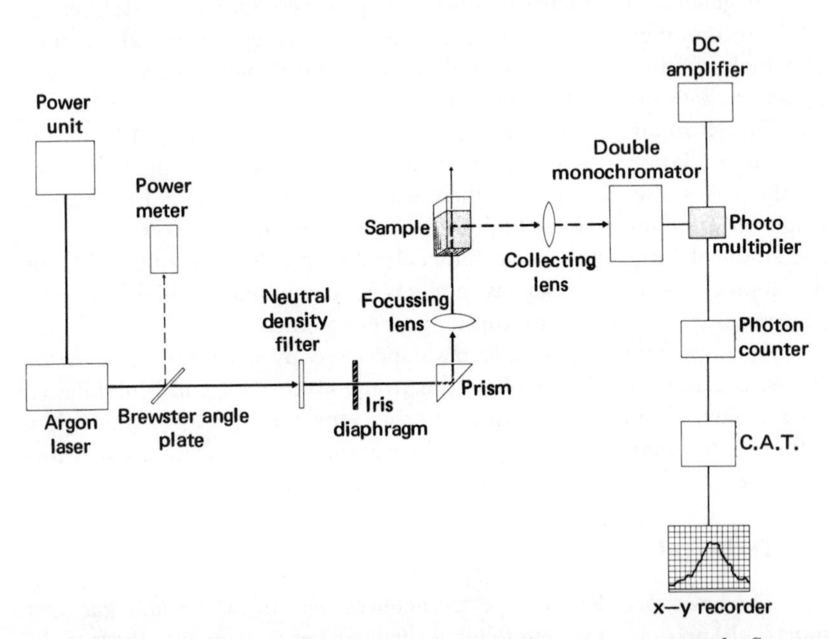

Fig. 4.8 Schematic representation of a laser Raman spectrograph. See text for details. [From Wallach et al. (83) by permission of the North Holland Publishing Co.]

The choice of laser sources is important, since these are expensive; they are, however, usually interchangeable. With samples absorbing near 500.0 nm, a krypton ion laser with lines at 568.2 and 647.1 is useful, except for resonance Raman experiments. More generally, the best choice is an argon ion laser operated at 514.5 nm or 488.0 nm with high-output power. This laser can be "single moded" permitting application of a recent technical advance, of great importance for the study of scattering systems such as membranes (namely, the iodine filter technique). This is applicable only to the 514.5 nm line of argon ion lasers. A 5.0 cm long cell filled with hot iodine vapor strongly absorbs the 514.5 nm Ar^+ line, yielding at 10^3 discrimination against Rayleigh-scattered light. By placing such a filter in the path of the scattered light, one can approach the incident frequency within 5–10 cm^{-1}, even using particulate samples.

The Ar^+ laser offers the additional advantage that detector efficiencies are great near 500 nm, as are Raman-scattering cross sections. These factors and high power levels are essential for work with dilute solutions. When background fluorescence is low, the signal-to-noise ratio of a Raman band approximates the square root of the product of laser power, sample concentration, and time of measurement. As an approximation, good spectra from 10 mg/ml peptide or protein solutions require about 1 w of 514.5 nm laser power at a scan rate of 25 cm^{-1}/min. Because of the small dimensions of laser beams, this can be achieved with very small sample volumes ($\sim$50 μl; i.e., $\sim$500 μg).

Fluorescence commonly complicates laser Raman spectroscopy, but can usually be avoided by careful sample and reagent purification. Nevertheless, some fluorescence may remain at levels that are initially identical whether irradiation is at 514.5 or 647.1 nm. Fluorescence declines most rapidly at high laser power and at low irradiating wavelength. The presence of fluorescence indicates absorption of light, some of which is dissipated as heat and can cause localized heating and sample damage. In our studies of oxygenated erythrocyte membrane suspensions, using 1.5 w of 488.0 nm laser power, we found the decrease in fluorescence paralleled by a decline in Na^+—K^+-ATPase activity. However, local heating could not be the sole source of enzyme damage, since in deoxygenated samples, the ATPase activity remained unimpaired (93). In computerized systems, derivative spectroscopy can eliminate much of the output interference due to broad fluorescent bands.

Because of fluorescence and Rayleigh scattering from particulate inhomogeneities and impurities, sample quality is more critical in Raman than infrared spectroscopy. All chemicals should be highly purified. Only distilled, deionized, and deoxygenated water should be used; if fluorescence is apparent, further purification may be necessary, e.g., centrifugation, gel permeation, and/or filtration through activated charcoal. Particulate samples should be filtered through the smallest aperture millipore filter permissible while filling the cell, using disposable micropipettes. Dust or gas bubbles on the inside walls of the curvettes cannot be tolerated. For measurements in D_2O the usual precautions must be followed to avoid contamination by H_2O. Cacodylate is a useful buffer, because it has a weak Raman spectrum with the exception of a line near 600 cm^{-1}.

For clear or moderately turbid solutions or suspensions, the laser beam is best focused vertically through the bottom window into the sample cell, and the scattered radiation collected from the side at 90° to the beam. Cells should be constructed of nonfluorescent fused quartz, with a square of cylindrical cross section and a flat, well-polished bottom. The top of the cell should be sealed with a transparent cap to eliminate dust and constructed so as to permit deoxygenation. The cell must be completely filled to avoid reflections from the curved air-liquid interface. For temperature measurements and for long-term sample stability, the cell should be mounted in a temperature-controlled jacket or holder. If a jacket is used, the circulating liquid should not pass over the entrance or collection window.

In normal operation, the monochromator collects only the radiation scattered from the cylindrical volume illuminated by the focused laser beam; this is only about 20–50 μm in diameter. Turbid samples diffuse the incident beam, as well as the Raman radiation, into a much larger volume, reducing the number of *signal* photons that reach the monochromator. At the same time, *noise* is increased because of the extensive Rayleigh scattering. This difficulty can be met by a number of techniques, including the use of iodine filters and appropriate computerization.

A significant biological advantage of laser Raman spectroscopy over IR absorption spectroscopy is that *water* interferes to a far lesser extent. Throughout most of the useful Raman spectrum, the scattering from water can be largely ignored. The scattering from the water O—H bending vibration in the amide I region, although significant in intensity, is relatively weak compared to its infrared absorption. In Raman spectroscopy, therefore, water background can be successfully overcome either by employing high concentrations of polypeptide or protein (e.g., 79), or by using computer-assisted difference techniques; the latter approach permits analysis of relatively low small polypeptide or protein concentrations in aqueous solution or suspension.

Polypeptides and Proteins

The Raman scattering of H_2O and D_2O (Table 4.8) serves as a useful background for the recent Raman analyses of poly-L-alanine,

Table 4.8 Raman Vibrational Spectroscopy of H_2O and D_2O

Vibration	H_2O	D_2O
O—X stretching	Very broad band with maximal scattering at:	
(ν_s)	$\nu = 3439$	$\nu = 2532$
	Less intense scattering occurs at:	
	$\nu = 3300$	$\nu = 2400$
	Minor components occur at:	
	$\nu = 3600$	$\nu = 2600$
	$\nu = 3500$	
Assocation (ν_A)	Very weak, broad band	
X—O—X′ bending	Relatively weak bands at:	
(ν_2)	$\nu = 1640$	$\nu = 1208$
Libration	Three weak bands between:	
(ν_L)	$\nu \sim 450–780$	$\nu \sim 375–550$
Hindered translation	Weak bands at	
(ν_T)	$\nu \sim 152–175$	$\nu \sim 152–175$
Hindered translation	Weak band at	
(ν_{T_2})	$\nu = 60$	

Note: ν = frequency of band maximum in cm^{-1}.

poly-L-proline, and polyglycine obtained in concentrated *aqueous* solution (83–86). These polypeptides yield a number of bands associated with the amide, skeletal, and side chain groups, and Table 4.9 lists some of the more prominent scattering bands associated with poly-L-alanine. Many of the Raman peaks show good correlation with IR absorption frequencies.

Using difference spectroscopy, high resolution Raman spectra have been obtained from dilute solutions of poly-L-lysine and poly-L-glutamate (83) in different conformations. Table 4.10 lists the major bands exhibited by the unordered, β-antiparallel, and α-helical forms of poly-L-lysine. The IR absorption bands are included for comparison. α-helical and unordered poly-L-glutamate exhibits Raman bands similar to those of poly-L-lysine in these two conformations. Not all bands have been assigned, but this technique appears capable of discriminating between α-helical, β-pleated sheet, and unordered polypeptide conformations. The great potential of a method which can unambiguously detect these three conformations in dilute aqueous solutions of polypeptides cannot be emphasized too strongly.

Amide II scattering is weak (79, 80, 83–86), but the amide III region is readily visible and is more complex than in the infrared. It has been assigned to bands near 1320 cm^{-1} in polypeptides (83–88), and to a conformation-sensitive region near 1250 cm^{-1} in proteins (79–81, 89).

Table 4.9 Infrared and Raman Frequencies for Poly-L-Alanine[a]

Raman[b]		Infrared	Calcd. Modes[c]		
cm^{-1}	RI	cm^{-1}	A	E	Approximate Group Mode
294	w				Skeletal deformation
314	w		310		Skeletal deformation
		325s		338	
375	s	368s	361	368	Skeletal deformation
444	vw	445m		474	Skeletal deformation
527	vs	528s	524		Skeletal deformation
		595	592		N—H out-of-plane bending C=O out-of-plane bending; (amide VI)
610	w	615w		612	
662	w	655m		655	N—H+C=O out-of-plane bending; (amide IV)
698	m	690w–sh		695	
756	m		762		N—H and skeletal modes; (amide V)
		770vw		778	
800	vw				
884	w	890s			
907	vs	907m–sh			Methyl-carbon stretch and skeletal modes
930	vw–sh				
964	m	960s			
1016	w				
1048	vw	1049vs			Methyl modes
1065	vw–sh				
1104	vs	1105s			Methyl-carbon stretch
1168	m	1167s			Methyl rocking
1243	w	1240w–sh			
1268	w	1268m			
1310	m	1300s			C—N stretching, N—H in-plane bending; (amide III)
1331	s	1328m–sh			C—H bending
1371	m	1376vs			Methyl symmetric deformation
1453	vs	1453vs			Methyl asymmetric deformation
1549	vw	1537vs, 1555w–sh			N—H in-plane bending C—N stretching; (amide II)
1654	vs	1651vs			C=O stretching (amide I)

[a] From Koenig and Sutton (79).
[b] RI = relative intensity.
[c] Potential energy distribution for calculated modes given by Miyazawa *et al.* (24).

S—S scattering is intense, and disulfide modes are easily discerned (79–81). The intensity of a line near 509 cm^{-1} is proportional to the number of S—S— bonds, and the ratio of this peak to the C—S—S line at 661 cm^{-1} has been related to the C—S—S—C dihedral angle (79, 81).

Table 4.10 Raman and Infrared Frequencies (1500–1700 cm^{-1}) of Poly-L-Lysine in Different Conformations[a]

Helix[b,c,d]		Antiparallel—β[b,c,d]		"Unordered"[b,c,d]	
Raman	IR	Raman	IR	Raman	IR
1517w	1516w }AII	1511w		1521m	
1537m	1535s }AII	1535w	1530s AII		1535 AII
				1547	
		1564m		1565	
1600m					
1617m		1617m			
		1631vs	1636s }AI		
1639s	1650vs }AI				
1647vs	1652m }AI		}AI	1653vs	1656 AI
		1672vs		1665vs	
			1685m }AI	1683vs	

[a] Wallach et al. (83).
[b] vs = very strong; s = strong; m = medium; w = weak.
[c] IR frequencies from Susi *et al.* (17).
[d] AI = amide I; AII = amide II.

In an important recent development, highly pertinent to the membrane field, Brown et al. (94) show that certain proteins exhibit low-frequency Raman bands (<50 cm^{-1}), which depend on protein conformation but not on sample state (solution, film, or crystal). Vibrations of such low frequency must reflect motions of all or very large portions of the molecules. The data suggest possible means for the detection of translational motions of proteins in membranes.*

Resonance Raman Spectroscopy of Protein-Associated Chromophores

The irradiation of substances absorbing light at laser wavelengths will lead to local heating, which one might expect to degrade some biological macromolecules. However, because of resonance enhancement at low sample concentrations, one can obtain resonance Raman (RR)

* The conformations of insulin and proinsulin have been considerably clarified by laser Raman spectroscopy (N.-T. Yu, C. S. Liu, and D. C. O'Shea, *J. Mol. Biol.* **70:** 117, 1972). Denaturation of insulin markedly alters the amide I and III regions, as well as scattering from S—S and C—S stretching, skeletal bending and skeletal stretching. The data suggest major conformational changes toward the β-structure upon insulin denaturation. The data from insulin allow easy interpretation because of the lack of tryptophan resonances.

scattering of laser radiation from the vibrational modes in the heme groups of aqueous solutions of oxy- and deoxyhemoglobin (95–97).

Brunner et al. (95) report large, reversible differences occurring between the spectra of oxy- and deoxyhemoglobin, in particular (a) the shift of a strong band at 1376 cm^{-1} in oxyhemoglobin to 1335 cm^{-1} in the deoxygenated species and (b) disappearance of bands at 573 cm^{-1}, 1505 cm^{-1}, and 1636 cm^{-1}; this is tentatively ascribed to the =C—N stretching vibrations in the four pyrrole units of the heme. Of great general importance is the fact that the Raman spectra did not deteriorate even after 2 hr of irradiation.

Strekas and Spiro (96, 97) also obtain prominent resonance Raman spectra from diverse hemoglobin derivatives and observe that the strong bands at 1638 cm^{-1} and 1589 cm^{-1} go out of resonance with deoxygenation. Presumably their 1638 cm^{-1} band corresponds to the 1636 cm^{-1} component reported by Brunner et al. (95).

Recent resonance Raman spectroscopic studies of cytochrome c (98, 99) appear to pave the way for Raman studies on membrane respiratory chains. They show dramatic differences in the RR bands arising from the heme chromophore, depending upon whether this is oxidized or reduced. These experiments were performed at very low concentrations to minimize light absorption effects. In general, oxidation reduces the intensity and complexity of the spectrum, but also induces the appearances of new bands which may reflect significant frequency shifts upon oxidation.

The Raman spectra of cytochrome c and of hemoglobin exhibit unusual polarization characteristics. Thus, several prominent bands show inverse polarization, i.e., the polarization plane of light scattered at right angles to the incident beam lies perpendicular to that of the incident light, when this is perpendicular to the direction of scattering. The depolarization ratio $p = I_{\perp}/I_{\parallel}$ approaches infinity in these cases (Fig. 4.9). In contrast to their usual prominence in Raman spectra, polarized bands are weak and few in the spectral region studied. The theoretical basis of this phenomenon is discussed by Spiro and Strekas (99).

Lipids

The Raman spectra of long-chain fatty acids, fatty acid esters, and some phosphatides have also been fairly well studied (100–103). A head group deformation vibration lies at 720 cm^{-1}, CH_2 deformation near 845 cm^{-1}, =C—H (out of plane) deformation at 967 cm^{-1}, C—C stretching near 1085 cm^{-1}, =C—H (in plane) deformation at 1261 cm^{-1}, CH_2 twisting at 1298 cm^{-1}, CH_2 wagging at 1370 cm^{-1}, CH_2 deformation at 1445 cm^{-1}, C—O (ester) stretching at 1731 cm^{-1}, C—D stretching near 2100

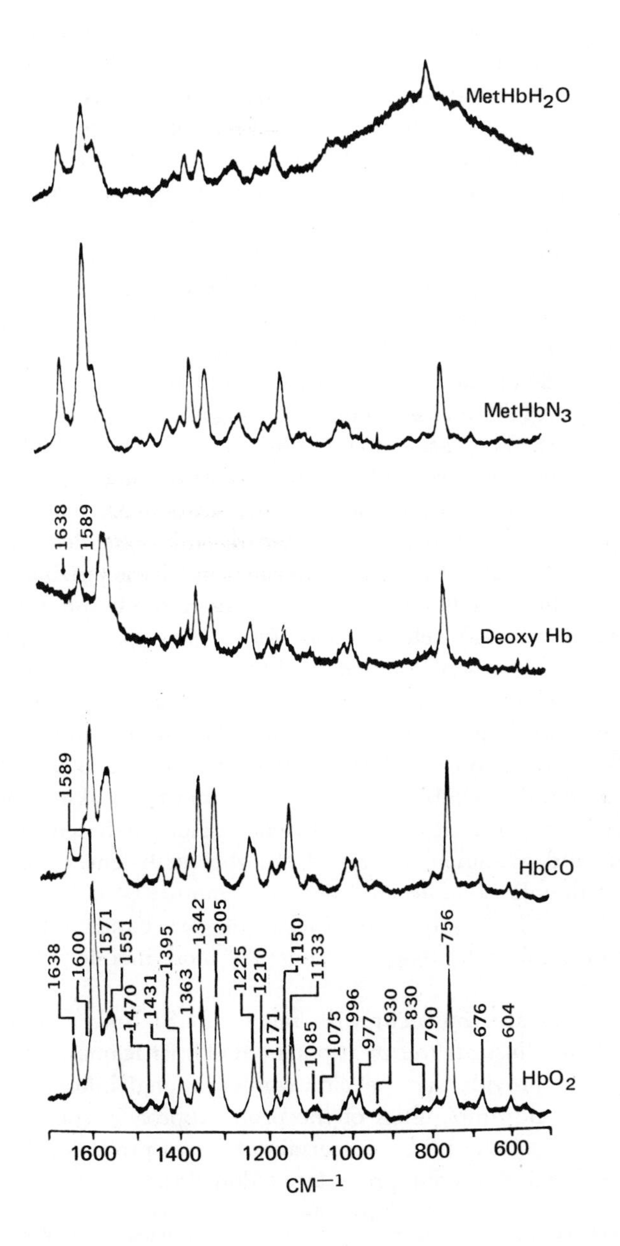

Fig. 4.9 Resonance Raman spectra of various hemoglobin derivatives. The large differences between deoxy- and oxyhemoglobin are particularly remarkable. [From Strekas and Spiro (96) by permission of the Elsevier Publishing Co.]

cm^{-1}, symmetric C—H stretching of CH_2 and CH_3 near 2850 and 2870 cm^{-1}, and asymmetric C—H stretching of CH_2 and CH_3 near 2930 and 2960 cm^{-1}. The bands between 1000 and 1200 cm^{-1} arise from skeletal vibrations where alternate carbons move in opposite directions along a hydrocarbon chain. In the crystalline (all *trans*) state, the band near 1080 cm^{-1} is weak relative to the 1125 cm^{-1} line. This is reversed upon melting due to the appearance of *gauche* isomers. In the crystalline state symmetric C—H stretching from CH_3 groups dominates, whereas in the liquid state symmetric C—H stretching of CH_2 groups becomes predominant.

In aqueous suspensions of micellar dipamitoyl lecithin, the intensity of bands near 1066 and 1130 cm^{-1} indicates the amount of all-*trans*-structure in the paraffin chains, and a broad band at 1089 cm^{-1} has been assigned to structures containing several *gauche* rotations of the melted paraffin (102). The ratios of the 1089/1128 cm^{-1} or 1089/1066 cm^{-1} bands provide a sensitive index of the extent of the ordering or paraffin chains and the cooperativity of thermal transitions.

Lippert and Peticolas (103) have recently extended their Raman studies into the field of model membranes by characterizing the Raman active vibrations in long-chain fatty acids and phosphatides sonically dispersed in water, as well as in solid form. They provide assignments for those Raman bands of long-chain fatty acids that serve to determine (a) all-*trans*-chain length and (b) the position and configuration of ethylenic bonds in homogeneous samples. The authors also show that the —C—C— stretch region of the hydrocarbon chains of free fatty acids and phosphatides exhibits two intense scattering peaks, which can be assigned to (a) the all-*trans*-methylene moiety between the carboxyl terminus and a double bond and (b) the methylene chain extending from the double bond to the methyl terminus; both bands vary with temperature. The relative intensities of the two bands indicate that the interior of a lipid multibilayer is generally more mobile than the surface region.

The data are consistent with much of ESR information discussed later, but also illustrate the great potential of laser Raman spectroscopy—its ability to provide very specific configurational information without the dangers and ambiguities of the probe approach, such as structural perturbations by the probe, uncertainties as to probe localization, and probe localization in nonrepresentative domains.*

* Laser-Raman spectroscopy of lecithin and lecithin-cholesterol mixtures gives considerable insight into lecithin-cholesterol interactions (R. Mendelsohn, *Biochim. Biophys. Acta* **290:** 15, 1972). The positions of the Raman lines near 1100 cm^{-1} and the intensity ratio 1125 cm^{-1}/1085 cm^{-1} are sensitive to hydrocarbon chain configuration and allow monitoring of phase transitions. By these criteria, the hydrocarbon chains of dipalmitoyl phosphatidylethanolamine are in predominantly *trans* configuration (crystalline) at room temperature, while those of egg phosphatidyl choline are in a liquid state.

Very intense *resonance Raman* (RR) spectra have been reported for several conjugated polyenes, including carotenoids (104, 105). In these studies, the absorption spectra and Raman spectra of the pigments remained stable for up to 1 hr of irradiation, and the spectra could also be obtained from living plant material (carrot roots, tomato). This work has been extended to include various vitamin A type molecules, permitting a correlation between isomer configuration and the RR spectra (106). Such studies clearly indicate the potential of RR spectroscopy in the study of membranes containing conjugated polyenes such as carotenoids and vitamin A. It also suggests that such substances, as well as the polyene antibiotics, might serve as useful Raman active probes.

Membranes

At this writing, only two publications on laser Raman spectroscopy of biomembranes have been published (107, 108); the first (107) concerns erythrocyte ghosts and describes four bands at 1110, 1340, 1420, and 1445 cm^{-1}, superimposed upon a broad fluorescence background. The 1110 cm^{-1} component is ascribed to fluidity of the hydrocarbon chains in the membranes. These data are suspect, because of the high fluorescence, which is most likely due to solution impurities and can be avoided by careful technique. Where there is fluorescence, there must be light absorption and some dissipation of absorbed energy as heat. At the levels of power concentrated into a small volume by a laser beam, this can produce very significant local heating and therewith artifactual fluidization of hydrocarbon chains.

In a highly significant report, Rimai et al. (108) extended their earlier resonance Raman work (RR) to the visual pigments of intact, dark-adapted bovine retinas at −70°C to −85°C; they employed an ion argon laser at 488.0 nm. As in their earlier studies on plant material (104–106), they find that the Raman spectrum from these organelles is

---

The insertion of cholesterol into egg lecithin liposomes reduces the spectral contribution of *gauche* isomers and increases that of the *trans* species. Cholesterol appears to prohibit the formation of some but not all *gauche* isomers, thereby reducing chain fluidity. Very likely the *gauche* isomers which are sterically blocked are those involving bends in the tightly packed, polar end of the chains. Insertion of cholesterol into a "crystalline" phospholipid array, loosens the tightly packed structure by permitting formation of *gauche* isomers at the apolar end of the hydrocarbon chains (cf. 102; J. E. Rothman and D. M. Engelman, *Nature New Biol.* **237:** 42, 1972).

The CH stretching vibrations (near 2900 cm^{-1}) of lipid Raman spectra provide another sensitive index for the configurations of hydrocarbon chains (K. Larsson, *Chem. Phys. Lipids* **10:** 165, 1973). Diverse packing arrangements can be identified in the solid state as can the increase in disorder upon transition to liquid-crystalline and micellar states. Excellent spectra can be obtained even in the presence of 90% by weight of water. A technique for the study of skin samples is described.

dominated by resonance-enhanced constituents; the spectra of non-resonant vibrations is attenuated by the absorbing chromophore. The spectra can, thus, be wholly attributed to the visual pigments. The major scattering peak at 1555 cm^{-1} is assigned to the ethylenic C=C stretching mode of retinal. The fact that the frequency is distinctly lower than that of free retinal or its basic Schiff base is thought to support the hypothesis that the pigment is bound as a *protonated* Schiff base, i.e., by a —$C=N^{+}H$— bond.*

References

1. Bellamy, L. J. *The Infrared Spectra of Complex Molecules.* New York: Wiley, 2nd ed., 1958.
2. Herzberg, H. *Infrared and Raman Spectra of Polyatomic Molecules.* Princeton, N.J.: Van Nostrand, 1945.
3. Wilson, E. B., Jr., Decius, J. C., and Cross, P. C. *Molecular Vibrations.* New York: McGraw-Hill, 1955.
4. Kendall, D. N. In *Applied Infrared Spectroscopy,* D. N. Kendall, ed. New York: Reinhold, 1966, ch. 1, p. 1.
5. Bellamy, L. J. In *Advances in Infrared Group Frequencies.* London: Methuen, 1968.
6. Kendall, D. N. In *Applied Infrared Spectroscopy,* D. N. Kendall, ed. New York: Reinhold, 1966, ch. 2, p. 31.
7. Higgs, P. W. *Proc. Roy. Soc.* (London) **A220:** 427, 1953.
8. McClure, W. O., and Edelman, G. M. *Biochemistry* **6:** 559, 1967.
9. Herscher, L. W. In *Applied Infrared Spectroscopy,* D. N. Kendall, ed. New York: Reinhold, 1966, p. 88.
10. Ibid.
11. Harrick, N. J. *Internal Reflection Spectroscopy.* New York: Interscience, 1967.
12. Kuyama, Y., and Shimanouchi, T. *Biopolymers* **6:** 1037, 1968.
13. Suzuki, S., Iwashita, S., Tsuboi, M., and Shimanouchi, T. In *Proc. Intern. Symp. Molecular Structure and Spectroscopy,* Tokyo: Science Council of Japan, 1962, paper A110.
14. Miyazawa, T. In *Poly-α-Amino Acids.* Biological Macromolecules Series 1, G. D. Fasman, ed. New York: Dekker, 1967, p. 69.
15. Townend, R., Kumosinski, T. F. Timasheff, S. N., Fasman, G. D., and Davidson, B. *Biochem. Biophys. Res. Commun.* **23:** 163, 1966.

* Resonance Raman spectroscopy has been employed to study the purple membrane pigment in "purple membranes" of *H. halobium* (R. Mendelsohn, *Nature* **243:** 22, 1973). The purple color arises from an unprotonated Schiff base, whose electron density is perturbed by an electrostatic interaction with protein. Mendelsohn suggests that the chromophore consists of a charge transfer complex between retinyllysine and a protein tryptophan.

By suitable selection of excitation wavelength, individual resonance Raman enhancement of chlorophyll a, chlorophyll b, and carotenoids in spinach chloroplasts can be produced (M. Lutz and J. Breton, *Biochem. Biophys. Res. Commun.* **53:** 413, 1973). The spectra give considerable insight into the associations of the chlorophylls with environmental molecules.

16. Blout, E. R., and Asadourian, A. *J. Amer. Chem. Soc.* **78:** 955, 1956.
17. Susi, H., Timasheff, S. N., and Stevens, L. *J. Biol. Chem.* **242:** 5460, 1967.
18. Miyazawa, T., and Blout, E. R. *J. Amer. Chem. Soc.* **83:** 712, 1961.
19. Timasheff, S. N., Susi, H., and Stevens, L. *J. Biol. Chem.* **242:** 5407, 1967.
20. Timasheff, S. N., and Susi, H. *J. Biol. Chem.* **241:** 249, 1966.
21. Krimm, S. *J. Mol. Biol.* **4:** 528, 1962.
22. Phillips, D. C. *Sci. American* **215** (5): 78, 1966.
23. Lipscomb, W. N. *Acc. Chem. Res.* **3:** 81, 1970.
24. Miyazawa, T., Masuda, Y., and Fukushima, K. *J. Polymer Sci.* **62:** 562, 1962.
25. Fukushima, F., and Miyazawa, T. Presented at the Ann. Meeting of the Chem. Soc. of Japan, Tokyo, 1964.
26. Itoh, K., Nakahara, T., Shimanouchi, T., Oya, M., Uno, K., and Iwakura, Y. *Biopolymers* **6:** 1759, 1968.
27. Malcolm, B. R. *Nature* **195:** 901, 1962.
28. Malcolm, B. R. *Biochem. J.* **110:** 733, 1968.
29. Malcolm, B. R. *Nature* **219:** 929, 1968.
30. Malcolm, B. R., *Nature* **227:** 1358, 1970.
31. Malcolm, B. R. *Biopolymers* **9:** 911, 1970.
32. Frommherz, P., Peters, J., Müldner, H. G., and Otting, W. *Biochim. Biophys. Acta* **274:** 644, 1972.
33. Fromherz, P. *Biochim. Biophys. Acta* **225:** 382, 1971.
34. Chapman, D., Kamat, V. B., and Levene, R. J. *Science* **160:** 314, 1968.
35. Maddy, A. H., and Malcolm, B. R. *Science* **150:** 1616, 1965.
36. Maddy, A. H., and Malcolm, B. R. *Science* **153:** 213, 1966.
37. Graham, J. M., and Wallach, D. F. H. *Biochim. Biophys. Acta* **241:** 180, 1971.
38. Wallach, D. F. H., and Zahler, P. H. *Proc. Natl. Acad. Sci.* (U.S.) **56:** 1552, 1966.
39. Wallach, D. F. H., and Zahler, P. H. *Biochim. Biophys. Acta* **150:** 186, 1968.
40. Jenkinson, T. J., Kamat, V. B., and Chapman, D. *Biochim. Biophys. Acta* **183:** 427, 1969.
41. Green, D. H., and Salton, M. R. J. *Biochim. Biophys. Acta* **211:** 139, 1970.
42. Choules, G. L., and Bjorklund, R. F. *Biochemistry* **9:** 4759, 1970.
43. Avruch, J., and Wallach, D. F. H. *Biochim. Biophys. Acta* **241:** 249, 1971.
44. Graham, J. M., and Wallach, D. F. H. *Biochim. Biophys. Acta* **193:** 225, 1969.
45. Chapman, D. In *Biological Membranes*, Vol. 2, D. Chapman and D. F. H. Wallach, eds. London: Academic Press, 1973, p. 91.
46. Blow, D. M. *Biochem. J.* **112:** 261, 1969.
47. Bertland, A. U. II, Johansen, J., and Ottensen, 91. *Federation Proc.* **27:** 596, 1968.
48. Bertland, A. U. II, and Kalckar, H. M. *Proc. Natl. Acad. Sci.* (U.S.) **61:** 629, 1968.
49. Kalckar, H. M., Bertland, A. U. II, Johansen, J., and Ottensen, M. In *The Role of Nucleotides for the Function and Conformation of Enzymes*, H. M. Kalckar, M. Klenow, A. Munch-Peterson, M. Ottensen, and J. H. Thaysen, eds. Copenhagen: Munksgaard, and New York: Academic, 1969, p. 247.
50. Bertland, A. U. II, Kalckar, H. M., and Graham, J. M. (unpublished results).

51. Chicheportiche, R., and Lazdunski, M. *FEBS Letters* **3:** 195, 1969.
52. Wallach, D. F. H., Graham, J. M., and Fernbach, B. R. *Arch. Biochem. Biophys.* **131:** 322, 1969.
53. Wallach, D. F. H., Gordon, A. S., Graham, J. M., and Fernbach, B. R. In *Physical Principles of Biological Membranes*, F. Snell, J. Wolken, G. J. Iverson, and J. Lam, eds. London: Gordon and Breach, 1970, p. 345.
54. Kendrew, J. C., Dickerson, R. E., Strandberg, B. E., Hart, R. G., Davies, D. R., Phillips, D. C., and Shore, V. C. *Nature* **105:** 422, 1966.
55. Kauzmann, W. *Advan. Protein Chem.* **14:** 1, 1959.
56. Nemethy, G., and Scheraga, H. A. *J. Chem. Phys.* **41:** 680, 1964.
57. Lynn, J., and Fasman, G. *Biochem. Biophys. Res. Commun.* **33:** 327, 1968.
58. Davidson, B., and Fasman, G. *Biochemistry* **6:** 1616, 1967.
59. Khalil, M. T. M., and Lauffer, M. A. *Biochemistry* **6:** 2474, 1967.
60. Maybury, R. H., and Katz, J. J. *Nature* **177:** 629, 1956.
61. Vakar, A. B., Pumpyanskii, A. Ya., and Semenova, L. V. *Appl. Biochem. Microbiol.* **1:** 1, 1965.
62. Gallagher, K. J. In *Hydrogen Bonding*, D. Hadzi, ed. London: Pergamon, 1959, p. 45.
63. Scheraga, H. A. In *Protein Structure*. New York and London: Academic, 1961.
64. Tomita, K., Rich, A., de Loze, C., and Blout, E. R. *J. Mol. Biol.* **4:** 83, 1962.
65. Calvin, M., Hermans, J., Jr., and Scheraga, H. A. *J. Am. Chem. Soc.* **81:** 5048, 1959.
66. Kreschek, G. C., Schneider, H., and Scheraga, H. A. *J. Phys. Chem.* **69:** 3132, 1965.
67. Taborsky, G. *J. Biol. Chem.* **245:** 1054, 1970.
68. Wallach, D. F. H., and Gordon, A. *Federation Proc.* **27:** 1263, 1968.
69. Wallach, D. F. H. *J. Gen. Physiol.* **54:** 3s, 1969.
70. Wallach, D. F. H., and Gordon, A. S. In *Regulatory Functions of Biological Membranes*, J. Järnefelt, ed. Amsterdam: Elsevier, 1968, p. 87.
71. Chapman, D. In *The Structure of Lipids*. London: Methuen, 1965, ch. 4.
72. Tschalmer, H. *J. Chem. Phys.* **22:** 1745, 1954.
73. Sheppard, N. In *Advances in Spectroscopy*, H. W. Thompson, ed. New York: Wiley-Interscience, 1959, p. 288.
74. May, L., Kamble. A. B., and Acosta, I. P. *J. Membrane Biol.* **2:** 192, 1970.
75. Chapman, D., and Kamat, V. B. (unpublished observations).
76. Wallach, D. F. H., and Zahler, P. H. (unpublished observations).
77. Papakostides, G., Zundd, G., and Mehl, E. *Biochim. Biophys. Acta* **288:** 277, 1972.
78. Lemkey-Johnston, N., and DeKirmenjian, H. *Exp. Brain Res.* **11:** 392, 1970.
79. Koenig, J. L., and Sutton, P. L. *Biopolymers* **8:** 167, 1969.
80. Smith, M., Walton, A. G., and Koenig, J. L. *Biopolymers* **8:** 29, 1969.
81. Lord, R. C., and Yu, N. T. *J. Mol. Biol.* **50:** 509, 1970.
82. Lord, R. C., and Yu, N. T. *J. Mol. Biol.* **51:** 203, 1970.
83. Wallach, D. F. H., Graham, J. M., and Oseroff, A. *FEBS Letters* **7:** 330, 1970.
84. Sutton, P., and Koenig, J. L. *Biopolymers* **9:** 615, 1970.
85. Koenig, J. L., and Sutton, P. L. *Biopolymers* **9:** 1229, 1970.
86. Koenig, J. L., and Sutton, P. L. *Biopolymers* **10:** 89, 1971.

87. Lord, R. C. *XXIII International Congr. of Pure and Applied Chemistry* **7:** 179, 1971.
88. Lewis, A., and Scheraga, H. A. *Macromolecules* **4:** 539, 1971.
89. Bellocq, A. M., Lord, R. C., and Mendelsohn, R. *Biochim. Biophys. Acta* **257:** 280–7, 29 Feb. 1972.
90. Woodward, L. A. In *Raman Spectroscopy Theory and Practice*, H. A. Szymanski, ed. New York: Plenum Press, 1967, p. 1.
91. Tobin, M. C. In *Laser Raman Spectroscopy*. New York: Wiley-Interscience, 1971.
92. Edsall, J. In *Proteins, Amino Acids and Peptides*, E. J. Cohn and J. Edsall, eds. New York: Reinhold, 1943, p. 9.
93. Oseroff, A., Graham, J. M., and Wallach, D. F. H. (unpublished).
94. Brown, V. G., Erfurth, S. C., Small, E. W., and Peticolas, W. L. *Proc. Natl. Acad. Sci.* (U.S.) **69:** 1467, 1972.
95. Brunner, H., Mayer, A., and Sussner, H. *J. Mol. Biol.* **70:** 153, 1972.
96. Strekas, T. C., and Spiro, T. F. *Biochim. Biophys. Acta* **263:** 830, 1972.
97. Spiro, T. F., and Strekas, T. C. *Proc. Natl. Acad. Sci.* (U.S.) **69:** 2622, 1972.
98. Strekas, T. C., and Spiro, T. G. *Biochim. Biophys. Acta* **278:** 188, 1972.
99. Spiro, T. G., and Strekas, T. C. *Proc. Natl. Acad. Sci.* (U.S.) **69:** 2622, 1972.
100. Yvernault, T. *Oleagineux* **1:** 189, 1946.
101. Bulkin, B. J., and Krishnamachari, J. *J. Amer. Chem. Soc.* **94:** 1109, 1972.
102. Lippert, J. L., and Peticolas, W. L. *Proc. Natl. Acad. Sci.* (U.S.) **68:** 1572, 1971.
103. Lippert, J. L., and Peticolas, W. L., *Biochim. Biophys. Acta.* **282:** 8, 1972.
104. Rimai, L., Kilponen, R. G., and Gill, D. J. *J. Amer. Chem. Soc.* **92:** 3824, 1970.
105. Gill, D. J., Kilponen, R. G., and Rimai, L. *Nature* **227:** 743, 1970.
106. Rimai, L., Gill, D. J., and Parsons, J. L. *J. Amer. Chem. Soc.* **93:** 1353, 1971.
107. Bulkin, B. J. *Biochim. Biophys. Acta* **274:** 649, 1972.
108. Rimai, L., Kilponen, R. G., and Gill, D. J. *Biochem. Biophys. Res. Commun.* **41:** 492, 1970.

CHAPTER 5

NUCLEAR MAGNETIC RESONANCE

Introduction

The nuclei of many atoms constitute spinning charges, whose oscillating electric fields induce localized *magnetic moments*, which can be oriented in an applied magnetic field. Absorption of electromagnetic radiation of appropriate frequency (range 1–220 MHz) can raise the potential energy of the nuclei, forcing realignment of their magnetic moments in the applied field. Nuclear magnetic resonance (NMR) utilizes this and closely related phenomena, essentially measuring the energy required for realignment. Its physical principles have been rigorously treated (e.g., 1–2), and excellent reviews are available on its application to biological molecules in general (e.g., 3) and to lipids (e.g., 4). Here we shall limit comment mainly to the utility of NMR in the study of model and biomembranes, stressing that this technique, like infrared spectroscopy and optical activity measurements, mirrors primarily certain properties of *intrinsic* membrane components. Specifically, NMR signals the *mobility* of various nuclei.

Theory

General

Nuclei with zero spin, e.g., ^{12}C, do not yield NMR spectra. However, many nuclei of immediate or potential interest exhibit a net nuclear spin I (Table 5.1). When placed in a constant magnetic field, such nuclei orient their magnetic moments either parallel or antiparallel to the spin axis, producing two energy levels. Thus, protons aligned parallel to a magnetic field of strength H_0 possess a potential energy $-\mu H_0$ (μ is the magnetic moment of the proton), and those oriented against the applied magnetic field have an energy of $+\mu H_0$, corresponding to the two values of the magnetic quantum number, $M = \pm\frac{1}{2}$.

A population of protons represents a Boltzmann distribution between

Table 5.1 Properties of Some Biologically Important Paramagnetic Nuclei

Isotope	Natural Abundance	Spin	NMR Frequency for 10 kgauss (MHz)
^{1}H	99.985	$\frac{1}{2}$	42.576
^{2}H	0.015	1	6.536
^{13}C	1.108	$\frac{1}{2}$	10.705
^{14}N	99.630	1	3.076
^{15}N	0.370	$\frac{1}{2}$	4.314
^{19}F	100.00	$\frac{1}{2}$	40.054
^{31}P	100.00	$\frac{1}{2}$	17.235

the two energy levels, the ratio of the proton numbers in the two levels being equal to $e^{2\mu H_0/kT}$, where T is the absolute temperature and k the Boltzmann constant. The low-energy state $(-\mu H_0)$ contains more protons than the higher-energy state $(+\mu H_0)$; however, because of the small energies involved, the excess in the low-energy population is only about 3–4 × 10^6 nuclei per mole.

The interaction energy E between a nucleus of spin I and a magnetic field H_0 can have $(I+1)$ values according to the equation

$$E = \frac{M\mu H}{I} \tag{5.1}$$

where μ is the nuclear magnetic moment, and M is a quantum number which can equal I, $(I-1)\ldots$, 0, $\ldots$, $-(I-1)$, $-I$. The selection rule for absorption demands that $\Delta M = 1$; hence, the absorption frequency for a given value of H_0 is $\nu_0 = H_0/hI$, where h = Planck's constant, and energy absorption occurs when

$$\nu_0 = \frac{\mu H_0}{hI} = \frac{\gamma H_0}{2\pi} \tag{5.2}$$

here γ is the gyromagnetic ratio that is characteristic for each nucleus. When frequency is maintained constant and H_0 varied, absorption occurs when the magnetic field experienced by the nucleus locally equals $\nu_0/hI\mu$. Thus, when protons are aligned in an external magnetic field and are exposed to electromagnetic radiation of the appropriate (radio) frequency ν_0, they are promoted from the lower-energy level $(-\mu H_0)$ to the higher-energy state $(+\mu H_0)$ when

$$h\nu_0 = 2\mu H_0 \tag{5.3}$$

Radiation of the same energy, ν_0, is emitted upon nuclear relaxation, i.e., when transitions occur in the reverse direction.

Since the nuclei in the relaxed state are in a slight excess, electromagnetic radiation of suitable frequency produces a measurable *net* absorption of energy. Because only nuclei with a net spin give rise to resonant absorption, only nuclei with odd values for the atomic number or atomic mass can be observed by the method. This eliminates, for example, the normal isotopes of carbon and oxygen.

Since different NMR-active nuclei absorb at different resonance frequencies, a molecule containing a number of such nuclei, exposed to electromagnetic radiation of varying frequency, should yield a number of frequency absorption maxima. Thus, in a field of 10^4 gauss, protons have a resonant frequency of 42.58 MHz (cycles/sec $\times 10^6$). The corresponding resonant frequencies of some other nuclei are given in Table 5.1. However, it is clear from the preceding that NMR data can be expressed either in frequency units, ν_0, or field units, H_0, and it is technically simpler to operate at a constant applied frequency (e.g., 40, 60, 100, 120, and 220 MHz) and vary the magnetic field, H_0. Then the resonant absorption is expressed *in terms of the magnetic field* (in gauss).

The Chemical Shift

The resonance frequency of a given nucleus depends on the magnetic field it actually experiences, but this does not follow the ideal equation, $h\nu_0 = \mu H_0/I$, in biologically important cases. Most often, the nuclei are screened by orbital electrons with a diamagnetic moment opposed to but proportional to H_0. Such shielding is further complicated by vicinal π electron systems (e.g., from aromatic amino acids) and through the influence of other nearby nuclei. Therefore, the field experienced by a particular nucleus depends not only upon H_0, but also upon the extent (positive or negative) to which the local environment modifies the applied field. In molecules, the various nuclei thus experience a field, H, *different* from that of the magnet as follows

$$H = H_0 (1 - \sigma') \tag{5.4}$$

where σ' is a dimensionless constant reflecting a nucleus' microenvironment. σ' differs for nuclei of atoms participating in different chemical bonds. The displacement of resonance due to such local chemical effects is called the *chemical shift* and reflects the different σ' values in diverse chemical configurations. σ' values are small, particularly in the case of protons, because these have fewer electrons in their immediate vicinity than other nuclei.

The values of shielding constants are generally chosen so that

$$\sigma = \sigma' \times 10^6 = \frac{H - H_{\text{ref}}}{H_{\text{ref}}} \times 10^6 \tag{5.5}$$

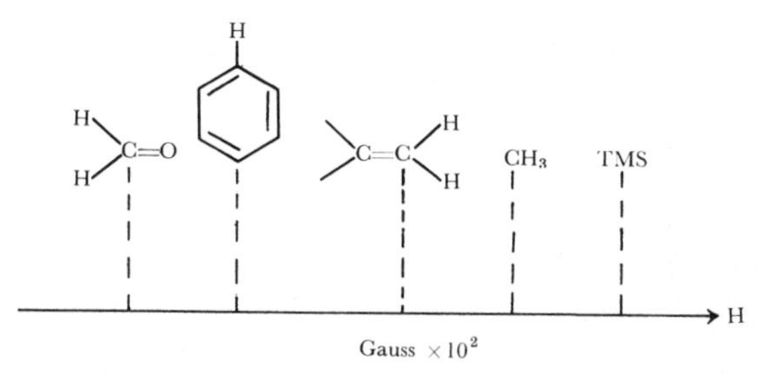

Fig. 5.1 Relative resonance positions for some biologically important protons.

where H_{ref} is the field where a proton, in a reference chemical array, resonates at the same frequency. The chemical shifts are then

$$\delta = \sigma_1 - \sigma_2 = \frac{H - H_{ref}}{H_{ref}} \times 10^6 \tag{5.6}$$

For protons in organic solvents, tetramethylsilane (TMS) is a convenient reference compound, the chemical shift being expressed as $\tau = 10 - (\sigma_1 + \sigma_2)$. Because chemical shifts are small in absolute magnitude, τ is expressed in parts per million (ppm), i.e., the shift in gauss per million gauss. In aqueous solvents, 2,2-dimethyl-2-sila-pentane-5-sulfonic acid sodium salt (DSS) is a common reference substance (Fig. 5.1). Both TMS and DDS absorb in regions where few other protons yield a signal.

Spin Coupling

As already noted, the field experienced by a given nucleus depends not only on H_0, but also on the nucleus' magnetic microenvironment. In the cases of liquids, or solids in solution, it appears that the quality of an instantaneous spin state is transmitted from one nucleus to the other via bonding electrons. This can occur even when electron spins are paired, because a nuclear magnetic moment slightly polarizes the bonding electrons magnetically, and this polarization can transmit the instantaneous spin state of one nucleus to another. As a result, one obtains a distribution of resonant frequencies about ν_0.

This process is normally countered by *spin lattice* relaxation. Here the energy of excited spins is transferred by a radiationless mechanism into the "lattice" of local nuclei in a time T_1 of 10^{-5}–10^4 sec. T_1 describes the lifetime of spin states. In the case of protons, these will be equally populated without the influence of an external field. When such is imposed instantaneously, the time required for the spin states to reach $1 - 1/e$ of their new equilibrium is T_1.

Spin lattice relaxation depends primarily on transitions between spin states induced by dipoles fluctuating near the resonant frequency of the nucleus under study. This dipole interaction varies as the inverse sixth power of distance; hence intramolecular contributions tend to dominate the phenomenon. Moreover, the larger the magnetic dipoles, the greater the relaxation effect.

At a given time, a population of randomly tumbling molecules includes components of rotational and translational motion with frequencies varying from zero to a maximum related inversely to the *correlation time* $1/\tau_c$. For a spherical molecule, the rotational correlation time varies with particle radius a (cm), viscosity η (poise), and absolute temperature, approximately according to the Stokes relation

$$\tau_c = \frac{8\pi\eta a^3}{6KT} \tag{5.7}$$

The correlation time relates to the intensity of resonance bands. Thus, an increase in local viscosity, changing correlation time from $1/\tau_1$ to $1/\tau_2$ raises the intensity of all components with frequencies $<1/\tau_2$ and decreases those with frequencies $>1/\tau_2$. This is because a nucleus relaxes most efficiently when the resonance frequencies $\nu_0 < 1/\tau_2$. Optimal relaxation occurs when $\nu_0 = \frac{1}{2}\pi\tau_2$. As a reference example, $1/\tau_c$ for water at 20°C is about 10^{-12} sec. The correlation times for ideal spheres of diverse radii, computed according to the Stokes relation, are given in Table 5.2.

"*Spin-spin*" *relaxation* involves a different process. Spin-coupled nuclei, with unhindered mobility, tumble rapidly, and the field imposed upon one nucleus by its neighbors is time averaged. However, when the motion of the coupled nuclei is constrained, e.g., by high local viscosity,

Table 5.2 Correlation Times of Differently Sized Ideal Spheres Calculated from the Stokes Equation

Radius	τ_c
50,000	1.25×10^{2}
10,000	1.00×10^{0}
5,000	1.25×10^{-1}
1,000	1.00×10^{-3}
500	1.25×10^{-4}
100	1.00×10^{-6}

Note: η=0.01 poises; T=293°K.

the field experienced by one nucleus varies with the motion of the surrounding nuclei. The nuclei will then precess out of phase with one another. This produces band broadening, particularly in solids and macromolecules.* As a principal consequence of this interaction, the local field H_e experienced by a nucleus differs from the laboratory field according to the relation

$$H_e = H_0 + H_d \tag{5.8}$$

where H_0 and H_d are the external field and the dipole field of vicinal nuclei, respectively. Nuclei with large magnetic dipoles clearly produce greater line broadening. The dipole interaction among nuclei is described by the relaxation time T_{2_d}, which is the time constant for the rate at which precessing nuclei get out of phase.

T_{2_d} increases with molecular motion, $1/\tau_c$, and rapid tumbling (e.g., $1/\tau_c \gg 2\pi\nu_0$) brings the intensities of low-frequency components to the point that $T_{2_d} = 2T_1$.

The total bandwidth is influenced not only by high-frequency motions, described by T_1, low-frequency motions, characterized by T_2, but also diverse magnetic inhomogeneities due to the sample, as well as the applied field, and described by T_{2_m}. The total bandwidth relates to the relaxation time T_2 as follows:

$$\frac{1}{T_2} = \pi\Delta\nu = \frac{1}{2T_1} + \frac{1}{T_{2_m}} + \frac{1}{T_{2_d}} \tag{5.10}$$

Experimental

Figure 5.2 illustrates the essentials of a typical NMR spectrometer. Most commonly, these instruments are operated at a fixed frequency (60 MHz to 220 MHz for protons) and sweep the external magnetic field continuously through the resonance areas; this is *continuous wave* NMR. The sensitivity is often increased by employing multiple sweeps, signal averaged by a computer of average transients (CAT). After n

* Bandwidth $\Delta\nu$ (Hz) is defined with respect to band shape as follows:

$$(\Delta\nu)^2 = A^2 \frac{2}{\pi} \text{Arctan} \frac{(\alpha \cdot \Delta\nu)}{\nu_c} \tag{5.9}$$

where A is the bandwidth in a rigid lattice, α a constant, and ν_c the correlation frequency of molecular motion. (ν_c reflects nuclear reorientation and in nonviscous media may exceed 100 Hz.) In a nonviscous environment, proton absorption bands may have a width $\Delta\nu \leq 1$ Hz, while in viscous media the band may become too wide to study. Often, rotational band narrowing can be estimated by the relation $\Delta\nu \simeq 2\pi\nu_0{}^2\tau_c$, where $\Delta\nu_0{}^2$ is in Hz^2.

sweeps, the signal has increased n-fold, but the noise by a factor of only $n^{-1/2}$, thus improving the signal-to-noise ratio.

A relatively new and important alternative approach is *pulsed* NMR. Here the sample is irradiated by short, intense pulses. The output is a free induction decay and is usually mathematically converted into the equivalent of constant wave spectroscopy by Fourier transformation (2). Pulsed NMR usually yields a spectrum in 1/100 the time required for continuous wave analysis. Even more significant perhaps is the fact that this technique can, in principle, avoid some of the interference due to

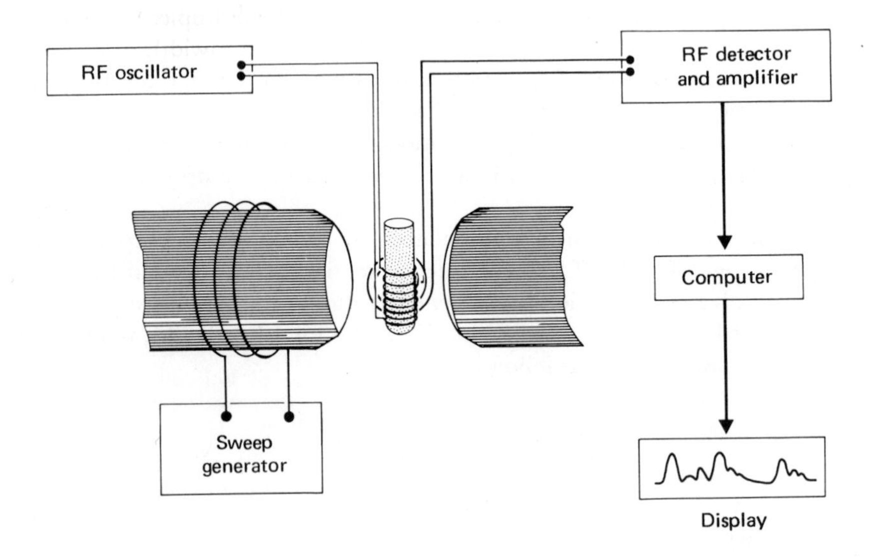

Fig. 5.2 Schematic of a typical NMR spectrometer. Essential details are given in text.

the solvent protons in H_2O solutions. In continuous wave spectroscopy of H_2O solutions or suspensions, the solvent protons essentially obscure the sample PMR spectrum by virtue of their abundance and their diverse resonances (spread over $\sim 6\tau$). However, H_2O protons relax far more rapidly than most other nuclei of interest. Accordingly, one should be able to introduce sufficient delay, between the end of the RF pulse and the onset of data collection, to allow relaxation of the H_2O protons and still resolve other resonance signals. This technique of delayed Fourier transform NMR spectroscopy is already being applied to the elimination of solvent interferences (5).

Comparison of the resonance bandwidths and intensities of a particular nucleus in an unhindered situation with identical nuclei in proteins, lipids, lipoproteins, or membranes provides a measure of the mobility of the group in the macromolecule. This realization has been most useful in the interpretation of NMR spectra from such substances.

NMR of Polypeptides and Proteins

The normal isotopes of carbon and oxygen possess zero net spins and NMR signals from ^{14}N are very broad; hence the NMR of proteins and polypeptides is due primarily to their protons.

In principle, these substances should yield rich proton magnetic resonance (PMR) spectra, because the protons of amino acid residues exist in numerous and diverse linkages (CH_3, CH_2, CH, COOH, NH_2, C_6H_5, etc.) and lie in environments of varying polarities. Moreover, the partial double bond character of the peptide C—N linkage causes short-range deshielding of peptide NH and α-CH protons (3), and their PMR spectra should vary with the array of the peptide linkages, i.e., with secondary structure. Indeed, this is occasionally observed, but not yet regularly enough for the conformational analysis of proteins in general. However, these effects are pertinent to this discussion, since the PMR spectra of lipoproteins and membranes may contain significant contributions from their peptide moieties.

Synthetic Polypeptides

Poly-L-alanine in chloroform solution is α-helical. Addition of trifluoroacetic acid causes the peptide to unfold and effects the displacements of the CH_3, CH, and NH proton peaks (Table 5.3); this is thought to be due to a decrease of the anisotropic shielding present in the α-helical conformation (6). Bradbury, Crane-Robinson, and Rattle (7) and Boublick at al. (8) also find shifts in the α-CH PMR of various polypeptides during conformational transitions. For example, the chemical shift of the α-CH protons in α-helical poly-L-arginine is 6.0τ, but 5.6τ in the unordered polymer. The chemical shifts of the α-CH protons of poly-L-lysine also increase with the coil-helix transition (8). Moreover, in α-helical poly-L-arginine, the α- and β-CH_2s exhibit broad PMR peaks, attesting to their hindered mobility. Solvent effects may contribute to the broadening (8), but in the case of poly-γ-benzyl-L-glutamate, changes in chemical shifts associated with coil-helix transitions most likely arise from factors other than solvent-polymer interactions (9).

Under favorable conditions, one should be able to determine the proportion of residues in helical, unordered, or other arrays from PMR spectra (7–9), but there are major obstacles to this goal. *First*, conformationally significant signals have so far only been obtained in mixtures of

Table 5.3 Chemical Shifts of Poly-L-Alanine in $CDCL_3$-TFA Mixtures (Shifts in parts per million from Tetramethylsilane)[a]

Wt % TFA	PLA Concn. (mg/ml)	$-b_0$	CH_3	CH	NH	TFA	CH—CH_3
9.76	15.2		1.557	4.283	7.933	11.000	2.726
18.16	26.4	341[b]	1.557	4.300	7.867	11.300	2.743
25.2	26.3		1.562	4.323	7.850	11.450	2.761
37.6	26.1		1.588	4.372	7.850	11.605	2.784
45.0	26.2		1.608	4.402	7.867		2.794
49.8	25.5	328.1					2.794
52.9	25.8	328.2	1.617	4.423	7.867		2.806
65.6	24.7		1.625	4.467	7.867	11.617	2.842
69.7	26.0	307.4					2.842
76.0	26.0		1.648	4.505	7.867	11.617	2.857
87.3	24.5	214.5	1.608	4.617	7.817	11.733	3.009
89.1	25.1	193.3	1.603	4.640	7.817	11.717	3.037
100.0	26.4	124.3	1.590	4.707	7.783		3.117

[a] From Stewart *et al.* (6). Chemical shift in δ ($\delta = 10 - \tau$). PLA = poly-L-alanine; TFA = trifluoroacetic acid.
[b] $-b_0$ measured on a solution containing 2.2 mg of PLA/ml.

trifluoroacetic acid and deuterated chloroform; less impressive effects have been obtained in aqueous solvents (7). *Second,* in the case of poly-γ-benzyl-L-glutamate of *low molecular weight* (<20,000), the resonance frequencies of the NH protons are 1.83τ for the α-helical and 2.05τ in the unordered form, while those of the CH protons are 6.0τ and 5.5τ in the α-helical and unordered forms, respectively; such differences in chemical shift are adequate to discriminate between the two conformations, but unfortunately disappear with higher degrees of polymerization (9). *Third,* the helix-coil transitions of poly-L-alanine and poly-L-leucine cause broadening of the NH and α-CH proton peaks at *different helicities* (7). *Fourth,* in several cases (poly-L-alanine, poly-γ-benzyl-L-aspartate, poly-γ-benzyl-L-glutamate, poly-L-glutamate, and poly-L-methionine), shifts in the α-CH proton line can differ tenfold, and only poly-L-alanine yields a defined N—H proton peak upon transition to the α-helical form (10). The various PMR manifestations of helix coil transitions in diverse polypeptides most likely arise from differences in (a) hydrogen bonding and (b) magnetic anisotropy of adjacent peptide bonds.

Proteins

In most proteins, protons can exist in so many different atomic and molecular environments that PMR spectra commonly yield poor resolution. As many as 25 scans may be required to produce identifiable peaks, and in many proteins, high resolution PMR spectra require denaturation. An important factor accounting for the low level of resolution with globular proteins is the tumbling rate of the whole macromolecule (11). The most important information obtained to date concerns the environment of certain specific residues upon denaturation (12). PMR studies thus show that certain tryptophyl, phenylalanyl, and tyrosyl residues of lysozyme are buried in the interior of the protein and become exposed during denaturation. Histones (13) and chymotrypsinogen (14) have also been extensively studied by PMR. Both exhibit proton peaks due to the methyl and methylene groups of alanine, leucine, isoleucine, threonine, and valine. Bandwidth measurements, which reflect molecular motion, indicate that in helical segments the mobilities of apolar residues is limited (13). The apolar groups in chymotrypsinogen yield only one band ($\Delta\nu = 0.2\tau$), and intensity changes occur only during unfolding of the protein. Significant improvement in the resolution of the PMR of lysozyme has been obtained at higher frequencies—100 MHz (15).

A promising technique for high resolution protein-PMR is that of Crespi and Katz (16), who incorporate ^{1}H-amino acids into fully deuterated proteins to aid elucidation of secondary and tertiary structure. They examined the phycocyanin of certain blue-green algae (*Phormidium luridum, Synechococcus lividus, Fremyella dipolsiphon*) and green algae (*Chlorella vulgaris*) after either biosynthetic or radiochemical incorporation. Biosynthetically incorporated ^{1}H-leucine produces a single peak at 1.29δ which sharpens upon denaturation, suggesting that most of the leucine residues are in a similar apolar environment. ^{1}H-alanine gives three peaks in contrast. The resonance of the CH_3 protons in ^{1}H-leucine incorporated into phycoerythrin by addition subsequent to isolation of the protein occurs at two positions—1.23δ and 0.6δ, indicating two different environments for the leucine residues. The sharpness of the peaks indicates high mobility.

The application of NMR techniques to the study of the secondary structure of, and conformational transitions in, globular proteins is still in its infancy, and membrane proteins have been studied hardly at all by this method; nevertheless, it is valuable to realize its potential. High resolution equipment (220 MHz) combined with the judicious use of pulse and relaxation methods, deuteration, ^{13}C enrichment, etc., can, without doubt, serve membrane biology profitably.

Moreover, while PMR is as yet not satisfactory for absolute conformational analysis, its ability to define the environment of certain amino acid residues will be of value in the study of membrane proteins. Perturbations of membrane structure, such as modification of the apolar environment within the membrane, should produce PMR changes which, under the right conditions, could yield important information also about membrane protein. Alterations in the lipid content of lipoprotein structures by phospholipase action, by lipid perturbants, by lipid exchange, or by lipid removal might change the environment of certain apolar or aromatic amino acid residues sufficiently to influence their PMR spectra.

Membrane proteins with paramagnetic substituents are immediately amenable to PMR study. For example, the PMR of ferrocytochrome c differs markedly from that of ferricytochrome c (3). Some anomalous high field peaks have been observed and may reflect the effect of the porphyrin ring current on adjacent amino acid residues; other differences have been attributed to conformational transitions (3). The PMR study of cytochromes in situ would appear a fruitful application of this technique to membrane biology.

NMR of Phosphatides

PMR

The phospholipid components of lipoprotein structures have been intensively investigated by PMR, with special emphasis on their physical state (liquid, liquid crystalline, crystalline, etc.), cholesterol-phosphatide interactions, lipid-protein interactions, as well as the interactions of various lipids with small polypeptides and other substances. Most studies have utilized PMR, but more and more workers are also investigating the NMR of ^{13}C-enriched or deuterated lipids.

As already noted, the field experienced by a highly mobile, spin-coupled nucleus will be time averaged, generating a narrow resonance band. Less mobile spin-coupled nuclei are exposed to a varying magnetic field made up of a constant component, the applied field, and a varying component, due to the motion of surrounding magnetic dipoles; these exhibit broad resonance bands. Broad band PMR can, thus, give information regarding the liquid/solid proportion of fats (17, 18). Figure 5.3(a) describes the absorption curve arising from a liquid/solid fat mixture. Figure 5.3(b) describes the derivative curve, which is often used to present NMR spectra; the broad absorption arises from the solid component, and the superimposed narrow band arises from the liquid phase (4).

Under certain conditions, one can resolve magnetic resonance due to specific lipid protons (Table 5.4). Thus, a CH_2 group next to an —OH has a different chemical shift from one vicinal to an —OCO—, i.e., 6.4τ, and 5.90τ, respectively (4). Aldehydic protons have low chemical shifts

(0.0–0.7τ). —NH proton resonance is hardly detectable (4), and the resonance peak of the normal nitrogen of lecithin choline is too broad to yield useful data.

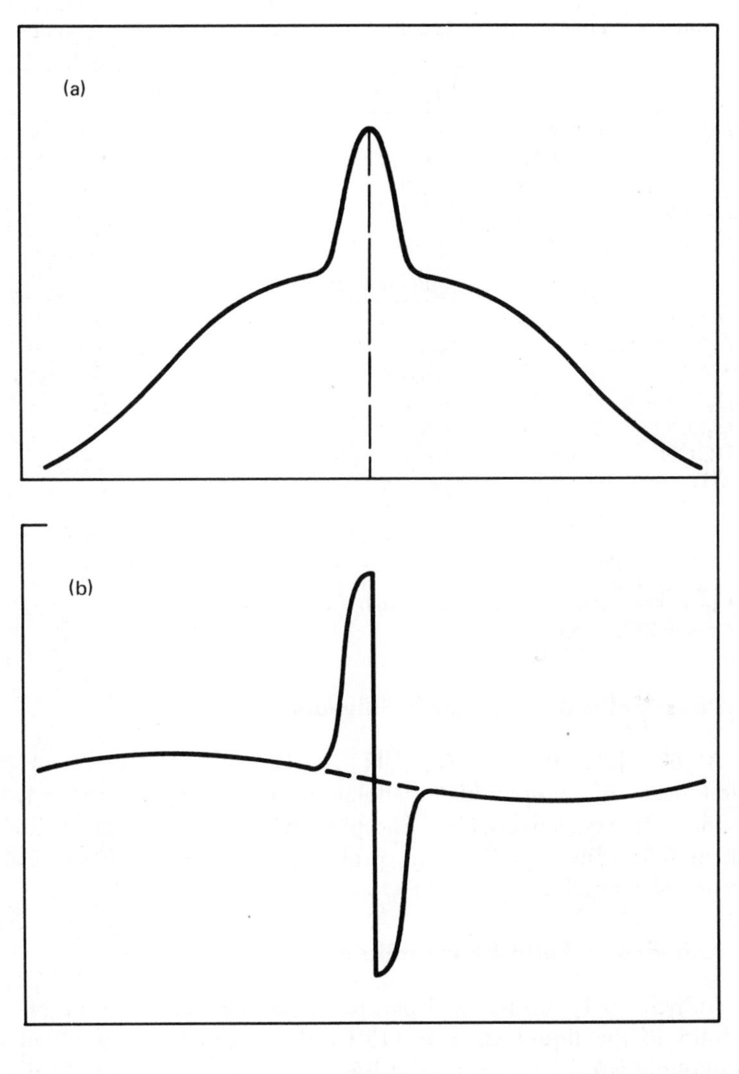

Fig. 5.3 (a) Typical "broad band" PMR resonance obtained from liquid/solid fat mixtures. The broad band arises from the solid component, the peak from the liquid. (b) First derivative of (a), bringing out the contribution of the liquid component (17, 18).

Table 5.4 Chemical Shifts in τ Values for Some Important Phosphatide Protons[a]

Proton	Dioleylphosphatidylethanolamine	Egg Lecithin[b]	Mixed Acyl Phosphatidylserine	O-Phosphoethanolamine[b]	O-Phosphocholine-Cl[b]
		Fatty Acyl Protons			
$—CH_3$	9.12	9.11	9.12		
$—CH_2—$	8.70	8.74	8.7		
$+C\underline{H}_2—CH$	7.95	8.0	7.95		
$—CH_2CO—$	7.69	7.71	7.71		
$—CH{=}CH—$	4.60	4.63	4.65		
		Polar Protons			
$—C\underline{H}_2NH_2$	6.70			6.71	
$—C\underline{H}_2N^+(CH_3)_3$	—	~6.1			6.3
$—N^+(CH_3)_3$	—	6.65[c]			6.75
$—CH_2OCO$	5.89	~6.1	5.9		
$—CH_2OPO$	5.67	5.7	5.9		
$—CHOCO$	~4.8	~4.8	~4.8		
$—N^+H_3$	1.67	—	—		

[a] From Chapman and Morrison (19). Unless otherwise stated, assignments are in $CDCl_3$ with τ of tetramethylsilane = 10.0.
[b] In D_2O; τ of tetramethylsilane = 10.0.
[c] τ = 6.77 in D_2O.

Phospholipids in Organic Solvents

At 60 MHz, the $—CH_3$, (CH_2), $—CH{=}CH—$, and $—CH_2CO$ proton peaks of phosphatidylethanolamine in $CDCl_3$ occur at 9.1, 8.7, 4.6, and 7.7τ, respectively (19). Phosphatidylcholine yields an additional peak at 6.6τ, due to $N^+(CH_3)_3$ protons; this is more evident in the presence of a small amount of D_2O (19).

Anhydrous, Solid Phospholipids

Anhydrous 1,2-dimyristoyl phosphatidylethanolamine changes from the solid to the liquid state at 115°C. When the temperature of this phospholipid is raised from that of liquid nitrogen toward the transition temperature, the width of the broad-band resonance peak gradually narrows from 15 to 0.09 gauss (20). At the transition temperature, the peak abruptly sharpens and intensifies, indicating a sudden increase of molecular motion. However, the PMR spectra give evidence for some molecular motion even at room temperatures, where this saturated phospholipid is in the solid state. When examined at 60 MHz and −180°C, anhydrous phosphatidylethanolamines exhibit "wide line" PMR spectra, with chain proton bandwidths near 15 gauss; these reduce to $\frac{1}{2}$ by heating to 75°C (20) or to $\frac{1}{4}$ in the presence of water (21). Lecithins behave analogously, but choline groups exhibit substantial mobility in comparison to the acyl chains or glycerol (22). Also, in the liquid crystalline state, the protons of the mobile choline group and the fatty acids yield bandwidths of 0.1 gauss, compared with 1 gauss for the protons of the immobile glycerol residue.

At higher fields (220 MHz), one can obtain high resolution PMR spectra from phospholipids in the smectic, liquid crystalline phase (23). Some lipids, such as sphingomyelin, give highly resolved spectra at about 60°C, with clearly resolved —CH_3, $(CH_2)_n$, and —$N^+(CH_3)_3$ bands (24). Moreover, these and related data further attest to the mobility of the choline group and point against a fixed position of this residue in a model lipid bilayer. This is particularly evident after elimination of the $(CH_2)_n$ signal by deuteration of the hydrocarbon chains, allowing clear resolution of the —$N^+(CH_3)_3$ band (25).

Hydrated Solid Phosphatides

At 25°C, in the presence of $<5\%$ water, egg yolk phosphatidylcholine yields a narrow band superimposed upon a broad component. The narrow band presumably represents mobile methyl and methylene protons of hydrocarbon chains. Upon transition to the liquid crystalline phase (~25°C), the PMR bands sharpen and stay so until ~90°C (26). At 90° a change from an anisotropic liquid crystalline state to an isotropic cubic phase occurs (27) with further resolution in the spectrum. Above 110°C, the spectrum becomes featureless due to a transition to another mesophase (21, 27). Similar data result with synthetic 1,2-dipalmitoyl phosphatidylcholine, but at different temperatures (28). It is important to note that the relative intensities of the $N^+(CH_3)_3$ and $(CH_2)_n$ bands depend markedly upon the lipid phase and the solvent environment of the choline group. Thus, addition of a small amount of D_2O to a $CDCl_3$ solution of egg yolk lecithin markedly intensifies these signals.

Phospholipids in Aqueous Dispersion

High resolution PMR spectra are readily obtained from phospholipids in organic solvents, in which atomic and molecular motion is unimpeded, but usually sonication in D_2O is required to produce spectra of comparable resolution at 60 MHz, with, for example, egg yolk lecithin or red cell membrane lipids (29–31), even though these relatively unsaturated phospholipids are rather liquid, even at room temperature in the absence of water. Between 30°C and 90°C, low intensity bands due to choline and methylene protons, superimposed on a broad component, are seen even with unsonicated preparations. In contrast, the PMR spectrum of a sonicated dispersion of egg yolk lecithin in D_2O resembles that of the solid phospholipid at 120°C, where it is in the cubic phase.

Sonication changes the broad-band spectrum for egg yolk lecithin in water to a narrow-band spectrum, resembling that of the phospholipid in chloroform. Because sonication causes free radical formation and can rupture covalent bonds, concern about its effects in these systems has been considerable. Indeed, it has been suggested that the procedure reduces the lipid phase to a liquid state (32, 33).

It appears, however, that a primary effect of sonication in pure liquid systems is the reduction of multilamellar particles into 200–800 A vesicles, bounded by one or two closed bilayer shells (34). Such size reduction would reduce anisotropic effects, allowing the magnetic fields imposed upon the protons of the small particles to be time averaged.

Finer et al. (35) have used 220 MHz PMR to clarify the effects of sonication on egg yolk lecithin suspensions. Their data support the view of Chapman et al. (34) and indicate that large multilayered phosphatide myelinics tumble too slowly to give narrow-band spectra. However, when such particles' motion is accelerated by sonic effects, multiple, frequent, highly energetic collisions result, fragmenting the myelinics into short-lived small phosphatide aggregates. These then coalesce to form spherules (190–300 A° diameter), each enclosed by a single, phosphatide bilayer shell (Fig. 5.4). In this shell, the phosphatide molecules appear to be organized as in the myelinics; packing appears such as to leave the choline and most methylene groups mobile, but immobilizes the glycerol ester residues and nearby methylenes.

It thus appears that simple, sonically induced, hydrodynamic effects bring out well-resolved PMR bands in model lecithin dispersions. Moreover, prolonged *homogenization* influences the PMR spectra of egg yolk lecithin at 33°C in a fashion similar to sonication (34, 35), but *homogenization* does not induce free radical formation or bond rupture. However, the well-recognized, more drastic effects of sonication, i.e., cavitation, free-radical formation, which may be unimportant here, must assuredly be evaluated in *lipoprotein* systems such as membranes. Thus, ultrasound acts as a vigorous protein denaturant, rapidly inactivating numerous enzymes, dissociating heme from hemoglobin-globin and altering many other proteins even at low irradiation levels (36). In this, membranes are unusually susceptible, even when thermal effects are excluded (36–38). Possible *additional* perturbations, due to *deuteration*, have already been discussed in the section on IR spectroscopy. Also supporting Chapman's contention are the X-ray studies of Wilkins et al. (39), showing that sonicated phosphatides maintain lamellar order and the data indicating that sonicated, dimyristoyl lecithin dispersions exhibit reversible broadening of the $[CH_2]_n$ signal at the established *bulk* crystallization temperature of this compound's hydrocarbon chains (40). Finally, one should note that spin-label studies of coarse and sonicated phosphatidylcholine show no significant difference in fatty acid chain mobility (41, 42).

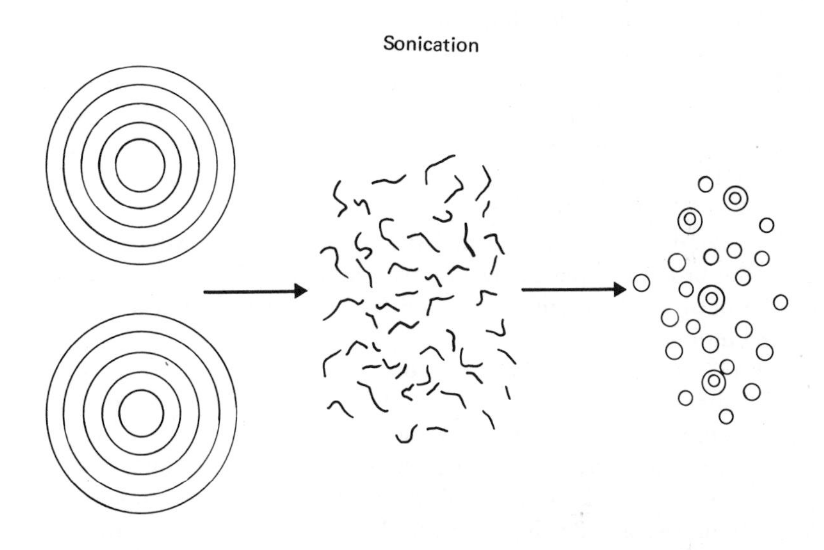

Fig. 5.4 Effects of sonication. Large particles, e.g., phosphatide myelin forms (*left*) tumble slowly and do not yield high resolution NMR spectra. Acceleration of such particles by sonic energy yields multiple, rapidly colliding, short-lived phosphatide aggregates (*center*). These then form 190–300 Å spherules, mostly enclosed by a single phosphatide bilayer (*right*) (34, 35). We stress that the sonic energy required to achieve this effect is known to destroy a number of proteins.

The most likely reason for the improved PMR of the small vesicles, thus, seems to be that they undergo rapid, isotropic reorientations which time-average residual anisotropic effects due to spin-coupled nuclei. However, an additional factor to consider is the role of association-dissociation equilibria between single molecules and micelles and between various micellar arrays. Thus, in a multilamellar myelin form, NMR signals may arise from the outer shell only. Moreover, a much lesser proportion of molecules would be in a free, rapidly tumbling state than in a small single-shelled vesicle with a large surface to volume ratio.

PMR Studies Using Paramagnetic Ions

Phosphatides sonicated in water, *liposome* dispersions, represent important models in membrane permeability studies, and NMR techniques are being applied to such systems in this light. Thus, Bergelson et al. (43), and V. F. Byshou et al. (44) have succeeded in differentiating the internal and external surfaces of lecithin vesicles in D_2O suspension, using changes of N^+ $(CH_3)_3$ PMR (100 MHz), induced by the paramagnetic ions Mn^{2+} and Eu^{3+} (Fig. 5.5).

Sonication of egg lecithin *in the presence* of Mn^{2+} drastically broadens the $N^+(CH_3)_3$ peak, presumably secondary to interaction of the Mn^{2+} with choline phosphates and due to a diminished T_2 of vicinal protons. Under these conditions, the Mn^{2+} would have access to both sides of the vesicle shell. In contrast, addition of Mn^{2+} *after* sonication yields two —$N^+(CH_3)_3$ components, namely a broad one and a superimposed sharp band at the same position, but with only 40% the intensity of the normal band. If it is assumed that the vesicle shells are Mn^{2+} impermeable, the broad component would arise from Mn^{2+}-complexed cholines on the outside of the membrane and the sharp band from free cholines on the inside.

Eu^{3+} creates consistent and more drastic effects. Here, addition of the ion after sonication splits the —$N^+(CH_3)_3$ signal into two components, one of normal position and bandwidth, and a second at higher field, which broadens increasingly with rising Eu^{3+} concentration. This component presumably arises from *external* choline-europium complexes (Fig. 5.5). Resonication, allowing access to the "inside," yields spectra indicat-

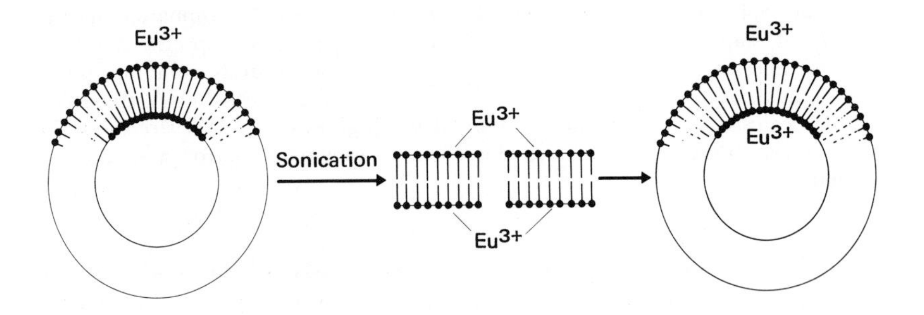

Fig. 5.5 Phosphatide bilayers resist permeation of lanthanide ions such as Eu^{3+}. Accordingly, such ions can ordinarily only influence the NMR of residues exposed at the external vesicular surface (left). Sonication (center) abolishes this permeability barrier, allowing the lanthanide ion to reach all phosphatide head groups. After cessation of sonication (right), the Eu^{3+} is bound at both external and internal surfaces of the bilayer. The lanthanide can, thus, act as a "sidedness" probe.

ing that the paramagnetic ion is now bound at both sides of the vesicle shell. The data also indicate that the rate constant for exchange of lecithin between inside and outside is less than 1 sec^{-1}. It also appears that about 1.5× as many lecithin molecules lie on the outside as the inside, a reasonable value for monolamellar vesicles of 200–300 A diameter. Levine et al. (45) have further explored the "sidedness" of lecithin bilayer vesicles with NMR techniques. They show that the inward and outward facing—$N^+(CH_3)_3$ groups of sonicated dipalmitoyl lecithin can be resolved at 100 MHz *above* the lamellar-liquid thermal transition. The residues facing inward are thought to account for the shoulder of the —$N^+(CH_3)_3$ resonance on the upfield side. This conclusion derives from the fact that the impermeant paramagnetic ions, Nd^{3+} and Eu^{3+}, when added to the bulk D_2O phase, separate and shift the two components. Nd^{3+} moves the main peak downfield and the lesser component upfield. Eu^{3+}, whose anisotropy is of opposite sign, causes the bands to shift in opposite direction. Although diamagnetic $UO_2{}^{2+}$ also resolves the two components, diamagnetic La^{3+} does not; Ca^{2+} has no effect. The spectra also suggest a broadening of the overlapping components at increasing field strength.

Chemical shift data indicate that the lecithin affinity for the tested ions follows the order $UO_2{}^{2+} > Eu^{3+} > Nd^{3+} > Ca^{2+}$. There also appears to be a rapid exchange of the ions between the various equivalent binding sites on the external vesicle surfaces. None of the ions substantially affects overall T_1 values, but the T_1 values of —$N^+(CH_3)_3$ groups directly liganded must be considerably reduced, since the exchange rate between sites is large, and the proportion of liganded phosphatides is small. All ions increase the temperature of the lamellar-liquid phase transition presumably by increasing phosphatide packing.

Clearly, the approaches using paramagnetic ions will aid in elucidating the sidedness of small, vesicular *biomembranes*.

^{23}Na NMR

Na^+ and K^+ transport constitute important membrane protein functions. It appears probable that these cations can be distinguished only when stripped of their hydration shells and that this occurs upon their reaction with "carriers" that are either mobile parts of membrane macromolecules or specific "ionophores" dissolved in apolar membrane regions.

The large nuclear quadrapole moment of ^{23}Na couples the nuclear spin to its environment, providing a NMR spin-lattice relaxation mechanism, T_1, which can be very strong, responds sensitively to the electronic environment of the sodium atom, and serves to monitor the formation of of Na^+-ligand complexes even when these are quite labile (46, 47). James and Noggle (48) have used this approach to study the interaction of $^{23}Na^+$ with solutions of phosphoserine and phosphoethanolamine as well as with aqueous dispersions of phosphatidylserine. Their data on T_1 as a function of pH indicate that the association of sodium with these substances is primarily ionic. Binding to phosphatidylserine is weak at physiologic pH, suggesting that this lipid is an unlikely sodium carrier. Although a negative result, the authors clearly pave the way to a novel approach to a major membrane problem and one probably more meaningful than ^{23}Na NMR studies previously reported for biomembranes (see below). They also point out possible extensions to K^+ transport using ^{39}K pulsed spectrometers with time-averaging computers.

^{13}C Resonance (CMR)

PMR studies reflect many structural properties of lipids in model and biomembranes, including transitions in lipid bi- and multilayers from an ordered state of the hydrocarbon chains at low temperatures to a less ordered one at higher temperatures; this effect depends on lipid unsaturation and the presence of cholesterol. However, PMR methods do not distinguish between diverse *populations* of lipids, characterized by rotational isomerization about given —C—C— bonds, and the degree and rate of interconversion between different configurations. This naturally limits interpretation of the PMR signals from biomembranes, particularly since multiple *cis*-unsaturations commonly prevail in these systems.

However, it appears that this dilemma can be resolved by ^{13}C NMR, even at the natural abundance of ^{13}C, since the ^{13}C resonance of alkane hydrocarbons shifts upfield by 5τ when a γ-methyl becomes juxtaposed to a methylene carbon through a *gauche* conformation of the intervening α—β bond. This configurational sensitivity of chemical shift allows analysis of the ratio of *gauche;trans* conformers and the susceptibility of this proportion to external factors (temperature, membrane perturbants, etc.). This is all the more feasible because ^{13}C NMR is much less sensitive to intermolecular interactions than ^1H NMR, and has thus been applied successfully by Batchelor et al. (49) in the case of lecithin liposomes.

These workers had previously shown that solvent effects on $(CH_2)_n$ are negligible and argue convincingly that temperature and electric field effects in a bilayer would also be small. They reason that the most important factor influencing methylene resonance shifts (^{13}C) involves rotational isomerization about the —C—C bond. Most important are the shifts near unsaturated carbon-carbon bonds. A *gauche* conformer about a 7–8 or 11–12 bond would shift the C9 and 10 resonances 1–2τ upfield.

Since the authors observe a large downfield shift for unsaturated ^{13}C resonance in egg yolk liposomes, they suggest that a *gauche* orientation about the 7–8 and 11–12 bonds of a 1-10 unsaturated system is energetically favored as the phase transition is approached. This is reasonable, since space-filling models (Fig. 5.6) show that a 7–8 or 11–12 *gauche* configuration tends to straighten out an oleate residue, allowing it to pack more tightly than in the *trans*-extended conformers.

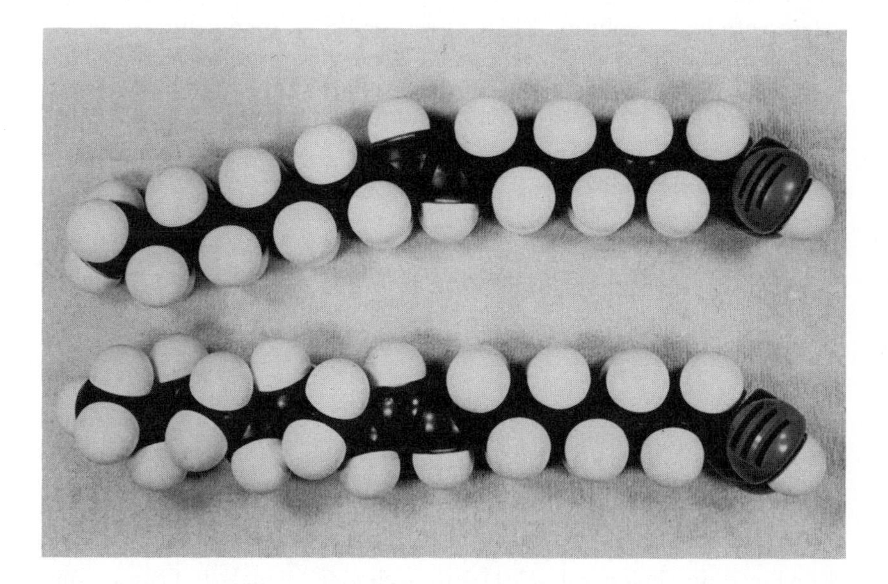

Fig. 5.6 Space-filling model showing oleic acid with a *trans*-, extended configuration (above), a terminal *gauche* configuration (below). The gauche configuration tends to "straighten out" the acyl chain (49). These data agree with laser Raman observations (Chapter 4).

In general, CMR adds an important potential to NMR studies of membranes and membrane components, since large chemical shifts (200τ) occur with ^{13}C, and its small gyromagnetic ratio makes it less prone to spin-spin band broadening.

Other CMR studies of lecithin dispersions (45) also reveal good resolution after sonication. The CH=CH signal appears particularly prominent and the choline resonance indicates high mobility.

Deuteron Magnetic Resonance

Nuclei with $I \geq 1$, as deuterium ($I = 1$), may exhibit splitting of their NMR lines. This is so in the case of dimyristoyl lecithin, where the fatty acid chains have been deuterated (25). Below its transition temperature, this material yields a diffuse chain spectrum due to overlapping CD_2 doublets, indicating that the various CD_2 groups in the fatty acid chain undergo different types of motion. However, just above the transition temperature, very significant band splitting occurs. The data are consistent with high resolution PMR as well as ESR experiments, suggesting that methylene groups near the polar head group are less mobile than those near the methyl terminals of the fatty acid chains.

Phosphatide-Cholesterol Interactions

Cholesterol, a significant component of many biomembranes, significantly alters phosphatide NMR in model systems.

PMR

Sonic codispersion of cholesterol with egg yolk or dimyristoyl lecithin significantly broadens the PMR bands of the latter, with the chain signals afflicted far more than the choline resonance (50). This reflects hindrance of the motion of the phosphatide hydrocarbon chains and thus is expectedly temperature dependent (51). In the case of dimyristoyl-lecithin:cholesterol (1:1), the choline signal indicates some mobility even below the chain melting temperature, but far more above the transition point, while the chain signals are severely restricted in both cases. However, the apolar ends of the fatty acid chains become more mobile as the temperature is raised toward 70°C.

Oldfield and Chapman (24) have progressed further in this area, examining sphingomyelin-cholesterol interactions at 220 MHz. At 20°C, below the transition temperature, $—N^+(CH_3)_3$ resonance, but no chain signals, is observed. At 40°C, broad chain signals are seen even after simple homogenization, and at 60°C the $—N^+(CH_3)_3$, $(CH_2)_n$, and $—CH_3$ bands are all well resolved.

In sphingomyelin-cholesterol mixtures (1:1), there is drastic broadening of the chain signals even above the transition temperature, indicating restrained motion of the acyl chains. The results also suggest greater immobilization of the methylene groups near the methyl terminals of the fatty acid residues. Compatible data have been obtained with unsonicated cholesterol-lecithin dispersions at the appropriate transition temperatures

(51, 21). The work of Darke et al. (51) indicates that when lecithin and cholesterol are sonically dispersed in water, a 1:1 molar complex with a lifetime in excess of 30 msec results. In the case of dimyristoyl-lecithin: cholesterol, above the transition temperature, all chain methylenes are less mobile than in pure phosphatide systems. It appears that the cholesterol —OH may be H-bonded to the lecithin phosphate, and also lies in close apolar association with the first 10 methylenes at the ester end of each acyl chain.

With cholesterol:phosphatide ratios <1, equimolar complexes are still formed but distributed in clusters. Darke et al. (51) suggest that this might occur in biomembranes (where the cholesterol:phosphatide ratio is often less than 1), *unless* membrane protein perturbs the complex. One should add that phosphatides other than lecithin may behave very differently.

Lee et al. (52) have recently examined the NMR of lecithin dispersions in D_2O, as well as cholesterol-lecithin interactions, relying primarily on spin-lattice (T_1) proton relaxation data. They are much concerned with arguments relying on spin diffusion in lipid bilayers, implying that proton relaxation times cannot give valid information about the motion of molecules in these systems. Indeed, they point out that work with ESR probes, ^{19}F-fluorostearic acid NMR probes (53), as well as ^{13}C-relaxation (54) are not consistent with spin-diffusion arguments. They examine both dipalmitoyl- and egg-lecithin in D_2O at 100 MHz. In sonicated suspension, the former undergoes dramatic, reversible PMR changes with temperature. Above 40°C, the $—N^+(CH_3)_3$, $(CH_2)_n$, and CH_3 resonances are well resolved, but at lower temperatures, the spectra become more and more featureless, apparently without change in particle size.

T_1 spin-lattice relaxation data for both dipalmitoyl- and egg-lecithin suggest a gradation of T_1 values for the protons of lecithin bilayers. They are consistent with ^{13}CMR results, but not with extensive spin diffusion. Other relaxation studies on sonicated egg-lecithin at 200 MHz (55) also suggest different T_1 values for various fatty acid protons. All in all, the data support others, suggesting that packing in phosphatide bilayers is maximal in the glycerol zone and that residue mobility increases toward the methyl terminals of the acyl chains and at the $—N^+(CH_3)_3$ residue.

At all temperatures studied, addition of cholesterol (in various proportions) to egg-lecithin increases the bandwidths and lowers the T_1 values of the various phosphatide proton resonances. This is distinct for the $—N^+(CH_3)_3$ group and dramatic for the chain signals. Part of the broadening in the latter region is due to contributions from cholesterol and therefore difficult to evaluate.

^{13}Carbon Magnetic Resonance

Keough et al. (56), measuring ^{13}C resonance, found all signals except the choline resonance to be broadened by cholesterol, although the $—CH_3$ band is rather little affected. The carbonyl carbons also appear immobilized, but, consistent with PMR studies, maximal hindrance appears approximately in the center of the acyl chain.

Interaction of Phosphatides with Biologically Active Small Molecules

Iodine, 2,4-Dinitrophenol, Tetraphenylboron

Jendresiak (57) explored the interaction of egg yolk phosphatidylcholine with iodine, 2,4-dinitrophenol, and tetraphenylboron, all of which increase the electrical conductivity of phosphatidylcholine bilayers. Using a 220 MHz PMR spectrometer, he observed band narrowing of the choline signal with all three substances and an upfield shift, with reference to tetramethylsilane, for iodine and tetraphenylboron. It appears that the micellar state of the phosphatide is not broken by these reagents, although structural rearrangements of some type seem unavoidable. The strength of interaction with phosphatidylcholine increases in the order dinitrophenol > iodine > tetraphenylboron. The data are taken to indicate that, *if* the substances studied act as ion "carriers," an interaction with the polar moiety of the phosphatide would be an initial step.*

Epinephrine

Attacking a model of biological importance, Hammes and Tallman (58) examined the interaction of sonically dispersed egg yolk phosphatidylcholine, brain phosphatidylserine, and brain phosphatidylethanolamine with L-epinephrine, in the hope of clarifying this substance's stimulation of plasma membrane adenyl cyclase. Measuring proton spin-spin relaxation times T_2 at 60 MHz, these were found to be unaffected in the cases of phosphatidylcholine and phosphatidylethanolamine. However, T_2 was decreased in the case of phosphatidylserine, as evidenced by differential line broadening. The data suggest that at pD 7.4, L-epinephrine bears a positive charge and can bind to the COO^- of phosphatidylserine. The phenoxyl groups of epinephrine may also H-bond to the phosphatide.

* The interaction of halide ions with phospholipid dispersions can be studied by PMR (G. L. Jendresiak, *Chem. Phys. Lipids* **9:** 133, 1972). Studies at 220 MHz show that the $-N^+(CH_3)_3$ proton signal for lecithin consists of two components, probably arising from residues at the exterior and interior of the vesicles. Diverse anions separate the contributions of these two topographical components in the lyotrophic order. The anion effect varies with diverse phosphatides and relate to phosphatide-halide interactions.

A 1:1 stoichiometry with a binding constant of 10^{-3} M^{-1} would appear reasonable.

Interactions of Phosphatides with Polypeptides and Proteins

This topic has not been systematically investigated by NMR, but the action of the cyclic peptide antibiotics valinomycin, alamethicin, and gramicidin-S on model phosphatide systems has been studied (59). This is because such compounds may induce or alter ion transport across biomembranes. All are surface-active compounds, and alamethicin contains 19 amino acid residues, while valinomycin consists of 6 amino acids and 6 hydroxy acids. Gramicidin-S, with 10 amino acid residues, does not alter ion transport across mitochondrial membranes.

Valinomycin and alamethicin, when combined with phosphatidylserine or phosphatidylcholine sonicated in D_2O, at the molar ratio $\geq$ 100:1, broaden the hydrocarbon chain signals so much that they become indistinguishable. Apparently, these substances inhibit the molecular motion of the fatty acid chains in a cooperative fashion (since such low peptide:phospholipid ratios are effective). In contrast, gramicidin-S does not significantly alter the lipid chain signals.

The binding of lysolecithin by bovine serum albumin weakens and broadens the phosphatide $(CH_2)_n$ PMR band in a dramatic fashion (60), signaling an apolar lipid-protein interaction. Similar data have been obtained with α_{S1}-, β, and κ caseins (61). Thus, when increasing amounts of lysolecithin are added to the caseins in D_2O, the $(CH_2)_n$ band progressively decreases intensity as its bandwidth increases, while the —$N^+(CH_3)_3$ signal remains unaffected. This points to a hydrophobic lipid-protein interaction, while leaves the polar group of the phosphatide entirely mobile. (Confirmatory data were obtained by following the ESR spectra of the spin label 12-nitroxide methyl stearate; see below.)

Importantly, PMR spectroscopy at 220 MHz shows that increasing concentrations of lysolecithin progressively sharpen aromatic amino acid resonances in a fashion similar to that produced by urea addition, but at lower concentration. This suggests that binding of this phosphatide tends to alter the quaternary, tertiary, and/or secondary structures of these proteins to an extent permitting enhanced mobility of Tyr, Trp, and Phe.

PMR of Serum Lipoproteins

Native Structure

At 100 MHz and 45°C human low-density serum lipoproteins (LDL_2) exhibit resonance bands at 4.6, 6.8, 8.7, and 9.1τ (62) which, by reference to lecithin standards, are assigned to —CH=CH—,

—$N^+(CH_3)_3$, —$(CH_2)_n$, and —CH_3 protons (Fig. 5.7). The phospholipid acyl chains must be highly mobile under these conditions, since the PMR spectra resemble those of egg yolk lecithin in sonicated dispersion or in organic solution. However, the cholesterol ester of LDL_2 appears relatively immobile, since the C_{18} and C_{19} methyl groups at 9.3τ emerge only at about 70°C. The PMR spectra of α-lipoprotein (HDL_2 and HDL_3) are similar (62–64). Resonance due to the cholesterol ring protons (9.3τ) is not resolved at 13°C, but the band is well defined at 50° (63). Interactions between cholesterol and phospholipid do not appear strong, because the $(CH_2)_n$ peak of the lipoproteins is much more intense than in codispersions of lecithin and cholesterol. This idea is supported by the ready exchange of cholesterol between serum lipoproteins and other cholesterol pools, e.g., erythrocyte membranes (65); however, the interaction of cholesterol with phosphatide acyl chains may depend on their unsaturation (66). Nevertheless, Steim, Edner, and Bargoot (60) found identical PMR spectra for serum lipoproteins and sonicated suspensions of lipo-

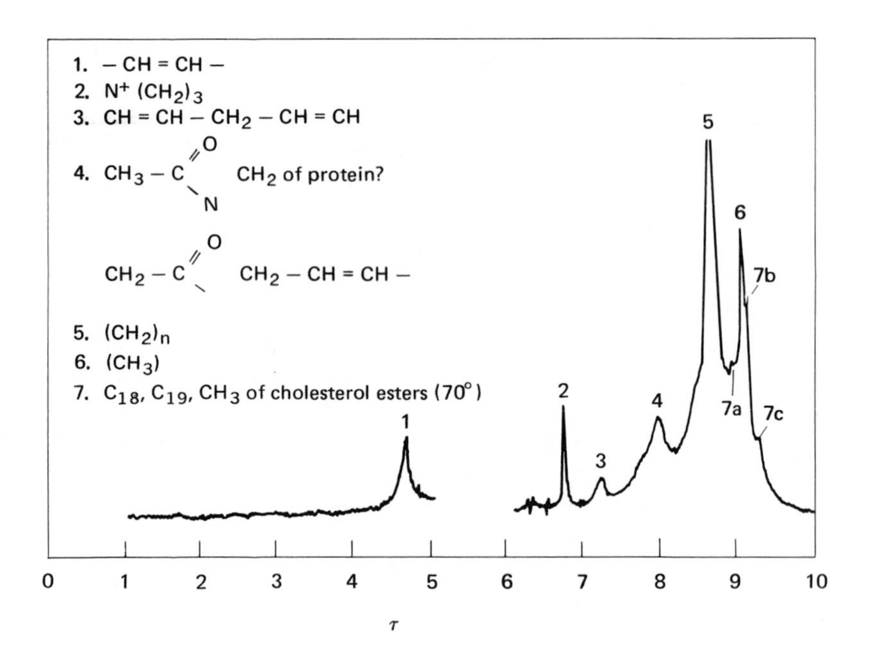

Fig. 5.7 100 MHz PMR of human low-density lipoprotein. Note the prominence of the $(CH_2)_n$, CH_3, and —CH=CH— signals, indicating a great mobility of the acyl side chains as in sonicated phosphatide dispersions; this at 45°C. The cholesterol ester signals upfield from the CH_3 band appear only above 70°, indicating greater immobilization of these compounds in the native proteins. The head groups are mobile at all temperatures (62). (Courtesy the authors and Elsevier Publishing Co.)

protein lipids in D_2O. They did not study the PMR of the lipoprotein phospholipid alone, but concluded from the $(CH_2)_n$, CH=CH, and steroid proton resonances that the lipid interactions in the lipoproteins were the same as in the sonicated lipid dispersion and that the cholesterol is "mobile" in both; they also argue that the lipids of serum lipoproteins are predominantly micellar and that they do not participate in important *apolar* interactions with the protein moiety. The greater mobility of cholesterol at higher temperatures parallels the increased mobility of certain amino acid residues, whose resonance near 3τ becomes apparent at 50°C (63).

The prominence and high resolution of the $—N^+(CH_3)_3$ proton peak of lipoproteins suggest that this group is not ionically linked to the protein. However, the chemical shift of the band is as in chloroform-methanol mixtures (62) suggesting that the lipoprotein cholines are not fully exposed to bulk water. Moreover, the choline peak in HDL_2 and HDL_3 is less intense than in sonicated phospholipid dispersions (62) and more like that of the phospholipid-cholesterol dispersions (60). Indeed, the intensity of the choline peak in HDL_2 and HDL_3 is lower than that of the $(CH_2)_n$ band (63), whereas the reverse is true in lecithin sonicates. This may reflect the contribution of protein resonances to the $(CH_2)_n$ band, or may be because of some restriction of choline mobility in the lipoproteins, or both.

Structural Perturbations of Serum Lipoproteins

Treatment of LDL_2 (62), HDL_2, and HDL_3 (63) with sodium dodecyl sulfate (SDS) brings out a resonance band at $2.5–2.8\tau$, attributable to the protons of aromatic amino acids. The CH=CH resonance band intensifies and sharpens concurrently, signaling increased mobility of unsaturated segments in the fatty acids; SDS itself does not contribute structure in this spectral region. The data have been taken to point to apolar binding of lipid to protein, but the emergence of the aromatic proton signal may be due solely to perturbation of hydrophobically stabilized secondary and tertiary structures in the protein moiety by SDS (7).

Heating of HDL_2 and HDL_3 to 50°C allows resolution proton signals due to aromatic residues, lysine ($7.0–7.1$ and $8.3–8.4\tau$), and glutamic acid ($7.9–8.0\tau$) (63) with 220 MHz instruments. Moreover, denaturation of HDL_2 by trifluoroacetic acid intensifies the $—CH_3$ proton signal near 9.0τ (62); unfolding of the protein is presumably involved, increasing the molecular motion of side chain methyl groups of amino acids and/or of cholesterol.

As noted, the $—N^+(CH_3)_3$ band of lipoproteins is sharp and intense, weighing against ion pairing between this group and anionic amino acid residues. This and other data suggest that lipid-protein interactions in serum lipoproteins are apolar. However, Steim et al. (60) contend that PMR data do not convincingly demonstrate apolar binding of lipids by the protein moiety of LDL_2. They reason that the $(CH_2)_n$ resonance of LDL_2 is too well resolved to be compatible with the type of apolar immobilization of fatty acid chains that they find in the hydrophobically stabilized complexes of bovine serum albumin and lysolecithin; in these the $(CH_2)_n$ band is weak and diffuse. However, in the studies of Steim et al. (60), the weight ratios protein:lysolecithin (6:1–20:1) far exceed those found in serum lipoproteins. Under their conditions, 63–85% of the total $(CH_2)_n$ signal could arise from amino acid side chain methylenes. This points up one difficulty in attempting estimates of group mobilities by PMR in heterogeneous systems; the spectra generally yield only composite values, and a highly mobile component might be obscured by an immobilized one, and vice versa. At present, there appears to be no clear solution to this problem.

Cellular Membranes

NMR studies on cellular membranes have focused on erythrocyte ghosts, generally isolated as in Dodge et al. (67), but some spectra have also been obtained from myelin, sarcoplasmic reticulum, liver bile fronts, as well as the membranes of *Halobacterium halobium* and *Acholeplasma laidlawii*. Until recently only PMR spectra could be obtained, and in general the signals recorded from membranes are thought to arise primarily from membrane lipids. As will be discussed, they depend strongly on variables such as instrumental frequency, temperature, sonication, etc.

Erythrocyte Membranes

PMR

At 60 MHz and $T = 40 \leq 40°C$, *unsonicated* erythrocyte membranes yield a rather featureless PMR spectrum; the broad absorption between 7–9τ presumably arises from lipid and protein hydrocarbon residues (68). At 220 MHz weak bands appear superimposed upon the broad component. These lie at 6.2–6.16, 6.7, 7.9, and 9.1τ (69).

In contrast, *sonicated* erythrocyte membranes yield sharp bands due to $—N^+(CH_3)_3$ protons (6.7τ), but diffuse signals due to $(CH_2)_n$ protons (8.8τ) and $—CH_3$ protons (9.1τ) (Fig. 5.8), at 60 MHz and 40°C. All intensify as the temperature is raised up to 120°C (68). Importantly, the

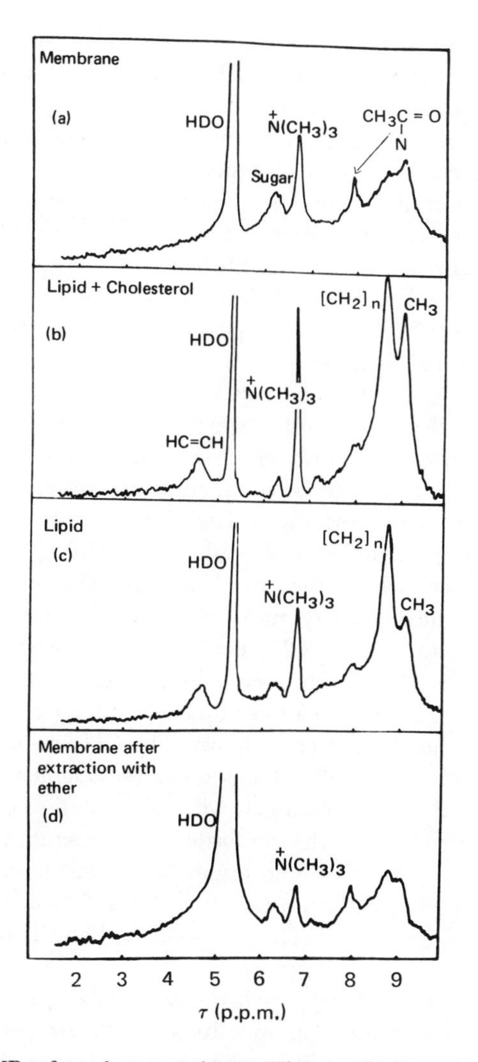

Fig. 5.8 PMR of erythrocyte ghosts. High resolution can be achieved at 60 MHz only after sonication. Some sharp resonances appear at 220 MHz without sonication (69). The figure represents the 60 MHz PRM of sonicated human erythrocytes at 40°C under various conditions. (a) The membrane spectrum has sharp choline and sugar signals, indicating unrestricted mobility. This is in marked contrast to the feeble or lacking acyl chain signals, which appear severely restricted in mobility in the membrane. (b) The PMR of the toal membrane lipid; all acyl chain signals are clearly resolved, including the CH=CH resonance band. The same is true for the cholesterol-free material (c). In contrast, when cholesterol (primarily) is extracted from ghosts with ether (d), the acyl chain signals remain damped (68). The $(CH_2)_n$ and (H_3) signals, as well as the —CH=CH— band, resolve well only at 120°C. (Courtesy of the authors and Elsevier Publishing Co.)

—$N^+(CH_3)_3$ band is very well resolved under these conditions compared with the broad, weak $(CH_2)_n$ and CH_3 signals. The latter, as well as the CH=CH band at 4.7τ, resolve well only at >120°C.

The diffuseness of the membrane hydrocarbon signals most likely derives from effects other than cholesterol-phosphatide interactions, since the PMR spectrum of the total erythrocyte membrane lipid (cholesterol plus phosphatides) sonically dispersed in D_2O displays sharp and intense $(CH_2)_n$ and CH_3 peaks (Fig. 5.8). Moreover, the erythrocyte membrane PMR is almost the same after cholesterol extraction as before, i.e., with weak and diffuse $(CH_2)_n$ and CH_3 peaks (68). These results suggest that cholesterol-phosphatide interactions are not primarily responsible for the peculiar features of the hydrocarbon PMR signals of erythrocyte ghosts. On the contrary, these data and others have led Chapman and associates (68) to suggest that the hydrocarbon chains of erythrocyte membrane phosphatides are constrained by interactions with *membrane proteins*.

Glaser et al. (70), working at 220 MHz, could not obtain a high-resolution PMR spectrum from unsonicated erythrocyte ghosts at 18°C. At 80°C, however, where the membrane protein is largely unfolded (according to optical activity measurements), a number of resonance bands appear, including a —CH_3 peak at 9.1τ; however, no significant $(CH_2)_n$ resonance is observed even under these conditions. The latter result is ascribed to lipid-lipid and/or lipid-protein interactions, restricting the mobility of phosphatide acyl chains. The CH_3 resonance is thought to arise from amino acid side chains only. Its emergence at 80°C, plus concurrent optical activity changes, without resolution of $(CH_2)_n$ resonance is taken to show that the conformation of membrane proteins can change without alteration of lipid organization. However, this argument does not explain why, if the protein is really unfolded, the contribution of amino acid residues to the $(CH_2)_n$ resonance band is not more prominent.

Glaser et al. (70) observe well-resolved $(CH_2)_n$ resonance band at low temperatures only after phospholipase C (*B. cereus*) treatment under conditions that effect about 70% phosphatide cleavage, but no changes in optical activity. They take this as additional evidence that the proteins and lipids of membranes can change state independently. What is undefined in this work is *how much* lipid or protein can change independently. Moreover, while the PMR spectrum of the enzyme-treated membranes resembles that of lipid extract, the 9.34τ resonance band due to cholesterol methyl protons is not resolved, suggesting again that lipid-lipid and/or lipid-protein interactions are more prominent in the membrane.

Steim et al. (60) propose that lipid hydrocarbon chains in the erythrocyte membrane are immobilized within a lipid bilayer, because in egg yolk lecithin the PMR spectrum shifts to narrow bands, concurrent with the temperature-dependent phase transition from a lamellar to a cubic phase. However, such correlations are ambiguous, particularly in view of the major differences between red cell lipids and egg yolk lecithins.

^{13}Carbon magnetic resonance

Metcalfe et al. (71) have shown better resolution with ^{13}C NMR than obtained with PMR and have demonstrated the high mobility of the $—N^+(CH_3)_3$ group even in intact erythrocyte ghosts. In these experiments, they examined 10% suspensions of *unsonicated* ghosts in D_2O at 28°. No ^{13}C enrichment was employed, but the $—N^+(CH_3)_3$ ^{13}C NMR was resolved, nevertheless.

^{23}Na NMR

The interaction of erythrocyte ghosts with sodium ion has been investigated by measuring the NMR bandwidth of ^{23}Na (72). As noted earlier, ^{23}Na has a large nuclear quadrupole, and under conditions of rapid chemical exchange of all quadrupolar Na^+ species, the observed bandwidth $\Delta\nu$ is a weighted average of all these forms:

$$\Delta\nu = (\Delta\nu)_{\text{free}} \cdot (x)_{\text{bound}} + \sum_i \cdot (\Delta\nu_i)_{\text{bound}} \cdot (x_i)_{\text{bound}} \tag{5.11}$$

where $(\Delta\nu)_{\text{free}}$ and $(\Delta\nu_i)_{\text{bound}}$ are the bandwidths of the free, hydrated, and bound sodium ions, respectively; $i =$ the possibility of multiple binding sites, and (x) the corresponding mole fractions.

In 0.16 M NaCl, the ^{23}Na bandwidth is 12 ± 1 Hz. It increases to 20 Hz by addition of normal erythrocyte ghosts and to 30.4 Hz upon addition of sodium dodecyl sulfate (to 0.03 M). The detergent alone did not effect broadening of the ^{23}Na resonance band. Treatment of the membranes with nonionic Triton X-100 or freeze thawing did not simulate the action of dodecyl sulfate. These preliminary studies are of considerable interest, indicating that erythrocyte membranes bind Na^+ and that further binding sites, presumably on membrane proteins, are generated by denaturing levels of dodecyl sulfate.

Myelin

In contrast to erythrocyte membranes, extraction of cholesterol from sonicated myelin preparations significantly sharpens the PMR bands $(CH_2)_n$ and $—CH_3$, but not those due to choline or the sugar moieties of galactolipids (73). The PMR spectrum of the total lipid extract of myelin dispersed in D_2O is essentially identical to that of myelin itself; here also, the $(CH_2)_n$ resolves well only after removal of cholesterol. Apparently, in myelin and myelin lipids, cholesterol is primarily responsible for hydrocarbon chain immobility. (It should be noted, however, that cholesterol methyl proton resonance, absent in myelin, is present in the lipid extract; this at least suggests some constraint of cholesterol mobility by membrane proteins even in myelin.) On the other hand, the $N^+(CH_3)_3$ group is freely "mobile," as evidenced by the narrowness of the line due to protons of this group.

Sciatic Nerve

The PMR spectrum (220 MHz) of an isolated rabbit sciatic nerve in deuterated Locke's solution, "doped" with 0.002 M Mn^{2+}, revealed resonance bands attributable to extracellular water, intracellular water, and phosphatides. The data suggest some fluidity of the phospholipid hydrocarbon chains. However, it is doubtful that the experiments yield detectable information about the axonal membrane, since most of the phosphatide in such preparations lies in the myelin and Schwann cell membranes (74).

Sarcoplasmic Reticulum

PMR

Davis and Inesi (75) have correlated the temperature-dependent Ca^+ efflux from sarcoplasmic vesicles with their PMR spectra (90 MHz). All experiments were carried out in deuterated buffers, which did not measurably interfere with the membranes' Ca^{2+} transport functions.

Interestingly, these membrane vesicles, even when *unsonicated*, yield relatively well-resolved —CH_3, $(CH_2)_n$, and HC=CH band at 20°C, but *no choline methyl signal.* This is in marked contrast to erythrocyte membranes. However, the choline methyl PMR appears when the temperature rises above 20°C and intensifies (together with —CH_3 resonance) up to a plateau between 40°C and 50°C. The temperature-dependent alterations of the —$N^+(CH_3)$ PMR is interesting, suggesting a transformation from a hindered to a mobile state between 25° and 40° (roughly paralleling the Ca^{2+} efflux rate) and an additional protein-dependent transition at higher temperatures.

The authors unfortunately do not evaluate the contribution of protein methyls and methylenes to their PMR spectra, assuming these signals to arise from lipids only. If this were the case, about 80% of the phosphatide acyl chains would be organized into a lamellar phase, with limited molecular motion, and 20% into a fluid region.

This system is clearly of great intrinsic interest and also illustrates the accessibility of *small vesicular* biomembrane fragments to PMR analysis.

^{13}C *NMR*

Detailed ^{13}C and ^{1}H NMR relaxation studies on the lipids of sarcoplasmic vesicles have now been reported by Robinson et al. (76). Although the natural abundance of ^{13}C is only $\sim 1\%$, useful spectra of sarcoplasmic membranes and their lipids were obtained without enrichment techniques.

Terminal $—CH_3$, $(CH_2)_n$, $—CH{=}CH—$, and choline methyls yield well-resolved resonance signals. The intensities of these bands in native membranes is equivalent to 60–90% of those found in appropriate amounts of sonicated sacroplasmic lipid extract in D_2O. This is in contrast to ^{1}H resonance, where the band intensities in the membranes account for only 25% of the resonance in lipid extracts. This is presumably due to the facts that the gyromagnetic ratio of ^{13}C is less than $\frac{1}{4}$ that of ^{1}H and that intermolecular proton-carbon distances are much smaller than intermolecular proton-proton separations.

The choline $(CH_3)_3$ T_1 values in membranes are appreciably shorter than those of lipid extracts, perhaps due to lipid-protein interactions of the head groups. While the chain signals of the lipids resemble those of the membranes, they lack single T_1 values in both cases, and the T_1 of the terminal methyl proton resonance is far less temperature sensitive in membranes than in lipid vesicles. The peculiarities of the chain signals could be explained by extreme heterogeneity of lipid composition and/or nonhomogeneous distribution of differently structured lipids, with slow diffusion between these separated domains. This, however, would suggest that the extremely rapid lateral diffusion of spin-labeled phosphatides in liposomes (p. 314) and sarcoplasmic vesicles (p. 332) may not represent the behavior of most of the phosphatides. This potential weakness of the probe approach is further discussed later.

It would appear from the present data that the ^{13}C relaxation measurements are still too ambiguous to account for the behavior of all lipids in sarcoplasmic membranes and may fail to sense the properties of those lipids that are in close hydrophobic interaction with proteins. However, the data are consonant with the view that substantial topologic domains of sarcoplasmic membranes essentially comprise lipid bilayer arrays.

Acholeplasma Laidlawii

Thermal studies (77–79) show that the purified membranes of *Acholeplasma laidlawii* and the lipids extracted therefrom undergo a poorly cooperative gel liquid crystal transition over 30°C. Moreover, the thermal transitions vary with lipid composition, which, in turn, is a function of growth and media composition. Steim (80) has observed analogous behavior in *E. coli* membranes, where the transition extends from 15°–

45°C for organisms grown at 37°. He contends (77) that such data indicate that more than 75% of the organisms' lipid is in a bilayer array, but Chapman and Urbina (79) point out that such quantitative estimates are unjustified.

When *Acholeplasma laidlawii* is grown in media containing perdeuterated palmitic and lauric acids, these are incorporated into membrane phosphatides. Subsequent deuteron NMR reveals the hydrocarbon chains to be much *less fluid* than found in the reference model, di-(perdeutero) myristoyl lecithin, in the liquid crystalline phase (81). The data do not allow judgments as to how much of the membrane lipid is rigidified; the authors suggest that such membranes are topologically heterogeneous and contain some more fluid domains.

Liver Membranes

Fischer and Jost (82) found that the PMR of L-epinephrine measured at 90 MHz in physiologically buffered D_2O changes drastically upon

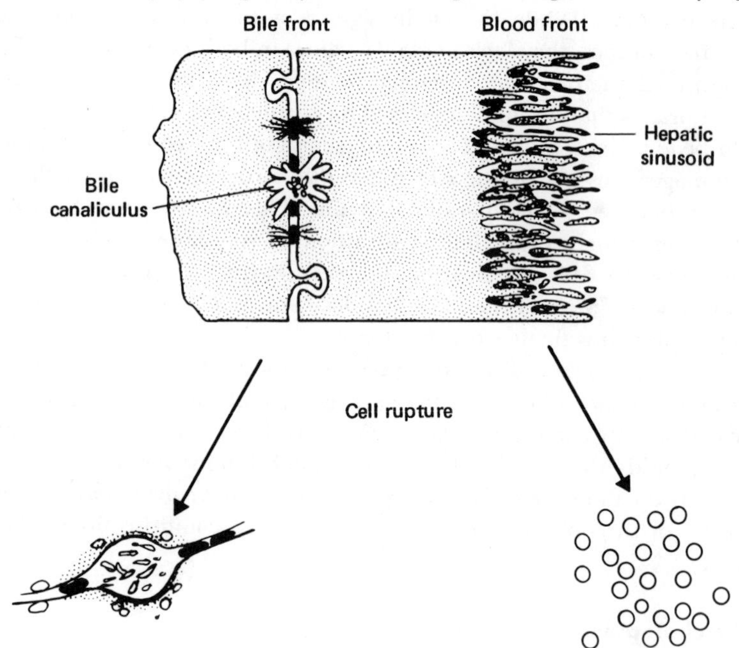

Fig. 5.9 The erythrocyte, being a single-cell type, represents a rather simple biological model. The hepatocyte, schematized here, is far more complex, because different portions of its surface are differentially specialized. We consider it critical that the spectroscopist recognize this anatomic fact. Liver "plasma" membranes, as studies in Hammes and Tallman (58), represent specialized "bile fronts" (lower left). Most of the hepatocyte surface vesiculates upon cell rupture (lower right). We discuss this matter thoroughly in Chapter 1.

addition of viable mouse liver cells. The data suggest that epinephrine is bound through its phenyl residue and distal to the side chain methylene. Dichloroisoproteronol, an inhibitor of epinephrine, competitively blocks the changes in the epinephrine PMR, while its own spectrum alters in a manner suggesting binding. Presumably, the binding occurs at the plasma membrane.

However, Hammes and Tallman (58) could not detect PMR evidence of epinephrine binding to isolated liver "plasma membranes." This is perhaps not surprising since their membrane isolate consists of "bile fronts," which represent only a small proportion of the hematocyte surface and one specialized for bile secretion. The epinephrine receptors are more likely to lie on the villi of the "blood front," which fragment into small vesicles upon cell disruption (Fig. 5.9).

The Contribution of Protein to Membrane PMR

The contribution of protein methylene and methyl groups to membrane $(CH_2)_n$ and CH_3 peaks requires clarification. Thus, the β- and α-CH_2 groups of poly-L-arginine yield prominent resonance bands near 8.2τ (8); in poly-L-methionine, their bands lie below 8.0τ (10). In histones the —CH_3 PMR bands of isoleucine, valine, and leucine lie at 9.12τ, but their location is clearly sensitive to microenvironment; for example, when adjacent to an electronegative atom, the band occurs at $\sim 6.8\tau$(13). The —CH_3 PMR of leucine, valine, and isoleucine residues has the same chemical shift as the CH_3 resonance band of sonicated erythrocyte ghosts.

The molecular weight of a typical phospholipid is about 760, and roughly 0.8% of its mass is due to hydrocarbon chain —CH_3 protons. If the proportions of leucine, isoleucine, and valine in erythrocyte membrane proteins are, as given by Rosenberg and Guidotti (83), 11, 5, and 7, respectively, approximately 1% of the protein mass derives from the —CH_3 protons of these residues. Thus, of the —CH_3 protons that might contribute to resonance near 9.2τ in the erythrocyte membrane, 19% comes from phospholipid, *56% from protein*, and 25% from cholesterol.

Quantitation is still difficult in PMR spectroscopy, and the real contribution of protein to membrane PMR spectra cannot now be specified. However, it appears that α-helical peptide segments would not contribute significantly, whereas unordered segments could yield strong methylene and methyl signals (7). If sonication of erythrocyte membranes causes some unfolding of α-helical segments, as is suggested by optical activity measurements, this might account for the concurrent resolution of the —CH_3 peak. [We have already commented on the established effects of sonication on proteins and here add that, in general, denatured proteins yield more highly resolved, intense PMR signals than the native ordered structures (13).]

Of the methylene groups due to the protein, only those distant from relatively electronegative atoms will yield PMR between 8.2 and 8.5τ. CH_2 protons adjacent to such atoms (i.e. near CO or —SH) exhibit lower tau values (13). Thus, the former should theoretically contribute to the erythrocyte membrane $(CH_2)_n$ peak at 8.7τ, but one cannot freely extrapolate from isolated proteins and synthetic polypeptides to membrane proteins.

Among the CH_2 protons contributing to the 8.2–8.5τ region are the β-CH_2's of methionine, leucine, glutamate, phenylalanine, tyrosine, and tryptophan, the γ-CH_2's of isoleucine, the β-, γ-, δ-, and ε-CH_2's of lysine, and the β-, γ-, and δ-CH_2's of arginine. Again, referring to the amino acid composition of erythrocyte ghosts, of the methylenes potentially capable of contributing to the 8.2–8.5τ region, roughly 70% are part of phospholipid and 30% of *protein.*

Protons of methylene groups adjacent to the electronegative side chain residues of serine, cysteine, methionine, glutamate, and aspartate will be almost the sole source of the absorbance in the 7.0 to 8.0τ region. Considering the total CH_2 and CH_3 proton absorbance from all sources in the 7.0–9.2 region, *protein and phospholipid should, in principle, contribute almost equally.* At least part of the sharpening and intensification of the $(CH_2)_n$ and CH_3 peaks observed with sonication, heat, or detergent treatment of erythrocyte membranes could be due to mobilization of *amino acid side chains* as a result of protein denaturation. Indeed, Kamat and Chapman (68) suggest that the 7.9τ peak observed in sonicated erythrocyte ghosts is due to the CH_2 groups of some amino acids—notably glutamate.

The protein contribution to the PMR of *Halobacterium halobium* membranes might be very substantial. In 4 M NaCl/D_2O, where the membranes are intact, the PMR spectrum is poorly resolved in the 8–9τ region (84). Disaggregation of the membrane by lowering ionic strength produces a well-resolved spectrum. Since 80% of this membrane is protein, we suspect that the resolution of the signals arises at least in part from increased freedom of amino acid side chains.

Effects of Detergents and Solvents

The effects of sodium dodecyl sulfate (SDS) and sodium deoxycholate (DOC), substances that are known to affect apolar bonds, provide clear evidence for apolar lipid-protein associations in membranes. Thus, SDS very markedly accentuates the $(CH_2)_n$ proton peak of erythrocyte membranes (Fig. 5.10) and bring out the CH=CH signal (68, 70). These

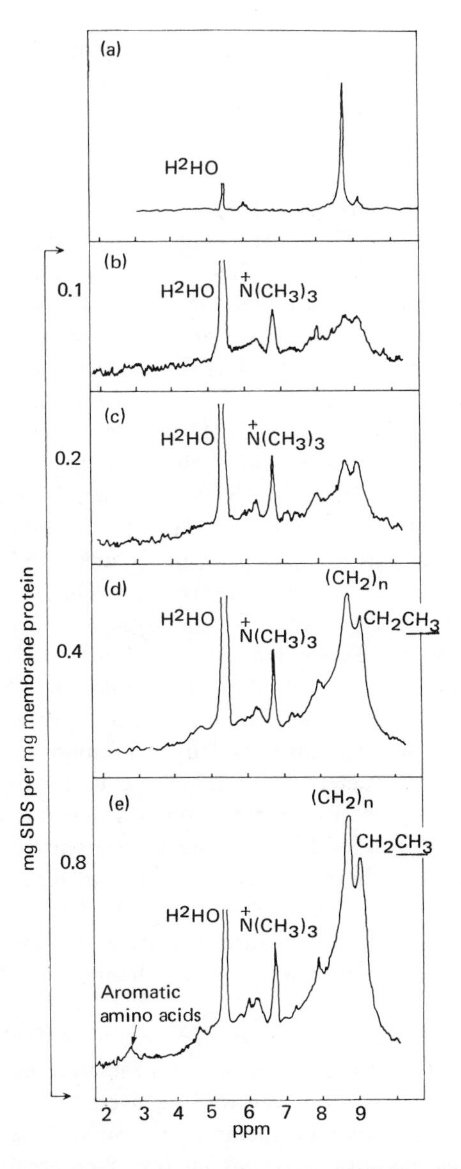

Fig. 5.10 The effect of sodium dodecyl sulfate (SDS) on the PMR of human erythrocyte ghosts. (a) The PMR spectrum of SDS; (b)–(d) with increasing SDS, the $(CH_2)_n$ and CH_3 signals become very prominent, far in excess of the contribution of SDS itself. In (d)–(e) the —CH=CH— resonance appears ($\sim$ 4.7τ), as the lipid structure becomes increasingly perturbed. In (e) the signal due to aromatic amino acid protons emerges, as membrane protein denatures. This fact is crucial to the interpretation of SDS effects on membrane optical activity. (At very low SDS concentrations, some membrane enzymes become activated.) [From Chapman et al. (68). Courtesy of *J. Mol. Biol.*]

effects begin to appear at 1 μmole/mg detergent membrane protein (68). At high SDS levels (about 2.5 μmoles/mg), the resonance due to proteins of aromatic amino acids becomes apparent (Fig. 5.10). The resolution of PMR signals due to aromatic CH=CH and $(CH_2)_n$ protons correlates with the progressive unfolding of membrane proteins by SDS, as shown by optical activity measurements (85).

The CH_2 groups of DOC are nearly all part of a cyclopentanoperhydrophenanthrene ring system. Their motion is, thus, constrained, and they do not contribute significantly to the PMR spectrum of DOC-treated membranes (31, 68). However, the methylene groups of SDS could add substantially to the $(CH_2)_n$ peak. With a 6:4 (w/w) protein: lipid ratio in erythrocyte ghosts (67), 1 mg protein is associated with about 0.7 mg lipid, of which about 0.47 mg would be phospholipid (67) and 0.28 mg would be due to phospholipid methylenes. Hence, with 1 μmole SDS/mg protein, 0.20 mg SDS methylene would be added to about 0.28 mg phosphatide methylene. How much the detergent methylenes actually contribute to the spectrum is unknown. However, SDS separates membrane lipids from membrane proteins, forming complexes with both to varying degrees. SDS protein/peptide complexes separate electrophoretically on SDS-laden polyacrylamide gels approximately according to molecular weight, and electrophoretic experiments show that the binding of SDS to membrane proteins depends upon the detergent concentration in the range used by Chapman et al. (68). Unfortunately, the PMR characteristics of SDS associated with specific proteins and with diverse lipids (in different phases) have not been studied. Accordingly, the effects of SDS on the membrane $(CH_2)_n$ signal are somewhat ambiguous; those of DOC are not, but this is less of a protein perturbant. Nevertheless, the emergence of CH=CH resonance is clearly due to increased mobility of unsaturated fatty acids, and the appearance of the aromatic PMR band surely signals changes in the conformations of membrane proteins.

In organic solvents, which are known to unfold proteins, e.g., trifluoroacetic acid, all the lipid signals and some amino acid resonances of erythrocyte membranes become well resolved (68). This can also be taken as evidence pointing to apolar lipid-protein interactions; studies with a helicogenic membrane solvent, such as 2-chloroethanol, might be additionally informative.

Conclusion

NMR has already contributed much valuable information to membrane biology and will assuredly add more. However, it is no panacea. Thus, the relevance of work on model systems and membrane isolates remains to be established here as in other spectroscopic methods. Moreover, theoretical problems of interpreting NMR of phospholipid and phosphatide-protein systems are formidable and obscure the significance of chemical shifts, as well as relaxation measurements. Important problems are (a) effects of sonication, (b) non-Lorentzian band shapes, (c) field dependence of bandwidths, (d) controversies about significance of spin-lattice (T_1) times and "spin-diffusion mechanisms," (e) resolution of contributions from protein, phosphatides, and cholesterol, and (f) the role of fatty acids of varying length and unsaturation. Moreover, the effects of diverse isolation techniques, sonication procedures, and the use of D_2O (in PMR) constitute unknown and potentially troublesome variables. Finally, one must always ask whether a given NMR signal is representative of all molecules bearing the reporter nucleus and to what extent the signals of potentially resonating nuclei are obscured.

References

1. Abragam, A. *The Principles of Nuclear Magnetism.* Oxford: Clarendon Press, 1961.
2. Pople, J. A., Schneider, W. B., and Bernstein, H. J. *High Resolution Nuclear Magnetic Resonance.* New York: McGraw-Hill, 1959.
3. Kowalsky, A., and Cohn, M. *Ann. Rev. Biochem.* **33:** 481, 1964.
4. Chapman, D. *The Structure of Lipids.* London: Methuen, 1965, p. 160.
5. Patt, L. S., and Sykes, B. D. *J. Chem. Phys.* **56:** 3182, 1972.
6. Stewart, W. E., Mandelkern, L., and Glick, R. E. *Biochemistry* **6:** 143, 1967.
7. Bradbury, E. M., Crane-Robinson, C., and Rattle, H. W. E. *Nature* **216:** 862, 1967.
8. Boublik, M., Bradbury, E. M., Crane-Robinson, C., and Rattle, H. W. E. *Eur. J. Biochem.* **12:** 258, 1970.
9. Bradbury, E. M., Crane-Robinson, C., Goldman, H., and Rattle, H. W. E. *Nature* **217:** 812, 1968.
10. Markley, J. L., Meadows, D. H., and Jardetzky, O. *J. Mol. Biol.* **27:** 25, 1967.
11. Bradbury, E. M., and Crane-Robinson, C. *Nature* **220:** 1079, 1968.
12. Cohen, J. S., and Jardetzky, O. *Proc. Natl. Acad. Sci.* (U.S.) **60:** 92, 1968.
13. Bradbury, E. M., Crane-Robinson, C., Goldman, H., Rattle, H. W. E., and Stephens, R. M. *J. Mol. Biol.* **29:** 507, 1967.
14. Hollis, D. P., McDonald, G., and Biltonen, R. L. *Proc. Natl. Acad. Sci.* (U.S.) **58:** 758, 1967.
15. Raferty, M. A., Dahlquist, F. W., Parsons, S. M., and Wolcott, R. G. *Proc. Natl. Acad. Sci.* (U.S.) **62:** 44, 1969.
16. Crespi, H. L., and Katz, J. J. *Nature* **224:** 560, 1969.
17. Chapman, D., Richards, R. E., and Yorke, R. W. *Nature* **183:** 44, 1959.
18. Chapman, D., Richards, R. E., and Yorke, R. W. *J. Amer. Oil Chem. Soc.* **5:** 243, 1960.
19. Chapman, D., and Morrison, A. *J. Biol. Chem.* **241:** 5044, 1966.

20. Chapman, D., and Salsbury, N. *Trans. Faraday Soc.* **62:** 2607, 1966.
21. Vesksli, A., Salisbury, N. J., and Chapman, D. *Biochim. Biophys. Acta* **183:** 434, 1969.
22. Salsbury, N. J., and Chapman, D. *Biochim. Biophys. Acta* **163:** 314, 1968.
23. Chan, S. I., Feigenson, G. W., and Seiter, C. H. A. *Nature* (Lond.) **231:** 110, 1971.
24. Oldfield, E., and Chapman, D. *FEBS Letters* **21:** 303, 1972.
25. Oldfield, E., Chapman, D., and Derbyshire, W. *FEBS Letters* **16:** 102, 1971.
26. Salsbury, N. J., and Harris, P. quoted by D. Chapman and G. H. Dodd, in *Structure and Function of Biological Membranes*, L. I. Rothfield, ed. New York: Academic Press, 1971, p. 28.
27. Small, D. M. *J. Lipid Res.* **8:** 551, 1967.
28. Chapman, D., Williams, R. M., and Ladbrooke, B. D. *Chem. Phys. Lipids* **1:** 445, 1967.
29. Chapman, D., and Penkett, S. A. *Nature* **211:** 1304, 1966.
30. Chapman, D., Fluck, D. J., Penkett, S. A., and Shipley, G. G. *Biochim. Biophys. Acta* **163:** 255, 1968.
31. Penkett, S. A., Flook, A. G., and Chapman, D. *Chem. Phys. Lipids* **2:** 273, 1968.
32. Steim, J. M. *Adv. in Chem. Ser.* **84:** 259, 1968.
33. Sheard, B. *Nature* (Lond.) **223:** 1057, 1969.
34. Chapman, D. *Ann. N.Y. Acad. Sci.* **195:** 179, 1972.
35. Finer, E. G. F., Flook, A. G., and Hauser, H. *FEBS Letters* **18:** 331, 1971.
36. Joly, M. *A Physicochemical Approach to the Denaturation of Proteins.* New York. Academic Press, 1965, pp. 9–16.
37. Grabar, P. *Biol. Med. Phys.* **3:** 191, 1953.
38. Hughes, D. E. In *Ultrasonic Energy*, E. Kelly, ed. Urbana, Ill.: Univ. of Illinois Press, 1965, p. 9.
39. Wilkins, M. H. F., Blaurock, A. E., and Engelman, D. M. *Nature* **230:** 72, 1971.
40. Chapman, D. In *Biological Membranes*, D. Chapman and D. F. H. Wallach, eds. New York: Academic Press, 1973, p. 91.
41. Hubbell, W. L., and McConnell, H. M. *Proc. Natl. Acad. Sci.* (U.S.) **61:** 12, 1968.
42. Barratt, M. D., Green, D. K., and Chapman, D. *Chem. Phys. Lipids* **3:** 140, 1969.
43. Bergelson, L. D., Barsukov, L. I., Dubrovina, N. I., and Bystrov, V. F. *Dokl. Akad. Nauk. SSSR*, **194:** 708, 1970.
44. Byshou, V. F. B., Dubrovina, N. I., Barsukow, L. J., and Bergelson, L. D. *Chem. Phys. Lipids* **6:** 343, 1971.
45. Levine, Y. K., Lee, A. G., Birdsall, N. J. M., Metcalfe, J. C., and Robinson, J. D. *Biochim. Biophys. Acta* **291:** 592, 1973.
46. James, T. L., and Noggle, J. H. *J. Am. Chem. Soc.* **91:** 3429, 1969.
47. James, T. L., and Noggle, J. H. *Proc. Natl. Acad. Sci.* (U.S.) **62:** 644, 1969.
48. James, T. L., and Noggle, J. H. *Anal. Biochem.* **49:** 208, 1972.
49. Batchelor, J. C., Prestegard, J. H., Cuschley, R. J., and Lipskey, S. R. *Biochem. Biophys. Res. Comm.* **48:** 70, 1972.
50. Chapman, D., and Penkett, S. A. *Nature* (Lond.) **211:** 1304, 1966.

51. Darke, A., Finer, E. G., Flook, A. G., and Phillips, M. C. *J. Mol. Biol.* **63:** 265, 1972.
52. Lee, A. G. L., Birdsall, N. J. M., Levine, Y. K., and Metcalfe, J. C. *Biochim. Biophys. Acta* **255:** 43, 1972.
53. Birdsall, N. J. M., Lee, A. G., Levine, Y. K., and Metcalfe, J. C. *Biochim. Biophys. Acta* **241:** 693, 1971.
54. Metcalfe, J. C., Birdsall, N. J. M., Fenney, J., Lee, A. G., Levine, Y. K., and Partington, P. *Nature* **233:** 199, 1971.
55. Horwitz, C. A. F., Horsley, W. J., and Klein, M. P. *Proc. Natl. Acad. Sci.* (U.S.) **69:** 590, 1972.
56. Keough, K. M., Oldfield, E., Chapman, D., and Beynon, P. *Chem. Phys. Lipids* (in press).
57. Jendresiak, G. L. J. *Chem. Phys. Lipids* **6:** 215, 1971.
58. Hammes, G. G. H., and Tallman, D. E. T. *Biochim. Biophys. Acta* **233:** 17, 1971.
59. Finer, E. G., Hauser, H., and Chapman, D. *Colloquium der Gesellschaft fur Physiol. Chem.* **20:** 368, 1969.
60. Steim, J. M., Edner, O. J., and Bargoot, F. G. *Science* **162:** 909, 1968.
61. Barratt, M. D., and Rayner, L. *Biochim. Biophys. Acta* **255:** 974, 1972.
62. Leslie, R. B., Chapman, D., and Scanu, A. M. *Chem. Phys. Lipids* **3:** 152, 1969.
63. Chapman, D., Leslie, R. B., Hirz, R., and Scanu, A. M. *Biochim. Biophys. Acta* **176:** 524, 1969.
64. Scanu, A. M., Reader, W., and Edelstein, C. *Biochim. Biophys. Acta* **160:** 32.
65. Ladbrooke, B. D., Williams, R. M., and Chapman, D. *Biochim. Biophys. Acta* **150:** 333, 1968.
66. Bruckdorfer, K. R., and Green, C. *Biochem. J.* **104:** 270, 1967.
67. Dodge, J. T., Mitchell, C., and Hanahan, D. J. *Arch. Biochem. Biophys.* **100:** 119, 1963.
68. Chapman, D., Kamat, V. B., deGier, J., and Penkett, S. A. *J. Mol. Biol.* **31:** 101, 1968.
69. Kamat, V. B., and Chapman, D. *Biochim. Biophys. Acta* **163:** 411, 1968.
70. Glaser, M., Simpkins, H., Singer, S. J., Sheetz, M., and Chan, S. J. *Proc. Natl. Acad. Sci.* (U.S.) **65:** 721, 1970.
71. Levine, Y. K., Birdsall, N. J. M., Lee, A. G., and Metcalfe, J. C. *Biochemistry* **11:** 416, 1972.
72. Magnuson, J. A., Shelton, D. S., and Magnuson, N. S. *Biochem. Biophys. Res. Comm.* **39:** 279, 1970.
73. Jenkinson, T. J., Kamat, V. B., and Chapman, D. *Biochim. Biophys. Acta* **183:** 427, 1969.
74. Dea, P., Chan, S. I., and Dea, F. J. *Science* **175:** 206–9, 1972.
75. Davis, D. G., and Inesi, G. *Biochim. Biophys. Acta* **241:** 1, 1971.
76. Robinson, J. D., Birdsall, N. J. M., Lee, A. G., and Metcalfe, J. C. *Biochemistry* **11:** 290, 1972.
77. Steim, J. M. In *Thermal Phase Transitions in Biomembranes* (in press).
78. Melchior, D. L., Morowitz, H. J., Sturtevant, J. M., and Tsong, T. Y. *Biochim. Biophys. Acta* **219:** 114, 1970.
79. Chapman, D., and Urbina, J. *FEBS Letters* **12:** 169, 1971.

80. Steim, J. M. In *Liquid Crystals and Ordered Fluids*, R. S. Porter and J. F. Johnson, eds. New York: Plenum Press, 1970.
81. Oldfield, E., Chapman, D., and Derbyshire, W. *Chem. Phys. Lipids* **9:** 69, 1972.
82. Fischer, J. J. F., and Jost, M. C. J. *Mol. Pharmacol.* **5:** 420, 1969.
83. Rosenberg, S. A., and Guidotti, G. In *Red Cell Membranes: Structure and Function*, G. A. Jamieson and T. J. Greenwalt, eds. Philadelphia: J. B. Lippincott, 1969, p. 93.
84. Chapman, D., and Kamat, V. B. In *Regulatory Functions of Biological Membranes*, J. Järnefelt, ed. Amsterdam/London/New York: Elsevier, 1968, B. B. A. Library, vol. 11, p. 99.
85. Wallach, D. F. H. *J. Gen. Physiol.* **54:** 3s, 1969.

CHAPTER 6

OPTICAL ACTIVITY

Introduction

The existence of conformation-dependent, disymmetric regions in all polypeptides has fostered attempts to quantify the secondary structure in soluble and membrane proteins by measurements of their optical activities: optical rotatory dispersion (ORD) and circular dichroism (CD). Optical activity has been widely reviewed (1–5); we, therefore, present only a few basics here before turning to membranes, whose optical activity derives almost exclusively from membrane proteins.

Electromagnetic radiation consists of sinusoidally propagated quanta generating electric (**E**) and magnetic (**M**) fields at right angles to each other and the direction of propagation. In *unpolarized* light, the **E** and **M** vectors are *randomly* distributed (Fig. 6.1), but purification of *monochromatic* light by absorptive or other *photoselective* techniques yields *plane* or *linearly polarized* light, where the **E** and **M** vectors show uniform

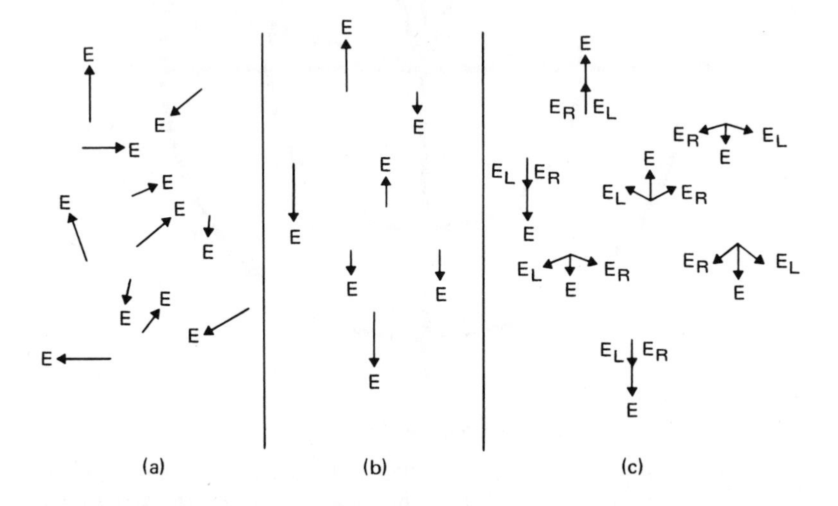

Fig. 6.1 Separation of unpolarized light (a) into *linearly polarized* light (b) and circularly polarized components (c). **E** is the electric vector. See text.

orientation (Fig. 6.1). Vector **E** of linearly polarized light is the resultant of two equal components rotating in phase but in opposite directions. A beam of linearly polarized light, thus, consists of *left* and *right circularly polarized* beams of equal frequency and intensity (Fig. 6.1); these can be experimentally resolved (6).

Optical Rotation or Circular Birefringence

The right and left circularly polarized components travel at unequal velocities through optically active media, i.e., the refractive index n_R (for the right circularized component) differs from n_L, changing the angles of E_R and E_L with p, the polarization plane of the incident light (Fig. 6.2); this is the origin of optical rotation. The rotation of the plane of polarization is given by

$$\alpha = \frac{1.8 \times 10^{10}}{\lambda} (n_L - n_R) \tag{6.1}$$

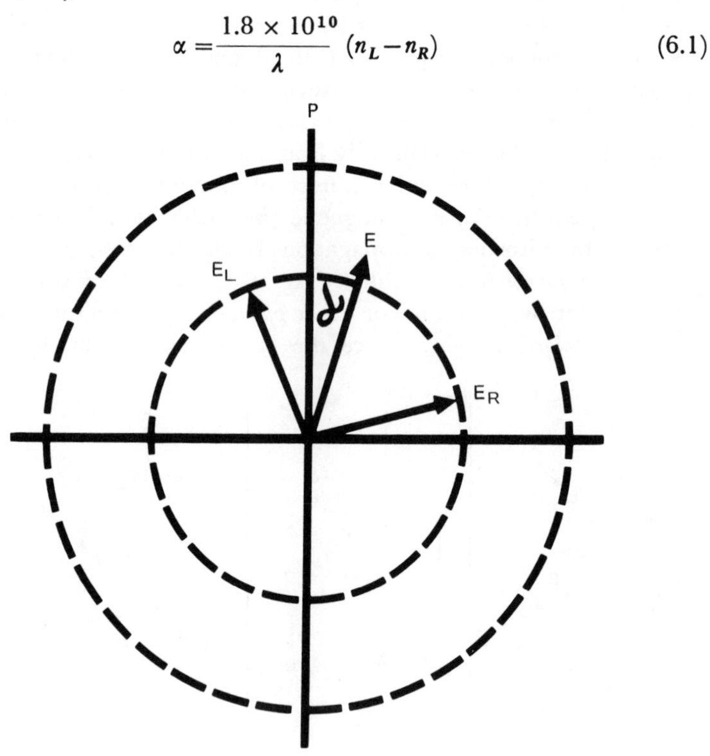

Fig. 6.2 Origin of optical rotation. The right and left circularly polarized components of linearly polarized light move at disparate velocities through optically active media. This changes the angles of these two components with respect to the polarization plane of the incident beam.

where α is the degrees of rotation per 10 cm and λ is in nanometers.

$$[\alpha] = \frac{\alpha}{c} \tag{6.2}$$

where c is the concentration of the optically active substance in grams per cubic centimeter of solution; the molar rotation $[m]$ is given by

$$[m] = \frac{\alpha M}{100c} = \frac{[\alpha] M}{100} \tag{6.3}$$

where M is the molecular weight of the optically active substance; $[m]$ is in deg $\cdot$ cm^2 $\cdot$ dmole^{-1}.

Circular Dichroism (CD)

Optically active substances absorb the left and right circularly polarized components of a linearly polarized beam to a different extent. The molar circular dichroism $\Delta\varepsilon$ for a solution of molar concentration M and a 1 cm light path is

$$\Delta\varepsilon = \varepsilon_L - \varepsilon_R = \frac{1}{[M]\ 10}\left(\log_{10}\frac{I_0}{I_L} - \log_{10}\frac{I_0}{I_R}\right) \tag{6.4}$$

However, CD is usually expressed as molar ellipticity θ, given by

$$\theta = 3300\ (\varepsilon_L - \varepsilon_R) \tag{6.5}$$

Variation of Optical Rotation and Circular Dichroism with Wavelength

In classical analysis, the response of molecular electrons to an electromagnetic field parallels the forced oscillation of a mechanical system; as the system's resonant frequency is approached and passed, it displays *dispersive* (in-phase) and *absorptive* (out-of-phase) responses. The change of n with λ, diagrammed in Fig. 6.3, is dispersive (also in optically inactive media). At the center of the absorption band, n traverses the value it approaches asymptotically at higher and lower wavelengths.

In optically active substances, the variation of $(n_L - n_R)$ with λ, i.e., optical rotatory dispersion (ORD), follows a typical dispersive pattern (Fig. 6.4); hence optical rotation is detectable far from the center of an optically active band. In contrast, the CD proportional to $(\varepsilon_L - \varepsilon_R)$ vs. λ is observed only in the vicinity of the absorption band (Fig. 6.4). With a single, optically active Gaussian band, the ORD extrema occur at $\pm$ the wavelengths where $\theta = 1/e\ \theta_{\max}$. For adjacent Gaussian bands, whose separation $(\lambda_1 - \lambda_2)$ is small compared with their bandwidth $\Delta\lambda$, their apparent separation will always be more than $\sqrt{2\Delta\lambda}$ (7).

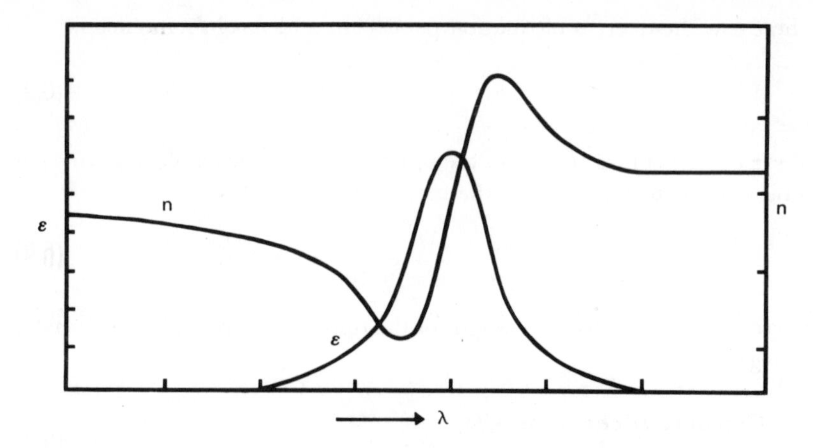

Fig. 6.3 Variation of refractive index n in the region of an absorption band. ε is optical absorption; λ is the wavelength. See text.

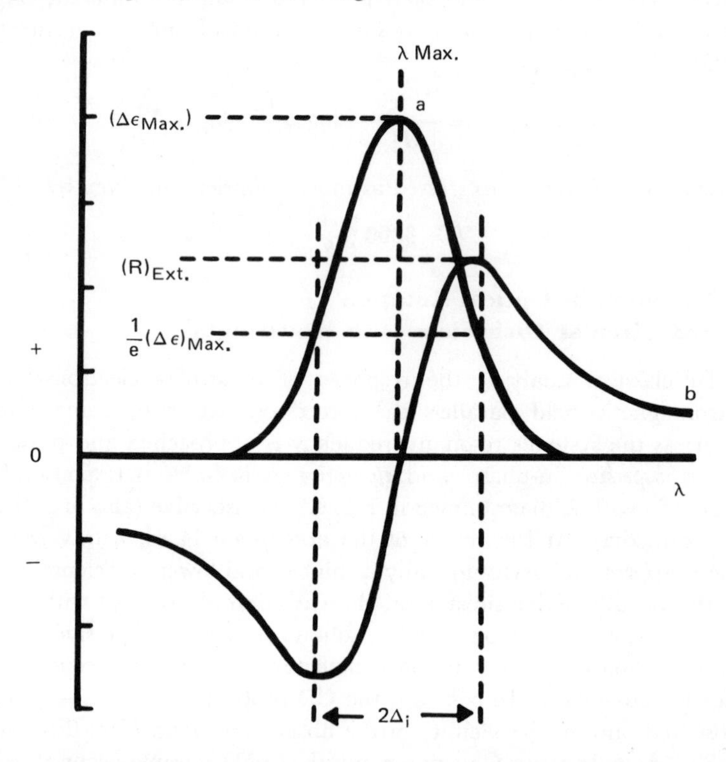

Fig. 6.4 Circular dichroism (curve a) and optical rotatory dispersion (curve b) of a Gaussian optically active band. The CD bands may be positive, as shown here, or negative. For Gaussian bands, the ORD extrema occur at $\frac{1}{2}$ bandwidth ($\triangle i$), where $\theta = 1/e \cdot \theta_{max}$.

Peptide Optical Activity of Polypeptides and Proteins

The optical activity of peptide bonds arises from (a) $\pi^0 - \pi^-$ transitions (i.e., promotion of electrons from the π^0 to antibonding π^- orbitals), responsible for the intense absorption near 190 nm and (b) $n - \pi^-$ transitions (i.e., excitation of nonbonding oxygen electrons to the π^- orbital), accounting for the weak band near 220 nm (Fig. 6.5).

In polypeptides, the optical activity of the peptide transitions is not a simple, summed contribution from all such linkages, but depends upon their spatial relationships; ORD and CD measurements have, thus, been useful in estimating the conformational proportions of certain soluble polypeptides, with the α-helical, β-, and "unordered" conformations yielding distinct ORD and CD spectra (4, 8).

The optical activity of solutions of "unordered" homopolypeptides is generally dominated by the $\pi^0 - \pi^-$ transition at 198 nm, accounting for the intense negative CD at 198 nm and the large negative ORD extremum at 205 nm. The small positive CD extremum at 223 nm is due to the $n - \pi^-$ transition. It is, however, unlikely that such generalities can be readily applied to constrained "unordered" regions in globular proteins (9–11). Thus, the model study of Fasman et al. (11) shows that when unordered polypeptides are immobilized in films, the negative CD band shifts to 200–205 nm, and a negative shoulder appears at 215–230 nm. These findings and similar data on partly unfolded proteins point up the influence of steric constraints of the spectra. We shall return to this in our comments on the optical activities of globular and membrane proteins (12–14).

Polypeptides in the right-handed α-helical conformation display two large negative CD bands near 222.5 nm and 208 nm and a positive one near 192 nm. The former two arise from the $\pi^0 - \pi^-$ transition, which is split into two components (at 206 nm and 192 nm). The 222.5 nm band

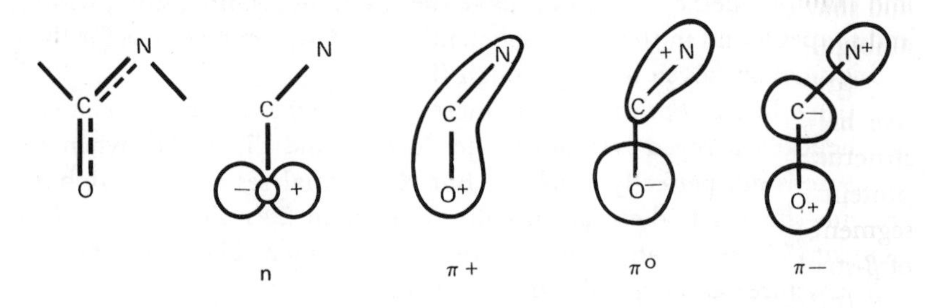

Fig. 6.5 Electronic transitions accounting for peptide optical activity. A single peptide linkage is at the extreme left. In polypeptides, the various transitions become electronically coupled and therewith dependent upon the secondary structure of the peptide chain.

is due primarily to the $n-\pi^-$ transition at 224 nm. (The CD band lies at slightly shorter wavelength than the transition, because of the overlapping negative $\pi^0-\pi^-$ component.) The corresponding ORD spectra are characterized by a large minimum at 233 nm, a crossover to positive rotation at 223 nm, a shoulder near 210 nm, and a maximum at 198 nm. The minimum is due primarily to the $n-\pi^-$ transition. There are appreciable if minor differences between the ORD and CD spectra of various soluble homopolypeptides, but globular proteins of known 2° structure can show substantial discrepancies (e.g., 12, 15–17).

The CD of β-structured poly-L-lysine shows a single negative band at 218 nm, attributable to the $n-\pi^-$ transition (18–19) and a positive extremum at 195 nm. However, recent studies on other β-structured polypeptides show that this conformation can yield very diverse optical activity spectra, whose extrema may overlap those of α-helical polypeptides (20–23).

It was initially hoped that the contribution of α-helical, β-structured, and unordered segments of globular proteins would sum, permitting conformational analysis of proteins in general by comparing their ORD and CD spectra with those of reference polypeptides and computer-generated "conformational mixtures" (12, 22–27). Despite limited success (25–27), this hope is receding, primarily for the following reasons.

(a) Numerous proteins (28–31), including membrane components (32–35), contain significant and sometimes functionally variable proportions of β-conformation; unknown proteins must, thus, be analyzed in terms of at least the three known conformations. This is difficult, even if a single synthetic polypeptide could serve as standard, since 7 optically active chromophores are concerned—these are the $n-\pi^-$ transitions of the three conformations, the $\pi-\pi^-$ transitions of the "unordered" and β-structures, the $\perp$-polarized helical $\pi-\pi^-$ transition, and the $\|$-polarized $\pi-\pi^-$ transition. Additional complications arise from band overlap and uncertainty whether the bands have the wavelengths, intensities, widths, and shapes found in the usual polypeptide models; some reasons for these ambiguities are given in (b) through (f).

(b) α-helical optical activity increases with the number of consecutive helically arrayed residues up to 20 (36) and diminishes when the structure is not perfectly α-helical. But X-ray analyses of many globular proteins show that there are usually fewer than 20 residues per helical segment (30, 36, 37) and that distortions are common. The optical activity of β-structures also depends on chain length.

(c) Even in synthetic polypeptides, optical activity is side chain dependent. This effect is slight in α-helices (23) and marked in β-structures (22).

(d) "Unordered" segments in proteins, if irregular, are usually fixed and structured and vary from protein to protein. They are not random arrays, i.e., irregularly distributed elements of diverse size and structure. Such regions also differ from the "random" states of poly-L-glutamate or poly-L-lysine, which have elements of local extended helix (9–11).

(e) X-ray studies (37) and fluorescence probe experiments (38) show clearly that many peptide linkages of globular proteins reside in an apolar, highly polarizable environment, unlike that of reference polypeptides in water solution. This microenvironment will tend to reduce the rotational strengths of the optically active bands and lower their resonant frequencies (12).

(f) Conformations other than the conventional right-handed α-helical, β-, and "unordered" structures may exist in proteins. For example, one must not forget about the left-handed α-helix; while this is intrinsically less stable than the right-handed helix, extrinsic structural forces may favor it in certain situations (39). Thus, poly-β-benzyl-L-aspartate, a water insoluble polymer, forms a *left-handed* α-helix in 1:99 dichloroacetic acid:chloroform. However, when *spread at an air-water interface, the sense of the helix shifts rapidly to right-handed.* This matter is further discussed in Chapter 4.

Realization of these fundamental problems has engendered a search for new analytical means, to extract unambiguous conformational information from ORD and CD (15–17). Saxena and Wetlaufer (15) address primarily the ambiguities due to so-called unordered regions, while Rosenkranz and Scholtan (16) attempt to use more suitable polypeptide standards. Both use proteins, whose conformation is known by X-ray crystallography, as absolute standards.

A rigorous recent study from Yang's laboratory (17) extends the use of structurally defined proteins as reference standards. The authors reason that the optical activity X (either rotation or ellipticity) at any wavelength can be taken as

$$X = f_H X_H + f_\beta X_\beta + f_U X_U \tag{6.6}$$

where f_H, f_β, and f_U are the fractions of peptide in helical, β-, and unordered conformations, respectively. The sum, fs, is unity, and each $f \geq 0$. They used five reference proteins, whose secondary structure has been determined by X-ray crystallography, namely, myoglobin, lysozyme, ribonuclease, papain, and lactate dehydrogenase. The Xs of these proteins at different wavelengths were fitted by a least-squares method to define the appropriate values for X_H, X_β, and X_U. These data were then used to compute the conformational proportions of other proteins whose structure had also been analyzed by X-ray.

They show that this approach represents a marked improvement over the use of synthetic polypeptides as conformational references. They also point out that CD and ORD spectra computed, using standard proteins as references, deviate markedly from those given by reference polypeptides. These directions will assuredly be furthe r explored.

The difficulties encountered in attempts at the conformational analysis of soluble globular proteins indubitably apply also to membrane proteins (13), but have been largely ignored in concern about light-scattering artifacts. We shall return to this, but first emphasize that (a) different membranes can vary greatly in their morphological, chemical, and physical properties, and (b) except in the case of some viral membranes, properties of membranes cannot be attributed to a predominant "structural protein," as suggested on occasions (40); nearly all membranes contain numerous different proteins.

Optical activity measurements on membranes, therefore, can at best only yield the "average" conformational proportions, which might mirror structural homologies among the proteins of a given membrane or diverse membranes (13, 41). Such information about the secondary structure of *membrane* proteins in situ would be of value, even at a time when X-ray analysis of crystallized soluble proteins is burgeoning. However, the general shortcomings of ORD and CD measurements contain serious and often hidden obstacles to the attainment of this quest. Moreover, this matter has become obfuscated by numerous publications, many reviewed by Urry (42), arguing that *all* deviation of membrane optical activity from that of synthetic polypeptide models is due to light scattering by the material. If this were so, membrane proteins would have an unusually high "helical content," a finding of general importance which should not be airily dismissed as by Stoeckenius and Engelman (43). [Indeed, some feel that corrected helicity reflects membrane function (e.g., 42)]. We shall return to this after reviewing the facts, but stress here the urgency for *multiple criteria* (e.g., ORD, plus CD, plus IR) and also for an understanding of the light-scattering properties of membranes (14, 44–46).

Membrane ORD Spectra

The ORD spectra of all membrane types studied so far show the following anomalies (Fig. 6.6):

(a) shapes close to that obtained with pure, right-handed α-helix,
(b) low amplitude, and
(c) displacement of the entire spectrum to longer wavelengths than observed with α-helix.

This is also true for membranes that lack IR evidence of β-structure under the conditions of optical activity measurement.

With mitochondria and their fragments, the situation is extreme (42, 47–50), but these structures *do* contain significant proportions of β-structured peptide, which, moreover, depends on their metabolic state (33, 47, 51, 52). Other membranes, particularly those of microorganisms

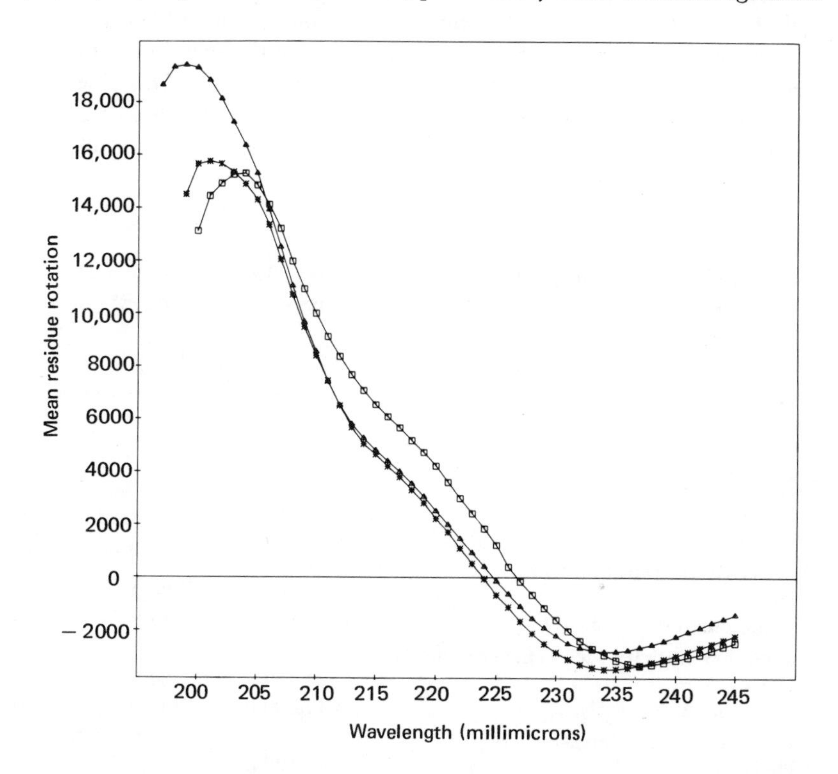

Fig. 6.6 ORD spectra of plasma membrane vesicles from Ehrlich ascites carcinoma. □——□ control; △——△ same with added lysolecithin *——* same with added phospholipase A. The controls exhibit a minimum at 236 nm. This shifts to 233 nm with phospholipase A or lysolecithin. [From Gordon et al. (13) courtesy of Elsevier Publishing Co.]

where electron transport components are built into the plasma membranes [e.g., *Acholeplasma laidlawii* (32), *Micrococcus lysodecticus* (53)] also give the IR signs of β-structure (32, 53), and Choules and Bjorklund (32) estimate the conformational proportion of the plasma membranes of *Acholeplasma laidlawii* as 36% β, 51% α, and 13% "unordered."

Membrane optical activity, differs from that of soluble lipoproteins (54) and of ionic complexes between acidic phosphatides and various proteins (55, 56)—this despite the fact that such complexes may be larger in size than some membrane fragments (see below).

A peculiarity of membrane ORD is the fact that, if the membrane protein were partly in right-handed α-helix and the rest unordered, the spectra should not have the "α-helical" shape (13, 40, 41). Conversely, if the proteins are primarily helical, they should show greater amplitude. This is also true for the CD spectra. The membrane ORD and CD spectra appear *as if* most of the optical activity of the membrane protein were masked with only right α-helix being expressed. This does not account for observed band shifts, particularly since, in mixtures of α-helix and unordered conformation, even small proportions of the latter displace the spectra to the blue.

Table 6.1 Rotatory Parameters That Yield a Good Fit for the CD and ORD Spectra of Ehrlich Ascites Carcinoma Membranes[a]

Parameter	Transition[b,c]		
	I	II	III
Bandwidth (nm)			
Helix	11.2	7.3	13.8
"Unordered"	10.0	10.3	8.5
Wavelength of transition (nm)			
Helix	197.5	206.9	222.7
"Unordered"	198.0	217.0	235.0
Maximum ellipticity ($deg \cdot cm^2 \cdot dmole^{-1}$)			
Helix	34,119.3	−16,234.5	13,913.0
"Unordered"	−12,000.0	3,240.0	−219.0
Rotational strength ($ergs\text{–}cm^3 \times 10^{40}$)			
Helix	23.8	−7.1	−10.6
"Unordered"	−7.5	1.9	−0.1
Conformation (%)			
Helix	58.0	58.0	58.0
"Unordered"	42.0	42.0	42.0

[a] This is only one of several equally "good fits" obtained with various rotatory parameters. In this solution, the *rotational strengths* of the various peptide transitions are in the same relative proportions found in poly-L-glutamic acid, myoglobin, hemoglobin, and lysozyme, but their absolute intensities are 15% less than those computed for lysozyme. From Straus et al. (12).

[b] For α-helix, I, II, and III, refer to the $\perp \pi^\circ - \pi^-$, $\parallel \pi^\circ - \pi^-$, and $n - \pi^-$ transitions, respectively. For "unordered" conformations, the numbers refer to the three bands specified in Carver et al. (23).

[c] Standard deviation between experimental curve and curve computed from parameters is 1.93%.

Since ORD and CD spectra have common origins and have a precisely defined relationship (2, 3), any explanation of the anomalous optical activity of membranes must account for the peculiarities of both. Lenard and Singer (40) did not study the relationships of amplitude, shape, and band position of their spectra in detail, but suggest that the red displacement of their membrane spectra might be due to the optical activity of α-helices, packed parallel but at a slight twist. This concept has also been invoked to explain the small red shift that accompanies aggregation of α-helical poly-L-glutamic acid (57), but in this case the spectral displacement is accompanied by increased optical activity. In any event, the small displacement of the $n-\pi^-$ CD maximum does not account for the observed "red shift" in the ORD; a bandwidth change must be involved also. Urry and associates (48) first argued that the displacement of the ORD minimum in membranes is due to a shift of the $n-\pi^-$ transition, giving as a reason the 3 nm displacement of the long wavelength CD band of mitochondria. But this could be correct only if the transition were quite isolated from the long wavelength $\pi^0-\pi^-$ transition, which is not the case, and if their CD spectra (as well as those of others) did not show red shifts also in the $\pi^0-\pi^-$ region; these facts cannot be explained by changes in only the position $n-\pi^-$ band. Wallach and associates, using extensive computer analyses, could find *no unique* plausible solution for the optical activities of plasma membranes over the entire peptide spectrum *any more than for globular proteins of known 2° structure* (12, 13). Thus, one set of optical activity parameters giving an excellent "fit" of both ORD and CD spectra of Ehrlich ascites carcinoma plasma membrane vesicles is given in Table 6.1, but other combinations of parameters also yield computed spectra matching the observed rather well.

Membrane CD Spectra

Numerous CD spectra have been reported for erythrocyte membranes under various conditions (13, 32, 40, 45, 58, 65)—plasma membranes from Ehrlich ascites carcinoma (13, 41), *B. subtilis* (40), *Acholeplasma laidlawii* (32), sarcotubular vesicles (66, 67), liver bile fronts (67), membranes of oxyntic cell microsomal vesicles (68), as well as mitochondria and submitochondrial vesicles (48–50).

The spectra reported for one membrane type vary considerably from laboratory to laboratory. Thus, Gordon and Holzwarth (45) report a CD minimum of 224 nm for erythrocyte ghosts, while Sonenberg (60) gives a value of 222 nm. Furthermore, one finds large differences among different membrane preparations, a fact that Urry suggests to reflect

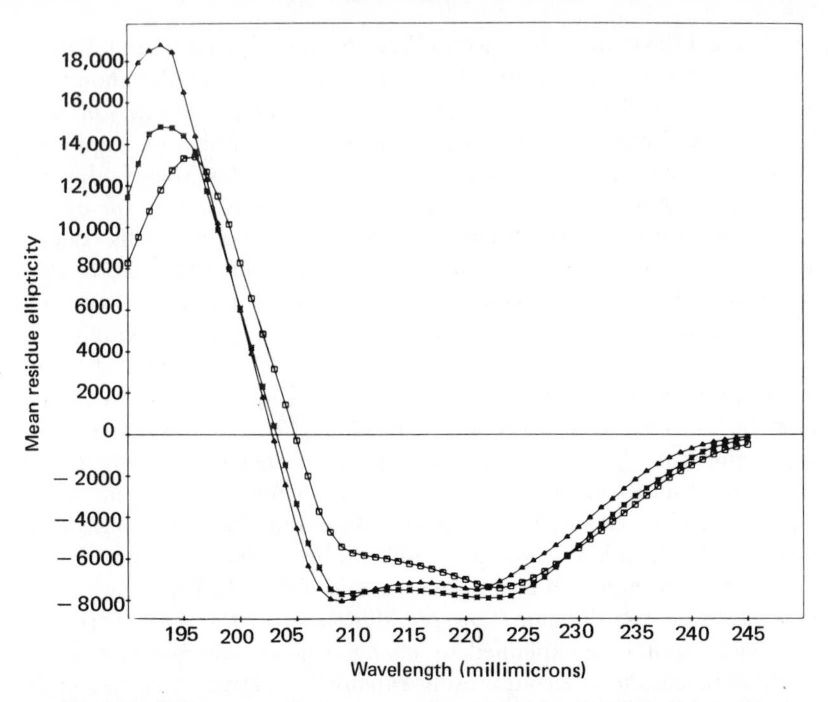

Fig. 6.7 CD spectra of plasma membrane vesicles from Ehrlich ascites carcinoma. □——□ control; △——△ lysolecithin added; *——* phospholipase A treated. [From Gordon et al. (13) courtesy of Elsevier Publishing Co.]

functional differences.* As in the case of ORD, most spectra exhibit a shape suggesting a mixture of "unordered" peptide and α-helix, but are of a lower amplitude than generally expected from polypeptides (Fig. 6.7).

However, mitochondria and mitochondrial fragments (e.g., 48–50) exhibit CD spectra of feeble amplitude with a major minimum at 223–225 nm, quite far to the red. We cannot explain this fact unambiguously; part of the phenomenon may arise from light-scattering artifacts, and part from the substantial proportion of β-structure in mitochondria (Chap. 4).

* The CD spectra of liver bile fronts obtained from hypophysectomized rats show greater ellipticities in the region of peptide absorption than those from normal animals or rats treated with bovine growth hormone (M. S. Rubin, N. I. Swislocki and M. Sonenberg, *Arch. Biochem. Biophys.* **157:** 243, 1973). Studies on the effects of growth hormone on previously isolated liver bile fronts support these findings (M. S. Rubin, N. I. Swislocki, and M. Sonenberg, *Arch. Biochem. Biophys.* **157:** 252, 1973). The membranes from hypophysectomized but not normal rats show a consistent increase in ellipticity upon addition of 10^{-9} M growth hormone. This effect is not accompanied by any light scattering changes, but growth hormone (10^{-17}–10^{-7}M) produces a decrease in tryptophan fluorescence and a small emission red shift. The CD changes, which resemble those previously reported for erythrocyte ghosts (60) are convincingly attributed to hormone-induced changes in membrane protein structure.

Moreover, the possibility of metabolic α- and/or "unordered" $\rightarrow \beta$-transitions must be considered very seriously in membrane optical activity, if only because they have been demonstrated by infrared spectroscopy (Chap. 4).*

In this connection, the work of Masotti et al. (68), as well as that of Singer and Morrison (65), appears of considerable interest. The first study concerns the effect of ATP on the CD spectrum of microsomal membrane vesicles, derived from dog gastric mucosa cells. These membranes bear an ATPase insensitive to Na^+ or K^+, but stimulated by HCO_3^- and inhibited by SCN^-. It is thought involved in acid secretion. Untreated vesicles exhibit nearly equivalent CD minima near 223 nm and 208 nm. Addition of 1 mM ATP shifts the 223 nm band to 224 nm, lowers its intensity, and shifts the crossover from 202 nm to 207 nm. Mg-AMP is without effect, but there are few other controls. Nevertheless, the authors compute that only part of the observed change could be attributed to light-scattering alterations and suggest the possibility of an $\alpha \rightarrow \beta$ transformation. This obvious possibility is unfortunately not tested by infrared spectroscopy, although ATP-induced shifts to β-structure have been detected by this method (Chap. 4).

The study of Singer and Morrison (65) is equally interesting. They solubilize human erythrocyte ghosts in *n*-pentanol (Chap. 3) to get an optically clear solution. When CD measurements are taken (in d H_2O), the data in Table 6.2 result. Upon comparing with typical

Table 6.2 CD Intensities and Band Positions of Erythrocyte Ghosts, Pentanol-Solubilized Erythrocyte Ghosts in d H_2O, and α-helical Poly-L-Lysine

Ghosts[a]		Pentanol-Solubilized Ghosts[b]		α-Helical Poly-L-Lysine[c]	
λ_{min} 223	θ 13,800	λ_{min} 219	θ 10,100	λ_{min} 222	θ 35,700
λ_{min} 209	θ 12,100	λ_{min} 209	θ 9,740	λ_{min} 208	θ 32,600
λ_{min} 195	θ 24,000	λ_{min} 192	θ 19,100	λ_{min} 191	θ 76,900

[a] From Gordon et al. (13).
[b] From Singer and Morrison (65).
[c] From Greenfield et al. (24).

Note: The wavelengths are given in nm, and θs (ellipticities) in deg · cm^2 · dmole^{-1}.

* Circular dichroism spectra of erythrocyte membrane films resemble those of ghost suspensions. The spectra do not change upon dehydration from 92 to 0% relative humidity. This finding is taken to indicate that membrane protein structure does not vary with water content (M. J. T. Schneider and A. S. Schneider, *J. Memb. Biol.* **9:** 127, 1972).

erythrocyte CD data as well as the CD of a helical synthetic polypeptide, one suspects that the solubilization step has induced some β-structuring. (The CD minimum of most β-structured peptides lies near 218 nm.) This suspicion is strengthened by the fact that the helicogenic solvent, 2-chloroethanol, intensifies the CD signal, shifting the major CD minimum from 219 nm to 221 nm. The authors urge for great caution in the interpretation of membrane optical activity and in the application of membrane solubilizers.

Turbidity Effects

Several authors (e.g., 42, 45, 46) have addressed the effects of the particulate nature of membranes upon their optical activity. Indeed, Urry and associates (42) and Schneider et al. (58) have stated that *all* of the unusual features of membrane optical activity arise from two turbidity artifacts, namely, "absorption statistics" and anomalous light scattering in regions of light absorption.

Absorption Statistics

Absorption statistics produce anomalously low amplitudes of absorption and optical activity in spectral regions, where particles absorb light more than would their chromophores in true solution. The terms "Duysens' effect," "optical sieving," "bunching effects," "concentration masking," and "absorption flattening" refer to the same phenomenon, which has been examined in detail by Holzwarth (69). Urry and Ji (70) first suggested that absorption statistics might influence membrane optical activity, but their empirical approach cannot be accepted to be reliable for optical activity measurements, particularly on membranes (14, 45, 46).

First, we fail to see how any correction can be applied unambiguously to a system whose conformational proportions are not precisely defined (14); *second*, in the case of *membrane shells*, whatever corrections are applied differ for absorption and optical activity spectra (71).

In the simplest model, one can imagine a membrane suspension to consist of identically sized, homogeneous, cubical particles, randomly distributed in a transparent solvent and constituting a fraction f of the total suspension volume (69). If one assumes that the absorptivity of the particles equals $a \cdot \text{cm}^{-1}$ the transmission, T, per cm accordingly equals e^{-aT}. If I_0 is the intensity of the light incident on the sample and l, the path length, is finite, I_0 can be considered as comprising a multitude of light bundles, each with intensity i_0. In traversing l cm of suspension, each bundle will pass through T cm of absorber, and its emergent intensity i will accordingly equal $i_0 e^{-aT}$. The total transmitted intensity I_{tot} will then equal

$$I_{tot} = I_0 \int_0^l P(T) e^{-aT} \, dT \qquad (6.7)$$

where $P(T)$ is the probability of a beam encountering the equivalent of T cm of chromophore (69).

However, I_{tot} does not vary linearly with the length of the absorbing path but *exponentially*. Thus, the absorbance A equals

$$A = \frac{I_0}{I_{tot}} = -\ln \int_0^l P(T) \cdot e^{-aT} \, dT \qquad (6.8)$$

In *true solutions* of small molecules, a ray of light encounters 10^9–10^{10} chromophores/cm^3, the random, statistical variation of T from average T remains insignificant, and $P(T) \cong 1$. However, in a *particle dispersion*, the number of chromophores encountered becomes much smaller, i.e., $< 10^8$/cm^3, and the distribution of $P(T)$ spreads out. Under these conditions, the solution absorbance A_{sol} and suspension absorbance A_{sus} deviate, and their ratio becomes

$$\frac{A_{sus}}{A_{sol}} = Q_A \qquad (6.9)$$

where Q_A is the "absorption flattening" quotient (69).

Q_A can be evaluated if $P(T)$ is known, which is easy in the simple (but unrealistic) cubic model used by Duysens. Here each light bundle, of $d \times d$ cross section, passes through a column of cross-sectional area d^2 and length l, which can be divided into $x = l/d$ cubes of volume d^3. The probability that one of these contains a cube is q, the volume concentration of absorbing material and the probability P_y that a path contains y particles equals

$$P_y = \frac{x!}{(x-y)!\,y!} q^y (1-q)^{x-y} \qquad (6.10)$$

P_y corresponds to $P(T)$, and $q \ll 1$. Then

$$Q_A = \frac{(1-e^{-ad})}{ad} \qquad (6.11)$$

i.e., Q_A is always less than unity.

Clearly, Q_A depends on particle size; thus, Q_A approaches 1.0 for $l/2000 = d$, but declines to 0.8 for $d = l/20$.

Absorption statistics lead to spectral *distortion*, since spectral domains of high absorptivity are affected more than other regions, i.e., absorption maxima become "flattened."

The Duysen model has been repeatedly applied to ORD (ϕ) and CD (θ). Taking the latter and equating θ_{sol} and θ_{sus} with the CD of membrane material in solution, but *conformationally identical* to the intact membrane protein ,θ_{sus} one can then define a CD absorption flattening quotient $Q_B = \theta_{sus}/\theta_{sol}$. Proceeding as before, the particles are expected to generate CD signals varying exponentially with particle number. The observed CD would then constitute an intensity-weighted average of the CD from each ray and

$$\theta_{sus} = \frac{\int_0^l P(T)\ [T\theta_{sol}/fl] \cdot e^{-(A_{sol}T/fl)} dT}{\int_0^l P(T)\ e^{-(A_{sol}T/fl.)}\ dT} \qquad (6.12)$$

for the cubic Duysen model and a binomial distribution

$$\theta_{sus} = \theta_{sol} \cdot e^{-(ad)} \qquad (6.13)$$

and

$$Q_B = e^{-(ad)} \qquad (6.14)$$

Typical values for cubes, spheres, and shells have been provided by Gordon and Holzwarth (71) and are presented in Table 6.3. We note

Table 6.3 Computed Absorption Flattening Coefficients Q_A and, CD Flattening Coefficients for Cubes, Spheres, and Shells of Varying Sizes[a,b]

	Edge or Radius (cm)	Mol Wt	Q_A	Q_B
Cube	1×10^{-7}	6×10^{2}	0.99	0.98
	1×10^{-6}	6×10^{5}	0.89	0.80
	1×10^{-5}	6×10^{8}	0.40	0.11
Sphere	1×10^{-7}	2.5×10^{3}	0.98	0.96
	1×10^{-6}	2.5×10^{6}	0.89	0.71
	1×10^{-5}	2.5×10^{9}	0.29	0.05
Shell	1×10^{-6}	2.5×10^{6}	0.84	0.71
	3.5×10^{-6}	6.9×10^{7}	0.70	0.47
	3.5×10^{-5}	9.0×10^{9}		0.34

[a,b] From Gordon and Holzwarth (71).

Note: Shell thickness = 10^{-6} cm, a/cm is taken to be $2.3 \times 10^{5} \cdot cm^{-1}$, and mean residue weight = 100.

that for cubical and spherical particles, deviations become substantial at mol wt $>10^6$. However, in the case of spherical shells, Q_A and Q_B are relatively insensitive to molecular weight. This points up the importance of *membrane thickness* in computing Q_A and Q_B for typical membrane shells. However, membrane thickness cannot be readily defined and varies markedly from one membrane to the next providing a source of ambiguity.

Light Scattering

Because of their substantial size, membranes and their fragments scatter light in a complex fashion, and one must suspect that the degree of scattering differs for left and right circularly polarized light. This matter has been dealt with rather voluminously, but semiempirically, by Urry and associates (e.g., 42) and more rigorously by Gordon and Holzwarth (45), Gordon (46), and Wallach et al. (14), all of whom utilized the Mie theory; this relies upon the exact solution of Maxwell's equations (72) and has been extended to isotropic, spherical shells (73). Here each particle is taken to be spherically symmetrical and is given in a spherical coordinate system $[v, \theta, \phi]$. Each incident wave vector is expanded into an infinite series of spherical vector wave functions. Similarly, the vector waves scattered in all directions are expanded as an infinite series on the same basis. The field of the scattered waves must always be finite, approach zero as r, the particle radius, approaches ∞, and must also satisfy the continuity requirements at the particle solvent boundaries; these depend upon the complex refractive indices m_1, m_2, and m_3 of the external medium, the shell, and the shell contents, respectively (45, 46).

For CD and ORD, one need only consider forward scattering, since scattering at other angles is not detected by the instruments. One can compute this for spherical membrane shells *provided* one knows their *diameter*, their *thickness*, and the *values of* m_1, m_2, *and* m_3 *for left and right circularly polarized light* (L, R). In most cases $m_1 = m_3 =$ the average refractive index value of the solvent (usually water) in the region of peptide absorption (1.4 for H_2O). To evaluate m_{2L} and m_{2R}, the membrane refractive indices for left and right circularly polarized light, one requires solution values for absorbance, ORD and CD. Gordon and Holzwarth (45) take the values obtained after addition of 0.1% sodium dodecyl sulfate (SDS), and Urry and associates (42) employ SDS, sonication, trifluoroethanol, etc., to achieve a reference state. Unfortunately, extensive evidence indicates that such agents may induce large *conformation* changes, thus vitiating their intent.

Nevertheless, assuming that SDS does not affect conformation, taking a reasonable membrane refractive index (for unpolarized light) of 1.7, and neglecting dispersion of this refractive index, one can compute m_{2L} and m_{2R} (45):

$$
\begin{aligned}
m_{2L} &= 1.7 - \left(\frac{iA_{sol}\lambda}{4\pi NlV}\right) + \frac{\lambda(\phi_{sol} - i\theta_{sol})}{360\, lNV} \\
m_{2R} &= 1.7 - \left(\frac{iA_{sol}\lambda}{4\pi NlV}\right) - \frac{\lambda(\phi_{sol} - i\theta_{sol})}{360\, lNV}
\end{aligned}
\qquad (6.15)
$$

where A_{sol} is absorbance of the "solution" (reference state) to the base e; ϕ_{sol} and θ_{sol} are the "solution" ORD and CD in degrees; l is the path length in cm; N is the number of particles/cm^3; and V, the volume of the shell (45). From the phase lag and absorptive loss in the forward direction and the total scattering in all directions, one can then proceed to evaluate the absorbance, ORD, and CD of the membrane particles. Unfortunately, such calculations rest on a multitude of uncertainties—shell diameters, diameter distribution, shell thickness, uniformity of shell refractive index, etc. They are nonetheless at least heuristically useful.

Effects in Membranes

The obvious possibility that light scattering might interfere with optical activity measurements on particles much larger than the wavelength of the irradiating light led Wallach and Zahler (41) to perform their original ORD measurements on *small* plasma membrane vesicles (diameter <1500 A, and comparable to the length of a typical, α-helical reference polypeptide poly-L-lysine, mol wt 100,000) after earlier studies (44) suggested that light-scattering artifacts would be small in such a case. These membranes exhibit (a) a distinct red shift of the negative "helical" ORD extremum, as well as of the ORD crossover; (b) a $n-\pi^-$ CD minimum (at 223.2 nm), which lies in the wavelength range found with helical synthetic polypeptides (13); (c) a general red shift of the ORD spectrum; (d) a red shift of $\pi-\pi^-$ region of the CD spectrum; and (e) low intensity but "α-helical" shape (13).

As noted, optical activity measurements on other membranes, all larger particles, differ considerably, but all suggest substantial "helicity" and share one common anomaly—red displacement of the ORD trough. However, one or more of the features often thought typical of membrane optical activity have also been recorded for peptides/proteins in true solution (12); indeed, cytochrome c, which contains little α-helix (74), yields a CD spectrum closely resembling that of many membranes (32).

Indubitably, light-scattering artifacts can be considerable with large membrane particles under certain conditions. As noted, Gordon and

Holzwarth (45) and Gordon (46) have treated this rigorously by application of Mie theory for the case of particles containing only α-helical peptide and Urry and associates (42) have developed mathematical techniques to "improve" membrane optical activity, which they apply in the hope of developing meaningful correlations between membrane optical activity and function.

Unfortunately, neither of these groups has thoroughly considered how the general ambiguities in protein optical activity measurements apply to membranes, and both continue on the assumption that synthetic polypeptides in α-helical and/or "random" conformations are suitable standards for proteins in general.

Additional ambiguity arises from the necessity to establish for each membrane a solution state in which the membrane proteins are conformationally *unchanged* from their native condition. Urry et al. (42) and others (e.g., 45, 71, 74) have attempted to attain reference states by diverse manipulations which, while indubitably reducing particle size, yield dubious results, since they are also well known to alter protein structure. Thus, ultrasound is a potent denaturant, rapidly inactivating numerous enzymes, dissociating heme from hemoglobin-globin (75) and altering many proteins even with low levels of irradiation (76). Membranes are particularly susceptible, and it appears that the major mechanisms involved are not thermal (77), but involve free radical formation and streaming (78). Other evidence to the contrary, sonication must thus be suspected of inducing protein denaturation even when done in the cold.

Use of the French press (64) is also not above suspicion. Thus, in its operating range (about 1000 kg/cm^2), serum albumin begins to aggregate (79) and actomyosin depolymerizes (76). Membranes seem even more sensitive, and exposure to about 300 kg/cm^2 produces erythrocyte leakiness to K^+ and later increased fragility (79), while exposure to about 1000 kg/cm^2 is reported to alter the electrical conductivity of membranes (80).

Sodium dodecyl sulfate (SDS) has repeatedly been used to generate a solution "reference state" [e.g., (42, 45)], but at the concentrations usually employed (about 1 mg/mg membrane protein), this detergent acts as a potent protein perturbant, generating the antiparallel β-conformation with poly-L-lysine (81), producing NMR evidence of unfolding of membrane proteins (82) and inactivating membrane enzymes (83, 84). Indeed, depending on the affinity of polypeptides for SDS, these tend to form rods saturated with bound detergent (85). Organic solvents (68) are equally suspect. Thus, 2-chloroethanol drastically alters protein helicity (40,41), while pentanol (65) and acidic solvents (41) generate β-structures in membrane proteins. To sum, the listed procedures introduce additional ambiguities and thus cannot be trusted to clarify the contribution of scattering artifact to membrane optical activity.

Accordingly, Wallach et al. have extended the Mie approach to define light scattering effects, which *would* appear with particles of given dimensions and containing only peptide in various proportions of the standard conformations found in synthetic reference polypeptides (14). They have adopted the approach developed by Gordon (46) and Gordon and Holzwarth (45), using their SCATCO computer program. These calculations assume that each membrane particle comprises an isotropic spherical shell of 7.5 nm thickness surrounding an isotropic core. Data are computed for particle radii of 75 nm and 3500 nm. Every particle is considered a discrete, independent scatterer. Multiple scattering is neglected. The core and medium are each given a fixed refractive index $m_1 = m_3 = 1.4$, corresponding to water near 200 nm. Although membranes have a density near 1.18, rather than the 1.5 of polypeptides, this discrepancy does not noticeably affect the calculation.

The complex refractive indices m_{2R} and m_{2L} of the membranes for right (R) and left (L) circularly polarized light at a specific vacuum wavelength λ of irradiating light were computed from established solution values of absorbance (86), ORD (87), and CD (88) for poly-L-lysine.

These computations were performed to determine what aspects of membrane CD and ORD might arise exclusively from light scattering. In this, a conformational standard more appropriate for proteins in general than a synthetic polypeptide would have been desirable. However, such a standard does not exist, and the CD, ORD, and absorbance data available for poly-L-lysine are sufficient to test the basic questions.

Gordon (46) and Gordon and Holzwarth (45) assume a shell density of 1.5 gm/cm^3, but measured values for membranes lie near 1.18 (89). However, selected computations at 197, 210, and 220 nm, for α-helix: unordered proportions of 1.00, 0.5, and 0.2, show that density variations from 1.15 to 1.50 change the computed ellipticities of 75 and 3500 nm particles by $\leq 1\%$.

The relative refractive index $n = 1.2$ assumed by Gordon (46) and Gordon and Holzwarth (45) matches experimental data (44). Gordon (46), Gordon and Holzwarth (45), and Wallach et al. (14) neglect dispersive effects on n and assume that the refractive index $= 1.2$ throughout the membrane shell. If the membranes comprise separated lipid and protein regions $\geq$ than the wavelength of light, the last assumption would introduce error. Clearly, this problem does not apply to the 75 nm particle .

Mie-scattering computations for poly-L-lysine at α-helix:unordered proportions from 1.00 to 0.20 clearly demonstrate that light-scattering

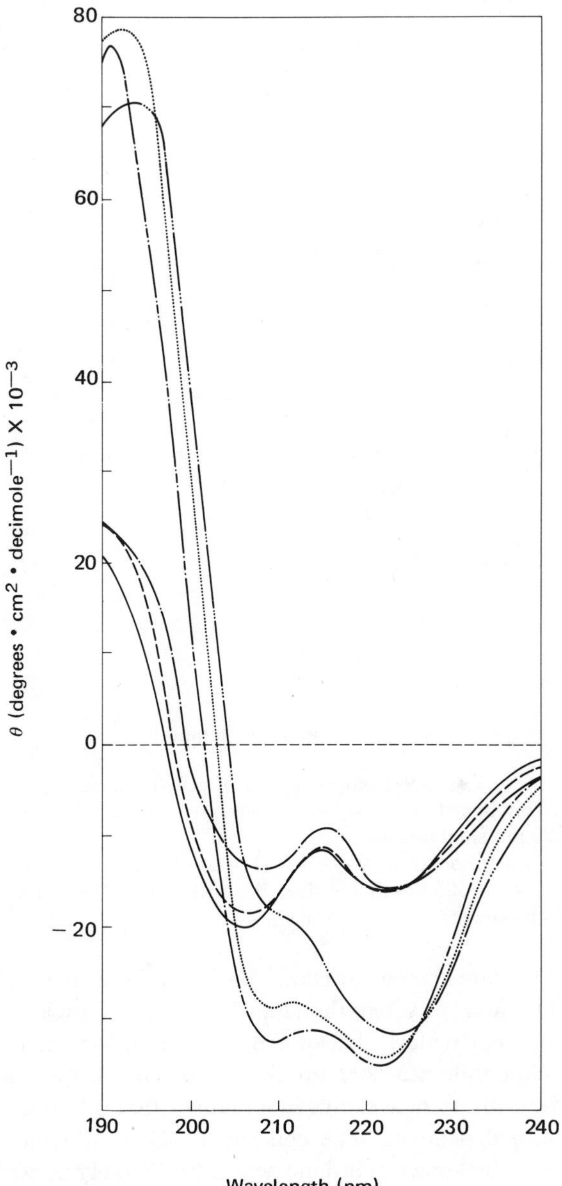

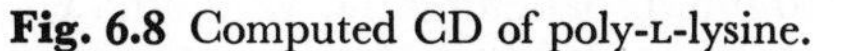
Fig. 6.8 Computed CD of poly-L-lysine.

100% α—helix, rest unordered: ——·—·—— solution; ·········· 75 nm shell radius; ——···—— 3500 nm shell radius.
50% α—helix: ———— solution; – – – – 75 nm shell radius; ——·—— 3500 nm shell radius. See text. [From (14), courtesy National Academy of Sciences.]

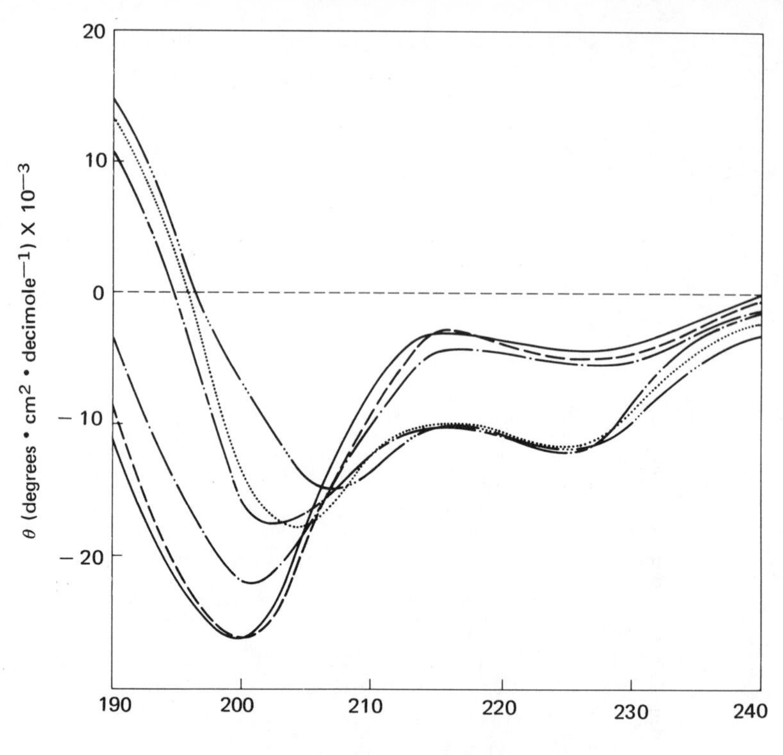

Fig. 6.9 Computed CD of poly-L-lysine.
20% α—helix, rest unordered: ——— solution; - - - - - 75 nm shell radius; ——·—— 3500 nm shell radius.
40% α—helix, rest unordered: ——·—·—— solution; ·········· 75 nm shell radius; ——···—— 3500 nm shell radius. See text. [From (14), courtesy National Academy of Sciences.]

distortions of CD are somewhat conformation sensitive near 222 nm, and strongly so at lower wavelengths (Figs. 6.8, 6.9). This is very marked below 215 nm, particularly because of the strong negative CD of the unordered polypeptide near 200 nm. As shown in Table 6.4, the anomalies vary markedly with particle radius in a manner that is both conformation and wavelength dependent. The computed CD of hypothetical 75 nm particles follows the known solution spectra very closely between 190 and 230 nm. However, the scattering effect of even such small vesicles becomes significant above 230 nm, broadening the "222 nm" CD band sufficiently to produce a shift of the "helical" ORD minimum from 233 to 235–236 nm, such as is seen in membranes.

The computations illustrated in Figs. 6.8 and 6.9 clearly show that *one cannot predict or "improve" light-scattering anomalies without knowing what*

Table 6.4 Predicted Ellipticities, θ, of Variously Sized Spherical Shells at Selected Wavelengths and Helicities

Helicity (%)[b]	Wavelength (nm)	θ at External Radius[a] (nm $\times$ 10^{-3})			
		~0[c]	0.05	0.1	1.0
100	197	44,300	57,040	59,800	66,050
100	210	−32,400	−30,210	−28,040	−21,830
100	220	−35,300	−33,310	−33,200	−30,900
50	197	10,000	6,280	8,100	12,250
50	210	−16,900	−17,000	−16,400	−14,420
50	220	−15,400	−14,850	−14,900	−14,220
20	197	−24,700	−24,100	−22,800	−19,800
20	210	−7,600	−9,100	−9,500	−9,960
20	220	−3,100	−3,800	−3,970	−4,200

[a] Shell thickness = 7.5 nm; θ in deg-cm^2-dmole^{-1}.
[b] Nonhelical peptide in unordered state.
[c] Solution.

conformations are present and in what proportion; but these data are exactly what one wishes to determine by optical activity measurements.

As far as membranes are concerned, one can obtain a good fit near 222 nm, knowing particle size, assuming suitable conformational proportions, and applying the Mie corrections (Figs. 6.10, 6.11). However, a fit near 222 nm invariably yields a major mismatch below 215 nm. This anomaly—a "good fit" near 222 nm, but not at shorter wavelengths—has been previously encountered with small, well-characterized, globular proteins (e.g., 12, 17); it is thus not unique to membranes and does not arise from light scattering.

We suspect that the discrepancies can be explained qualitatively as follows. (a) Globular and membrane proteins tend to give low "helical" signals (near 222, 210, and 195 nm) for reasons given earlier, i.e., short helical segments, distortions from perfect helicity and apolar environments. (b) In native proteins, peptide linkages not in α-helical or β-structured disposition comprise diverse, irregular, nonextended, but nonrandom arrays, each likely to yield a distinctive CD spectrum other than that of unordered poly-L-lysine. The summed contribution of such segments is likely to be diffuse. Thus, synthetic polypeptide standards are likely to underestimate overall "helicity" and exaggerate the proportion of "unordered" peptide segments. For these reasons, we tend to agree with others that membrane proteins may be more "helical" than

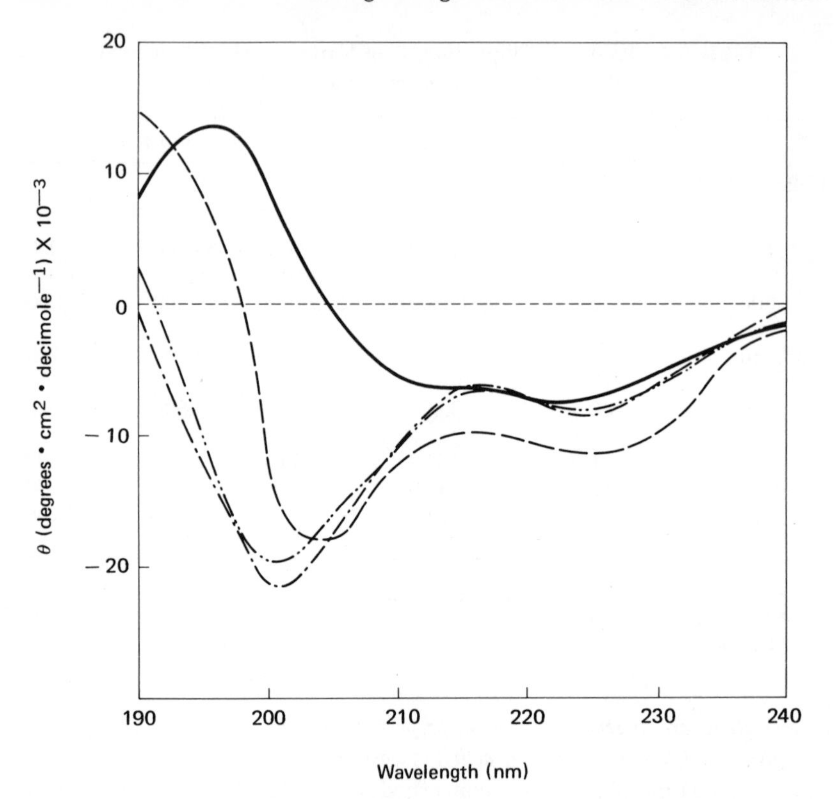

Fig. 6.10 CD of EAC plasma membranes compared with computed CD of poly-L-lysine.
——— membranes; ——·—— pol-L-lysine, 30% α—helix, rest unordered, solution; ——··—— poly-L-lysine, 30% α—helix, rest unordered, 75 nm shell radius; - - - - pol-L-lysine, 40% α—helix, rest unordered, 75 nm shell radius. See text. [From (14), courtesy National Academy of Sciences.]

indicated by their optical activity. However, we see no way by which "helicity" can, at present, be qualified unambiguously by optical activity measurements. For example, the lesser CD amplitude of one membrane type, compared with another, could reflect differences in overall "helicity" or in the proportion of *long* helical segments.*

In general, the CD band near 222 nm is rather broad and its position

* In their studies on the effects of growth hormone on liver bile fronts, Rubin *et al.* (M. S. Rubin, N. I. Swislocki, and M. Sonenberg, *Arch. Biochem. Biophys.* **157:** 252, 1973) invoke similar arguments to explain the CD and fluorescence changes observed. They also point out that sonication of liver bile fronts, as well as treatment with phospholipase A, destroy the characteristic CD response to growth hormone, alter the fluorescence properties of the membranes, and inactivate their Na^+–K^+-sensitive ATPase, presumably by changing the secondary and tertiary structures of sensitive membrane proteins.

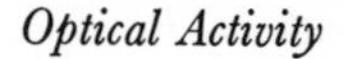

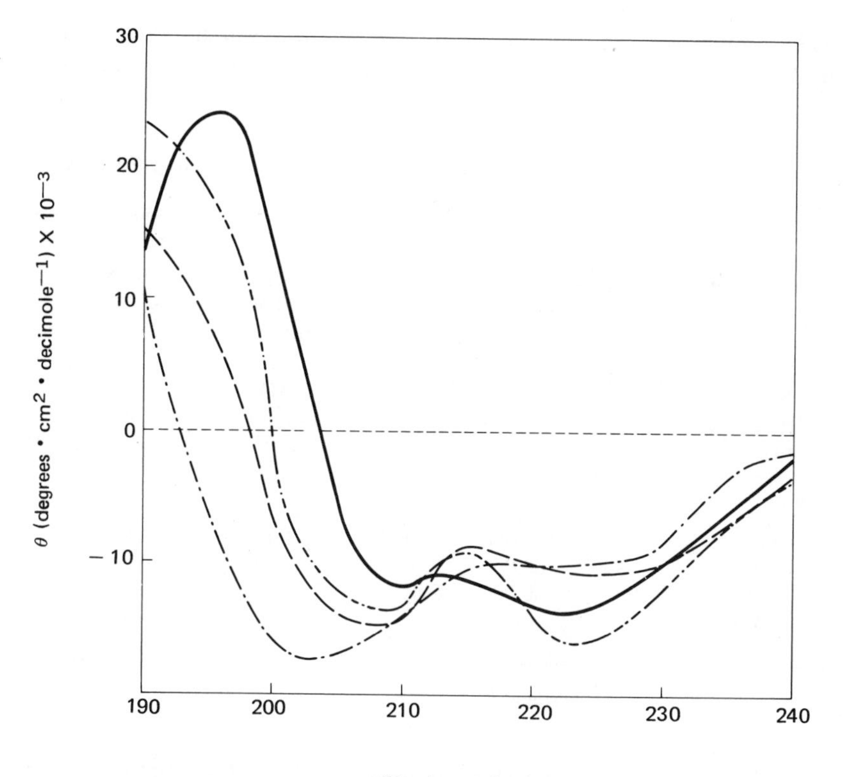

Fig. 6.11 CD of erythrocyte ghosts compared with computed CD of poly-L-lysine.
——— ghosts; - - - - poly-L-lysine, 40% α—helix, rest unordered, solution; ——·—— poly-L-lysine, 40% α—helix, rest unordered, 3500 nm shell radius; ——·—— 50% α—helix, rest unordered, 3500 nm shell radius. See text. [From (14), courtesy National Academy of Sciences.]

and amplitude are difficult to assign precisely. (This is true even with purely helical polypeptides.) However, turbidity effects can account for small red shifts of this minimum (Figs. 6.8 and 6.9), particularly at high helicity. Moreover, they can broaden the "222 nm band" sufficiently to account for the long observed "red shift" of the ORD minimum of membranes.

Other mechanisms can displace the ORD minimum. Particularly important are the established uncertainties in the optical activities of β-structured peptides which may explain the behavior of galactose-4-epimerase. This is an oligomeric protein, whose infrared spectra clearly signal the presence of considerable β-structured peptide (31). Importantly, at low ionic strength, the protein dissociates into monomers, as evident in the analytical ultracentrifuge (Table 6.5), and this process is associated with an *ORD red shift* of 6.0–7.0 nm. This appears due to broadening of the "222 nm" band, and might reflect a change in the proportion and/or type of β-structure. *An ORD red shift, thus, cannot necessarily be assumed unique to membranes or attributed to light scattering.*

Table 6.5 Molecular Weight and Changes in Optical Rotatory Dispersion of Galactose-4-Epimerase with Variations of Ionic Strength and Temperature

Solvent (*Tris*–HCL M pH 7.5)	Temp. (°C)	Mol Wt	$S_{20,w}$[a]	ORD Trough (nm)	$[m']$ ($deg \cdot cm^2 \cdot dmole^{-1}$)
0.1	25	240,000	10.0, 6.3	230	—
0.01	25	120,000	6.3	230	4200
0.001	9	60,000	2.8, 4.3	233	4540
0.001	25	60,000	2.8, 4.3	236	4000
0.001	40	—	—	237	3700

[a] From Bertland and Bertland (91).

Mie computations, however, indicate that the ORD minimum need not shift when large membrane fragments are comminuted into small vesicles by a nondenaturing process. Thus, while erythrocyte ghosts exhibit an ORD trough near 236 nm and ORD crossover at 226.5 nm and the first CD minimum at 223.2 nm, endocytotic disruption (90) of the ghosts into vesicles with diameters ≤ 100 nm, by exposure to low $\Gamma/2$, reduces light scattering, but does not measurably influence the optical activity parameters under study (Table 6.6). Exocytotic disruption of the ghosts into normally oriented vesicles (90) by gentle shear in low $\Gamma/2$ media containing Mg^{2+} also produces no significant change in optical activity (Table 6.6). The two vesicle types appear identical in composition, except that during endocytosis, some of the high-molecular-weight component "spectrin" is extracted. Conformational changes are not excluded even here, but the persistent ORD "red shift" can be explained by the fact, illustrated by our computations, that the scattering-induced widening of the "222 nm" CD band changes negligibly as vesicle radius is decreased from 3500 to 75 nm. This, of course, means that *one cannot unambiguously "correct" membrane* optical activity by reducing particle size (e.g., 67).

It thus appears that the ORD red shift may be due wholly or in

Table 6.6 Optical Activity Parameters of Erythrocyte Ghost as Well as Normally Oriented and Inside-Out Vesicles Derived Therefrom

Sample	ORD Trough (nm)	$[m]$ deg · cm^2 · dmole^{-1}	ORD Crossover (nm)[a]	First CD Minimum (nm)
Erythrocyte ghost	236	6200	226	223.2
Normally oriented vesicles	236	6150	228	223.2
Inside-out vesicles (IO)	235	6540	228	223.2
IO vesicles spectrin	235	4800	228	223.0
Spectrin	233	4000	223	222.5

[a] At $<50\%$ helicity, the position of the ORD crossover in the membranes is not explained by Mie scattering.

part to the large size of membrane particles. However, the other features of membrane optical activity cannot be attributed to light scattering alone. Rather, the interpretation of membrane optical activity suffers from the same uncertainties that plague the use of CD and ORD measurements in the conformational analysis of globular proteins.

Epilogue

The substantial literature on membrane optical activity in many ways mirrors the ambiguities and controversy that plague CD and ORD measurements of proteins in general. In the case of the membrane isolates studied so far, additional complications arise from absorption statistics artifacts and light scattering. In the case of small vesicles, these may cause little more than minor flattening and broadening of the CD bands; with large membrane fragments, more severe artifacts appear.

Despite these severe obstacles, measurements of membrane optical activity have yielded some important structural information, much of it when the method was first introduced into the field. *First,* most membranes contain more than 30% of their peptide linkages in a right-handed α-helical conformation; this is greater than what is found in most globular proteins. *Second,* the proportion of α-helix can often be markedly enhanced by the generally helicogenic solvent 2-chloroethanol. *Third*, membrane optical activity in the accessible spectral regions generally arises predominantly from peptide chromophores, with negligible or minor contributions from amino acid side chains, lipids, carbohydrate, and RNA. However, we note that the first measurement of membrane optical activity by Ke (92) revealed a striking, anomalous optical activity of membrane-associated chlorophyll in chloroplasts.

We anticipate that further informative application of optical activity techniques to membrane structural analysis will require further basic developments as well as some major changes of strategy.

First, as we have stressed before (13, 14) and reemphasize here, synthetic polypeptides cannot be taken as reliable reference substances for conformational analyses of proteins (membrane or soluble). However, the rapid progress in protein X-ray crystallography and the utilization of X-ray analyzed proteins as optical activity standards for soluble proteins (15–17) offers promise also for the membrane field. In this connection, it would appear prudent to join optical activity measurements with other techniques, in particular infrared spectroscopy and, in the future, laser Raman and NMR spectroscopy.

Second, one must account adequately for possible artifacts arising from the particulate nature of the membranes. This requires much more than the sample computations presented here and in Urry (42) and Gordon and Holzwarth (45). Indeed, Mie calculations will need to be performed for particles of diverse sizes, geometries, densities, and refractive indices, and these computations coupled to conformational analyses using proteins as reference substances (17).

With this information available in tables, and knowing particle size, geometry, etc., reasonable conformational analyses may be possible on membrane fragments of all types. The required computations comprise a Herculean task and have not been undertaken. Fortunately, in the case of very small particles, the artifacts are also small, and an approach such as described in Chen et al. (17) may give a reasonable conformational analysis without resorting to turbidity corrections.

Third, optical activity measurements in the regions of peptide absorption require excessive time for kinetic measurements. A large element in this difficulty derives from the low intensity of light sources currently used in this spectral region. Tunable lasers, together with frequency-doubling or quadrupling devices, may be of use here, particularly if the detection system is joined to a small computer.

The lack of high-intensity light sources for the far ultraviolet introduces additional complications, because so many metabolic substrates, cofactors, inhibitors, etc., absorb strongly in this spectral region and cannot be used satisfactorily at their biologically active levels.

In summary, we feel a need for major developments *outside* the membrane field, before optical activity measurements can add much new information to the domain of membrane structure and function.

References

1. Yang, J. T. In *Poly-α-Amino Acids*, G. D. Fasman, ed. New York: Marcel Dekker, 1967, vol. I, p. 239.
2. Beychok, S. In *Poly-α-Amino Acids*, G. D. Fasman, ed. New York: Marcel Dekker, 1967, vol. I, p. 293.
3. Beychok, S. *Ann. Rev. Biochem.* **37:** 437, 1968.
4. Gratzer, W. V., and Coburn, D. A. *Nature* **222:** 426, 1969.
5. Bovey, F. A. In *Polymer Conformation and Configuration.* New York: Academic Press, 1969, ch. IV.
6. Imbert, C. *Phys. Lett.* **31:** 337, 1970.
7. Wellman, K. M., Laur, P. H. A., Briggs, W. S., Moskowitz, A., and Djerassi, C. *J. Am. Chem. Soc.* **87:** 66, 1965.
8. Schellman, J. A., and Schellman, C. In *The Proteins*, H. Neurath, ed. New York and London: Academic Press, 1964, 2nd ed., ch. 7.
9. Tiffany, M. L., and Krimm, S. *Biopolymers* **8:** 347, 1969.
10. Dearborn, D. G., and Wetlaufer, D. B. *Biochem. Biophys. Res. Commun.* **39:** 314, 1970.
11. Fasman, G. D., Hoving, H., and Timasheff, S. N. *Biochemistry* **9:** 3316, 1970.
12. Straus, J. H., Gordon, A. S., and Wallach, D. F. H. *Eur. J. Biochem.* **11:** 201, 1969.
13. Gordon, A. S., Wallach, D. F. H., and Straus, J. H. *Biochim. Biophys. Acta* **183:** 405, 1969.
14. Wallach, D. F. H., Lowe, D., and Bertland, A. U. *Proc. Natl. Acad. Sci.* **70**: 3235, 1973.
15. Saxena, V. P., and Wetlaufer, D. B. *Proc. Natl. Acad. Sci.* (U.S.) **68:** 969, 1971,
16. Rosenkranz, H., and Scholtan, W. *Hoppe Seyler Ztschr. Physiol. Chem.* **352:** 896, 1971.
17. Chen, Yee-Hsiung, Yang, J. T., and Martinez, H. M. *Biochemistry* **11:** 4120, 1972.
18. Pysh, E. *Proc. Natl. Acad. Sci.* (U.S.) **56:** 825, 1966.
19. Rosenheck, K., and Sommer, B. *J. Chem. Phys.* **46:** 532, 1967.
20. Davidson, B., Tooney, N., and Fasman, G. D. *Biochem. Biophys. Res. Commun.* **23:** 156, 1967.
21. Fasman, G. D., and Potter, J. *Biochem. Biophys. Res. Commun.* **27:** 209, 1967.
22. Stevens, L., Townend, R., Timasheff, S., Gasman, G. D., and Potter, J. *Biochemistry* **7:** 3717, 1968.
23. Carver, J. P., Shechter, E., and Blout, E. R. *J. Am. Chem. Soc.* **88:** 2550, 1960.
24. Greenfield, N., Davidson, B., and Fasman, G. D. *Biochemistry* **6:** 1630, 1967.
25. Timasheff, S., Susi, H., Townend, R., Gobunoff, M. J., and Kumosinki, T. F. *Conform. Biopolym. Pap. Int. Symp.*, 1967, p. 133.
26. Greenfield, N., and Fasman, G. D. *Biochemistry* **8:** 4108, 1969.
27. Pflumm, N. N., and Beychok, S. *J. Biol. Chem.* **244:** 3973, 1969.

28. Blake, C. C. F., Koenig, D. F., Mair, G. A., North, A. C. T., and Sarma, V. R. *Nature* **206:** 757, 1965.
29. Birktoft, J. J., Blow, D. M., Henderson, R., and Steitz, T. A. *Phil. Transact. Roy. Soc.* (London), Ser. B. **257:** 67, 1970.
30. Phillips, D. C. *Proc. Natl. Acad. Sci.* (U.S.) **57:** 484, 1967.
31. Bertland, A. U., and Kalckar, H. M. *Proc. Natl. Acad. Sci.* (U.S.) **61:** 629, 1968.
32. Choules, G. L., and Bjorklund, R. *Biochemistry* **9:** 4759, 1970.
33. Graham, J. M., and Wallach, D. F. H. *Biochim. Biophys. Acta* **241:** 180, 1971.
34. Avruch, J., and Wallach, D. F. H. *Biochim. Biophys. Acta* **241:** 249, 1971.
35. Yaron, A., Katchalski, E., Berger, A., Fasman, G. D., and Sober, H. A. *Biopolymers* **10:** 1107, 1971.
36. Perutz, M. *J. Mol. Biol.* **13:** 646, 1965.
37. Perutz, M. F., Muirhead, H., Cox, J. M., and Goaman, L. C. *Nature* (London) **219:** 131, 1968.
38. Stryer, L. *J. Mol. Biol.* **13:** 842, 1965.
39. Malcolm, B. R. *Biopolymers* **9:** 911, 1970.
40. Lenard, J., and Singer, S. J. *Proc. Natl. Acad. Sci.* (U.S.) **56:** 1828, 1966.
41. Wallach, D. F. H., and Zahler, P. H. *Proc. Natl. Acad. Sci.* (U.S.) **56:** 1552, 1966.
42. Urry, D. *Biochim. Biophys. Acta* **265:** 115, 1972.
43. Stoeckenius, W., and Engelman, D. M. *J. Cell Biol.* **42:** 613, 1969.
44. Wallach, D. F. H., Kamat, V. B., and Gail, M. H. *J. Cell Biol.* **30:** 601, 1966.
45. Gordon, D. J., and Holzwarth, G. *Proc. Natl. Acad. Sci.* (U.S.) **68:** 2365, 1971.
46. Gordon, D. J. *Biochemistry* **11:** 413, 1972.
47. Wallach, D. F. H., Graham, J. M., and Fernbach, B. *Arch. Biochem. Biophys.* **131:** 322, 1969.
48. Urry, D. W., Mednieks, M., and Bejnarowicz, E. *Proc. Natl. Acad. Sci.* (U.S.) **57:** 1043, 1967.
49. Steim, J. M., and Fleischer, S. *Proc. Natl. Acad. Sci.* (U.S.) **58:** 1292, 1967.
50. Wrigglesworth, J. M., and Packer, L. *Arch. Biochem. Biophys.* **128:** 790, 1968.
51. Graham, J. M., and Wallach, D. F. H. *Biochim. Biophys. Acta* **193:** 225, 1969.
52. Wallach, D. F. H. *Chem. Phys. Lipids* **8:** 341, 1972.
53. Green, D. H., and Salton, M. R. J. Personal communication (to be published).
54. Scanu, A., and Hirz, R. *Proc. Natl. Acad. Sci.* (U.S.) **59:** 1968, 1968.
55. Ulmer, D. D., Valee, B. L., Gorchein, A., and Neuberger, A. *Nature* **206:** 825, 1965.
56. Gulik-Krzywicki, T., Schechter, E., Luzzati, V., and Faure, M. *Nature* **223:** 1116, 1969.
57. Cassim, Y., and Yang, J. T. *Biochem. Biophys. Res. Commun.* **26:** 58, 1967.
58. Schneider, A. S., Schneider, M-J., and Rosenheck, K. *Proc. Natl. Acad. Sci.* (U.S.) **66:** 613, 1969.
59. Lenard, J., and Singer, S. J. *Science* **159:** 738, 1968.
60. Sonenberg, M. *Biochem. Biophys. Res. Commun.* **36:** 450, 1969.
61. Wallach, D. F. H. *J. Genl. Physiol.* **54:** 35, 1969.
62. Urry, D. W., Masotti, L., and Krivacic, J. *Biochem. Biophys. Res. Commun.* **41:** 521, 1970.
63. Glaser, M., Simpkins, H., Singer, S. J., Sheetz, M., and Chan, S. I. *Proc. Natl. Acad. Sci.* (U.S.) **65:** 721, 1970.

64. Glaser, M., and Singer, S. J. *Biochemistry* **10:** 1780, 1971.
65. Singer, J. A., and Morrison, M. *Biochim. Biophys. Acta* **274:** 64, 1972.
66. Mommaerts, W. F. H. *Proc. Natl. Acad. Sci.* (U.S.) **58:** 2476, 1967.
67. Masotti, L., Urry, D. W., Krivacic, J. R., and Long, M. M. *Biochim. Biophys. Acta* **266:** 7, 1972.
68. Masotti, L., Urry, D. W., Krivacic, J. R., and Long, M. M. *Biochim. Biophys. Acta* **255:** 420, 1972.
69. Holzwarth, G., in *Membrane Molecular Biology*, C. F. Fox and A. Keith, eds. Sinauer Associates, Stamford, Conn., 1972, p. 228.
70. Urry, D. W., and Ji, T. H. *Arch. Biochem. Biophys.* **128:** 802, 1968.
71. Gordon, D. J., and Holzwarth, G. *Arch. Biochem. Biophys.* **142:** 481, 1971.
72. Van den Hulst, H. C. In *Light Scattering by Small Particles*. New York: Wiley, 1957.
73. Aden, A. L., and Kerker, M. J. *J. Appl. Phys.* **22:** 1242, 1951.
74. Dickerson, R. E., Takano, T., Eisenberg, D., Kallai, O. B., Samson, L., Cooper, A., and Margoliash, E. *J. Biol. Chem.* **246:** 1511, 1971.
75. Weissler, A. J. *J. Accoust. Soc. Am.* **32:** 1208, 1960.
76. Joly, M., *A Physicochemical Approach to the Denaturation of Proteins*. New York: Academic Press, 1965, pp. 9–16.
77. Graber, P. *Biol. Med. Physics* **3:** 191, 1953.
78. Hughes, D. E. In *Ultrasonic Energy*, E. Kelly, ed. Urbana, Ill.: University of Illinois Press, 1965, p. 9.
79. Yamoto, H. *Oyakama Igakkai Zasshi* **64:** 874, 1952.
80. Mitayake, T. *Oyakama Igakkai Zasshi* **69:** 461, 1957.
81. Sarkar, P. K., and Doty, P. *Proc. Natl. Acad. Sci.* (U.S.) **55:** 981, 1966.
82. Chapman, D., Kamat, V. B., DeGier, J., and Penckett, S. A. *J. Mol. Biol.* **31:** 10, 1968.
83. Wallach, D. F. H. *J. Genl. Physiol.* **54:** 35, 1969.
84. Dunnick, J. K., Marinetti, G. V., and Greenland, P. *Biochim. Biophys. Acta* **266:** 864, 1972.
85. Reynolds, J. A., and Tanford, C. *J. Biol. Chem.* **245:** 5161, 1970.
86. McDiarmid, R. Ph.D. Thesis, Harvard Univ., Cambridge, Mass., 1965.
87. Greenfield, N., and Fasman, G. D. *Biochemistry* **8:** 4108, 1969.
88. Greenfield, N., and Fasman, G. D. *Biochemistry* **6:** 1030, 1967.
89. Wallach, D. F. H., and Kamat, V. B. *Proc. Natl. Acad. Sci.* (U.S.) **52:** 721, 1964.
90. Steck, T. L., Weinstein, R. S., Straus, J. H., and Wallach, D. F. H. *Science* **168:** 255, 1970.
91. Bertland, L. H., and Bertland, A. U. *Biochemistry* **10:** 3145, 1971.
92. Ke, B. *Nature* **208:** 573, 1965.

CHAPTER 7

FLUORESCENCE, FLUORESCENT PROBES, AND OPTICALLY ABSORBING PROBES

Fluorescence

Excitable molecules can be promoted into an upper *singlet electronic state*,* such as S_2 of Fig. 7.1 by absorption of energy.

The relationship between the wavelength λ (or frequency ν) of light-producing electronic excitation and E, the produced change in potential energy, is

$$E = h\nu = \frac{hc}{\lambda} \tag{7.1}$$

where E is in kcal/mole or electron volts, h is Planck's constant, and c is the velocity of light. Excitation is followed by a rapid (10^{-12} sec) non-radiative conversion to the lowest excited *singlet* state, S_1, which typically has a lifetime of about 10^{-9} sec. Unless the excited molecule undergoes photodecomposition (at a rate k_{ph}), it can return from S_1 to S_0 directly by five competing mechanisms:

(a) fluorescence, i.e., photon emission (rate constant k_e),
(b) resonance transfer of the excitation energy to another chromophore (k_t),
(c) radiationless internal conversion (k_i),
(d) collisional transfer of energy to other solute or solvent molecules by *external conversion* (k_{ext}), and
(e) intersystem crossing (k_{cr}).

These various processes can be summarized as follows.

* One must distinguish electronic state from *electronic orbital*. The former refers to *all* electrons, while an orbital describes the volume in which there is a 99.9% probability of finding an electron.

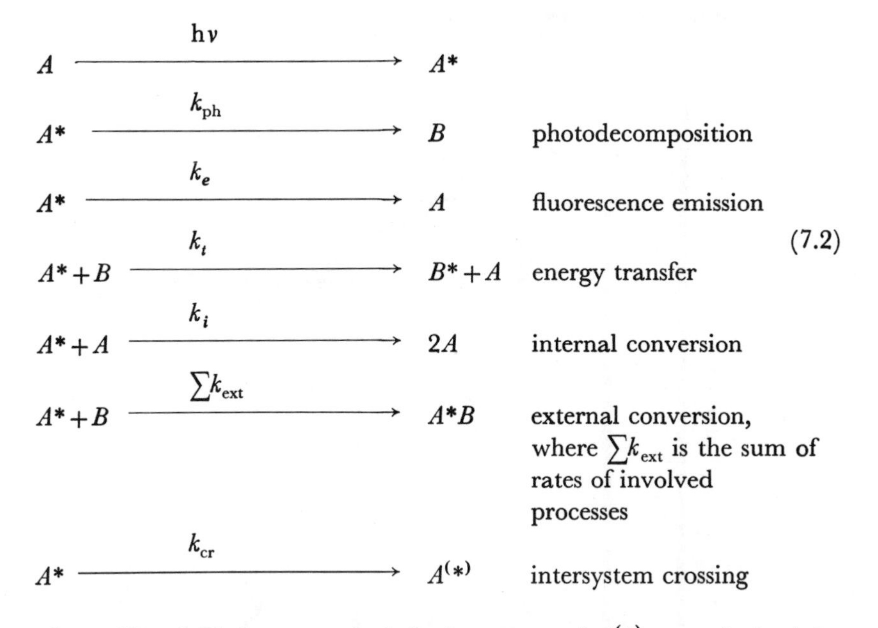

where A^* and B^* denote excited singlet states and $A^{(*)}$ an excited triplet state. The rates of these processes are sensitive to the physical and chemical environment of the fluorophore. For this reason, the fluorescence characteristics of intrinsic protein fluorophores (phenylalanine, tyrosine, tryptophan) and of adsorbed or convalently linked extrinsic fluorophores depend upon the architecture of the protein.

In some systems, the S_1 state (paired electron spins) may occasionally transfer to the triplet state T (spins unpaired, rate constant, k_{cr}). Photodecomposition may also occur from this state, but commonly the molecule returns to the ground state S_0 by (a) photon emission, here phosphorescence, (b) energy transfer, and (c) radiationless dissipation of energy. $T \rightarrow S_0$ transitions occur with low probability, because of the different electron spins of the two states. Hence T has a lifetime of 10^{-1} to 10^{-3} sec compared to 10^{-9} sec for S_0. The long lifetimes of T make energy dissipation by collision with other molecules probable, and phosphorescence is, thus, rarely observed in other than rigid media. This fact suggests that it may be a useful tool in the study of membranes.

Fluorescence measurements involve several different experimental techniques.

(a) *Fluorescence excitation spectra* represent the intensity of fluorescence as a function of the wavelength of the incident light. When appropriately corrected for instrumental factors and in cases of pure solutions of single fluorophores, the excitation spectra correspond to absorption spectra. They also show the same response to solvent characteristics as do ab-

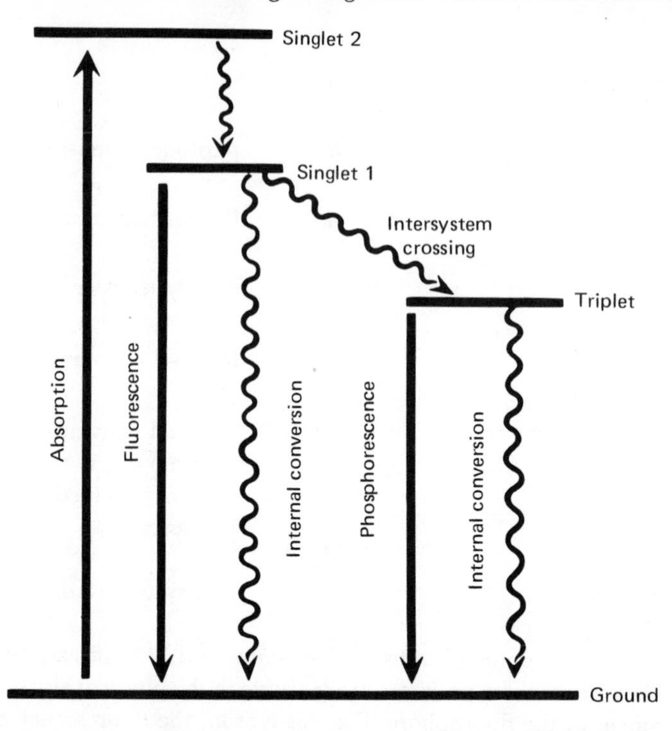

Fig. 7.1 Schematization of some basic phenomena involved in fluorescence and phosphorescence. The ordinate represents potential energy (in electron volts). We depict four electronic energy states: the ground state, a high-energy singlet state (2), the lowest-energy singlet state (1), and the lowest-energy triplet state. We represent quantum absorption and emission by solid arrows and nonradiative energy dissipation by ⇝. Nonradiative energy transfer is not shown. We give further details in the text.

sorption spectra. However, the sensitivity of fluorescence measurements is about 100 times that of absorbance spectroscopy.

(b) *Fluorescence emission spectra* give fluorescence intensity as a function of emission wavelength and represent the $S_1 \rightarrow S_0$ transition. Fluorophores in the excited state generally interact more with vicinal solvent molecules than do those in the ground state. Hence emission spectra reflect the environment of a fluorophore in a distinctive and sensitive fashion.

(c) *The quantum yield of fluorescence* is defined as Q = quanta emitted/quanta absorbed. The *maximum* value of Q is *one*, a condition found in several organic dyes. In terms of the above-noted rate constants,

$$Q = \frac{k_e}{k_e + k_{\text{ph}} + k_t + k_i + \sum k_{\text{ext}} + k_{\text{cr}}} \tag{7.3}$$

Because of nonradiative energy losses between absorption and emission, the *energy* yield of fluorescence is always less than one. The quantum yield depends upon the amount of internal conversion and "fluorescence quenching" or external conversion. The various external conversion mechanisms involve the rates and types of reaction of the excited molecule with solvent or other solute molecules; the quantum yield may, therefore, reflect the condition of the environment of the fluorophore. Also, Q clearly depends upon the duration of the excited state.

(d) *Electronic states* can be broadly categorized into *singlet* and *triplet* states (Fig. 7.1). In the former, *all* electrons have paired spins; in the latter *all but two* have paired spins.

In organic molecules, the important singlet transitions are $n-\pi^*$ and $\pi-\pi^*$. Both may generate optical activity as noted elsewhere. In a $n-\pi^*$ transition, an electron in a nonbonding orbital is promoted to an excited (π^*) bonding orbital. In a $\pi-\pi^*$ transition, a ground state π electron is excited to a vacant π orbital of higher energy (π^*).

An important, intrinsic aspect of fluorescence is the fact that the molecular geometry of an excited state can differ dramatically from the ground state. For example, the C—H bond angle of excited acetylene is 120°, compared with 180° in the ground state. Also critical is the nature of the lowest excited state, S_1, because this determines the fluorescent behavior, e.g., Q, of a molecule. Molecules, where the lowest singlet excited state corresponds to an $n-\pi^*$ energy difference, often have low values of Q, because of the small energy difference between their S_0 and the lowest triplet state. This, as well as the long lifetimes of such excited states, facilitates singlet $\rightarrow$ triplet conversion.

(e) *The lifetime of the excited state* τ_0 is defined as the time necessary for the fluorescence intensity to fall to $1/e$ of its initial value. Here one assumes that fluorescence is a first-order decay process, described by

$$A^* = A_0{}^* e^{-t/\tau_0} \tag{7.4}$$

where $A_0{}^*$ is the number of excited molecules at the instant excitation ceases, and A^* is the number remaining at any time, t, later. It requires about 10^{-15} sec to absorb or emit a photon. The relaxation time of small molecules in organic solvents is near 10^{-12} sec. In contrast, the natural lifetimes τ_0 for $n-\pi^*$ and $\pi-\pi^*$ transitions are 10^{-7} and 10^{-9} sec, respectively. The actual proportion of molecules in the excited state at any one instance depends on many variables, but is always small—about 10^{-13}.

(f) *Polarization of fluorescence*—irradiation of a randomly oriented population of fluorophores with plane-polarized exciting light preferentially activates those molecules, whose transition dipoles are parallel to the electric vector of the incident light, in a "photoselection" process.

Unless the dipoles reorient during the lifetime of the excited state, the emitted light has the same polarization plane as the exciting light (cf. infrared polarization, p. 147). Usually, rotational diffusion of excited fluorophores tends to randomize the orientations of the emission dipoles, depending upon (1) the lifetime of the excited state, (2) the viscosity of the "solvent" near the fluorophore, and (3) in the case of protein-bound fluorophores, the flexibility of the macromolecule and its rotational diffusion.

Fluorescence polarization p is defined as follows:

$$p = \frac{I_{\parallel} - I_{\perp}}{I_{\parallel} + I_{\perp}} \tag{7.5}$$

where $I_{\parallel}$ and $I_{\perp}$ are the intensities of light emitted with its polarization plane parallel ($\parallel$) and perpendicular ($\perp$) to that of the exciting beam. Fluorescence polarization measurements, particularly with nanosecond excitation pulses, can give considerable insight into processes (a) to (c).

Intrinsic Fluorescence

Biological membranes exhibit an inherent fluorescence due to the aromatic amino acids of membrane proteins (Table 7.1). The fluorescence yields of the amino acids in proteins is generally lower than that of the

Table 7.1 Excitation and Emission Maxima of Free Aromatic Amino Acids

	λ max (nm)		
Amino Acid	Excitation	Emission	Q[a]
Tryptophan	278 220 196	350	0.20
Tyrosine	274 222 192	303	0.21
Phenylalanine	258 206 187	282	0.04

[a] Quantum yield.

free compounds, primarily because of collisional quenching by certain other side chains. Moreover, in most globular proteins, the emitted fluorescence arises primarily from tryptophan, and energy transfer from tyrosine to tryptophan is common. The excitation spectra of aromatic amino acids, like their absorption spectra, are solvent sensitive and may thus exhibit typical "denaturation blue shifts." The emission spectrum of tryptophan shifts to the blue with decreasing environmental polarity, but that of tyrosine is relatively solvent insensitive.

Intrinsic fluorescence has been little employed in membrane studies outside of energy-transfer experiments and deserves greater attention, since aromatic amino acids can act as "intrinsic probes."

Wallach and Zahler (1) reported the fluorescence spectra of Ehrlich ascites tumor plasma membranes (Table 7.2), which resemble those of most globular proteins. The emission maximum of the tryptophans is shifted 15 nm to the blue, indicating that they lie in apolar regions. This is not perturbed by lysolecithin, and it is considered to reflect the amino acid environment of the fluorophore. The effect of 2-chloroethanol is important, since concurrent optical activity studies indicate that this solvent markedly raises the proportion of α-helix. This conformational change broadens the tryptophan emission peak to the red, suggesting that some of these residues have been transferred to a more polar environment. This is consistent with the broadening of the excitation band to the blue. Tyrosine fluorescence is also detectable under these conditions, perhaps due to diminished tyrosine $\rightarrow$ tryptophan energy transfer.

Initial studies on the intrinsic fluorescence of erythrocyte ghosts at neutral pH (2) yielded the same pattern as the tumor membranes

Table 7.2 Fluorescence Excitation and Emission Maxima of Ehrlich Ascites Carcinoma Plasma Membranes Under Diverse Solvent Conditions[a]

	λ_{max}(nm)	
Conditions	Excitation	Emission
pH 8.2, 20 mM Tris HCl	277–278	335
pH 8.2, 20 mM Tris HCL +lysolecithin	277–278	335
0.1 N HCl	277–278	335
2-chloroethanol: water (9:1; v/v)	272–277	335–342, 310[b]

[a] From (1).
[b] Shoulder.

described. However, Sonenberg (3), using more advanced techniques, discerned distinct tyrosine fluorescence at 303 nm. Importantly, he found that pure *growth hormone decreased tryptophan* fluorescence by about 20% at a concentration of only 70 *molecules/ghost*. The effect was maximal upon excitation at 225 and 282 nm, but slight with excitation at 295 nm. Fluorescence polarization decreased concurrently from 0.34 to 0.25. The effect is species specific and restricted to pHs near neutrality.

Occurring with such minute amounts of hormone and producing a concurrent change in circular dichroism (4), the phenomenon suggests a cooperative change of membrane protein conformation. The fact that the spectral changes are minimal at 292 nm, where tryptophan alone is excited, suggests diminished tyrosine → tryptophan energy transfer.

Optically Absorbing Probes

The absorption spectra of various dyes change when they are incorporated into micelles of amphipathic molecules. This fact has long been used to measure critical micelle concentrations (5). Indeed, the altered fluorescence of various dyes has been used for the same purpose (6). More recently, Traüble has observed a large increase in the extinction coefficient of bromthymol blue upon its interaction with dispersed lecithin (7). As expected, distinct absorption changes accompany temperature-induced phase transitions. However, one must note that bromthymol blue is as likely to perturb the structure of a phosphatide bilayer as is the fluorescent dye ANS (8) and that the data obtained are, thus, ambiguous.

Fluorescent Probes*

Solution Studies

Role of polarity

Long-standing evidence indicates that the spectral properties of many fluorescent substances vary with their local solvent environment, a fact extensively used to study the monomer → micelle transitions of amphiphilic molecules in water (6).

A more recent pharmacological study (9) shows the fluorescence yield of *berberine* to be solvent dependent. The drug is virtually nonfluorescent in water, but solution in a homologous series of alcohols

* A new review presenting some interesting insights into the application of fluorescence probes to membrane studies has appeared recently (G. R. Radda and G. Vanderkooi, *Biochim. Biophys. Acta* **265:** 509, 1972).

engenders a progressive rise of its fluorescence with increasing chain length of the alcohol. Despite their pharmacological interest, substances such as berberine have not been widely employed as fluorescent probes, and much greater interest has been focused on naphthalene derivatives such as 1-anilino-8-naphthalene sulphonate (ANS), 2-*p*-toluidinylnaphthalene-6-sulphonate (TNS) and their isomers.

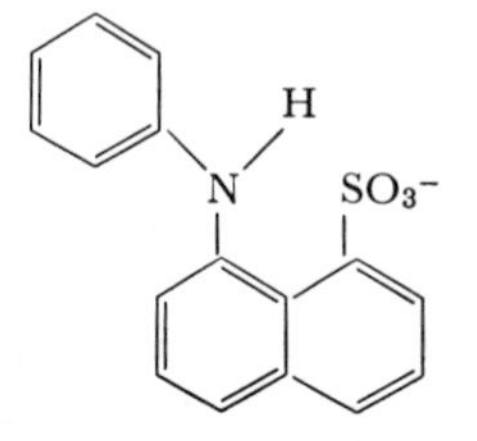

1-*anilino*-8-*naphthalene sulphonate*

The fluorescence efficiency Q and the position λ_{max} of the emission maximum of ANS and its isomers vary with solvent composition. In water $Q = 0.004$, and λ_{max} is at 515 nm. Q rises to 0.15 in ethylene glycol, 0.22 in methanol, 0.37 in ethanol, 0.46 in *N*-propanol, 0.56 in *N*-butanol, and 0.63 in *N*-octanol; while λ_{max} is 484, 476, 468, 466, and 464 nm, respectively. The fluorescence of ANS in ethanol-water mixtures follows the above pattern (Table 7.3).

Table 7.3 Fluorescence of ANS in Ethanol-Water Mixtures[a]

Volume Percent Ethanol	Emission Maximum (nm)	Absolute Quantum Yield
100	468	0.37
90	477	0.24
80	479	0.13
70	485	0.089
60	489	0.062
50	492	0.039
40	497	0.024
30	500	0.013
20	505	0.007
10	510	0.004
0	515	0.004

[a] From Stryer (10).

Between 100 and 80% ethanol, Q decreases by 68%, but λ_{max} shifts only 23%. It appears that Q is more sensitive to small amounts of polar solvent than is λ_{max}. Q signals local deactivation, whereas λ_{max} reflects the overall dipolar character of the solvent (10, 11). Concordantly, λ_{max} is the same in H_2O and D_2O, but Q_{D_2O} is three times greater than Q_{H_2O} (11).

The major factors affecting ANS fluorescence are the viscosity, polarity, and polarizability of its microenvironment. Increasing viscosity presumably reduces deactivation of the excited state occurring through the relative motion of the phenyl and naphthyl rings (12). This effect clearly also occurs when the probe is bound to a macromolecule.

A most important determinant of ANS fluorescence is the *solvation of the excited* state, which depends on the microscopic dielectric constant of the solvent. The long-wavelength absorption band of ANS is a $\pi-\pi^*$ transition, in which the excited state is more polar than the ground state, i.e., polar solvents interact more readily with the excited molecules. This decreases their potential energy and therewith the energy difference between excited and ground state. The emission wavelength is, thus, shifted to the red. Also, much energy is dissipated by nonradiative means, accounting for the low quantum yield in polar solvents.

The fluorescence of ANS in 8:2 dioxane-water mixtures is not affected by the chlorides of potassium, sodium, cesium, calcium, magnesium, and lanthanum, although these salts alter the fluorescence yield of ANS added to aqueous dispersions of *phosphatides* (13).

TNS behaves rather like ANS (14): Q increases and λ_{max} shifts blue with decreasing polarity in a series of primary alcohols. In water, 20% sucrose, 60% sucrose, and 20% polysucrose (Ficoll), TNS yields Q values of 0.0008, 0.005, 0.015, and 0.085, respectively, and λ_{max} lies at 500 nm, 500 nm, 486 nm, and 465 nm, respectively. The effects of high levels of sucrose and polysucrose are attributable to their viscosities. Frozen solutions of TNS have high fluorescence yields. The spectral properties of many fluorophores, other than those already discussed, also respond to environmental polarity, but these substances have been less well studied, e.g., ethidium bromide (15, 16), 2-(*N*-methylanilino-)naphthalene-6-sulfonate (MNS) (17).

Efforts to design fluorescent probes for membranes continue (e.g., 18). Important new substances include 12-(9-anthroyl)-stearate (AS), dansylphosphatidylethanolamine (DPE), *N*-nonadecylnaphthyl-2-amino-6-sulfonate (ONS) (Fig. 3.5). These can be codispersed with egg yolk lecithin, their mobility measured by nanosecond fluorescence, and their environment inferred from their spectra. The latter indicate that the fluorophores of AS, DPE, ONS lie in the apolar core, glycerol region, and polar faces, respectively, of the lipid micelles. The lifetimes of all probes is ~12n sec.

The rational design of significant probes is welcomed and offers many potential advantages. However, previously noted perturbations introduced by such substances cast a shadow on their ability in evaluating the structure of *biomembranes*.

The above studies, showing the dependence of the fluorescent properties of various organic substances on the dielectric and viscous properties of the solvent, underlie most applications of fluorescent probes to biological molecules. However, other fluorescence phenomena—polarization, energy transfer, "excimer" fluorescence—have potentially greater utility and specificity.

Volume probes

So called "excimer" fluorescence requires a high local concentration of mobile fluorophores; substances capable of "excimer" fluorescence can, thus, act as "volume indicators" (17). Many organic substances exhibit such fluorescence, e.g., aromatic hydrocarbons (19, 20), ketones (21), nucleosides, nucleotides, and nucleic acids (22–24), tryptophan, tyrosine, and phenylalanine (25). The report of Keleti (25) clearly demonstrates the typical pattern of "excimer" fluorescence. At low concentrations (<50 μg/ml), tyrosine excites between 283–290 nm and emits between 305 and 312 nm. At higher concentrations, self-quenching becomes apparent and normal tyrosine fluorescence is no longer detectable at 10 mg/ml. However, concentrated solutions of tyrosine (1–10 mg/ml) can be excited at 330–340 nm to emit at 405–420 nm. "Excimer" fluorescence of tryptophan is detected with concentrations of 1–5 mg/ml and is characterized by excitation at 385 nm and emission at 450 nm. Brocklehurst et al. (17) have introduced pyrene-3-sulphonate as an "excimer"-forming fluorescent probe for membrane studies.

Interaction of anilinonaphthalene sulfonates with peptide and related ionophores

The formation of aqueous alkali cation complexes by ionophores such as valinomycin, nigericin, and alamethicin markedly enhances ANS and TNS fluorescence (26) with concurrent displacements of emission wavelength to the blue. Only alamecithin-enhanced fluorescence in the absence of the cations and a number of ionophores, including the synthetic "crown" type, yielded no response with cations. In the responsive situations, fluorescence intensity varied as a function of cation concentration, at a given antibiotic and fluorophore level. From this relative cation affinities were computed for alamecithin ($Na^+ \cong K^+$), nigericin ($K^+ > Rb^+ > Na^+ > Cs^+$), and valinomycin ($Rb^+ > K^+ > Cs^+$). The reaction of the complexed ionophores with ANS and TNS also increased the fluorescence lifetime and polarization of the fluorophores. Application of this approach to biological systems should yield very useful information, particularly since fluorescence methods can discern very rapid reactions.

Interactions of anilinonaphthalene sulfonates with other small molecules

Naphthalene sulfonates exhibit fluorescent enhancement and blue shifts of emission not only upon binding to micelles macromolecules or ionophores, but also upon interaction with certain small molecules. This has been shown for ANS and lysolecithin, below its critical micelle concentration (12), and for *N*-methyl, *N*-phenyl-2-aminonaphthalene-6-sulfonate, which forms a 1:1 complex with the cyclic sugar, cycloheptaamylose (27). This fact suggests caution in attributing all fluorescence spectral changes of a membrane system solely to its aggregated lipids and/or proteins. This is true even for polarization studies, which cannot necessarily distinguish between a probe in a very fluid environment and one bound to a rapidly tumbling small molecule.

Lipid Systems

Mono- and multilayer experiments

The most fundamental applications of fluorescence techniques to membrane-related systems include the studies of Kuhn and associates on the transfer of electronic excitation energy between chromophores fixed in lipid mono- or multilayers (28). In many of their experiments, these workers used ionic cyanine dyes, conjugated with long hydrocarbon chains, as both energy donors and acceptors; A and B, shown below, form a typical donor-acceptor pair.

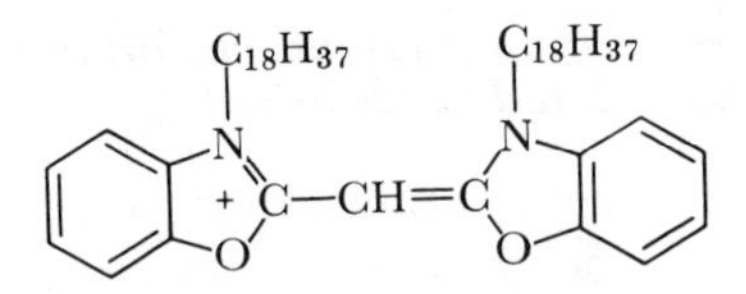

A: Donor—blue fluorescence (λ_{max} = 450 nm)

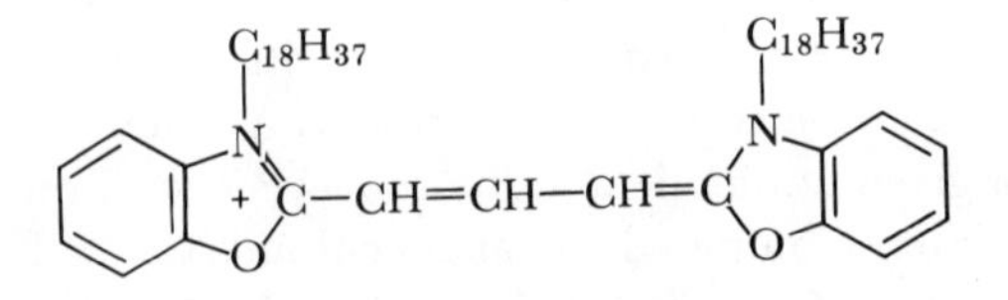

B: Acceptor—yellow fluorescence (λ_{max} = 520 nm)

These substances were chosen because they readily form monolayers at air-water interfaces, whence they can be deposited on solid surfaces as mono- or multilayers by the Blodgett technique (29). By suitable variations of this method, donor and acceptor layers can be deposited alternatingly and with varying spacings in a multilayer system. The dyes were generally used in conjunction with a 20-carbon cadmium or barium soap (cadmium or barium arachidate), serving to dilute the chromophores in a given monolayer and insuring the spacing between monolayers. The donor and acceptor layers could be deposited so that the two chromophore types were (a) contiguous, (b) separated by the length of one hydrocarbon chain (about 27 A), (c) separated by the length of two hydrocarbon chains, and (d) separated by greater distances by the use of spacer layers of barium or cadmium arachidate (Fig. 7.2).

By covering a monolayer of A with n monomolecular layers of barium or cadmium arachidate ($n = 2, 4, 6, \ldots$), before applying a layer of B,

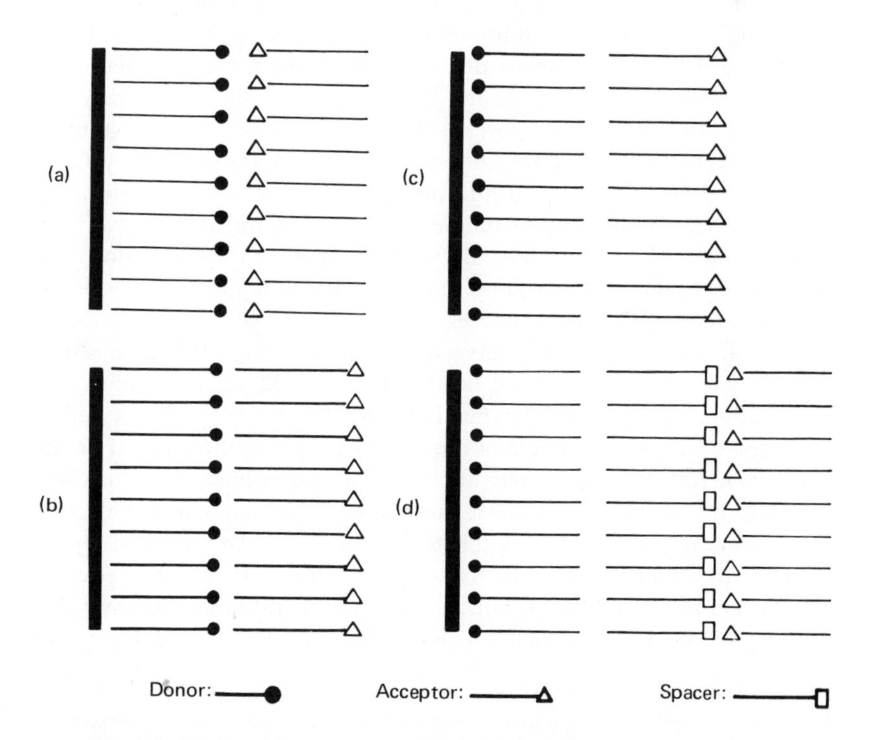

Fig. 7.2 Schematic of the experiments on energy transfer in lipid multilayers utilized by Kuhn and associates. The heavy line represents the glass substrate for the multilayers. In (a), donor and acceptor are closely spaced; in (d), they are far apart; (b) and (c) represent intermediate situations. See text for details.

Kuhn et al. (28) determined the efficiency of energy transfer from A to B as a function of the distance between the chromophore layers. With $n = 40$ (distance between A and B = 27 nm), excitation of A yields the fluorescence of A only. However, at lesser distances, there is increasing emission at 520 nm, due to B, and, when $n = 2$, the fluorescence of A is almost fully quenched by radiationless transfer of excitation energy to B.

The experiments of Kuhn et al. have validated and extended Förster's mathematical formulations for energy transfer (30) and continue to elucidate the properties of very thin films and their interactions with electromagnetic radiation. Frommherz, working in Kuhn's laboratory, recently extended this work to model studies of lipid-protein interactions (31). He developed new methods, permitting the adsorption of hemoglobin and several enzymes into lipid monolayers at air-water interfaces, and the deposition of the pure lipid-protein film complexes upon solid surfaces. The technique permits accurate determination of the number of protein molecules adsorbed per unit area of film surface. Interestingly, under the conditions employed, only one protein layer was adsorbed and this without denaturation of the protein, penetration of the lipid layer by apolar amino acid residues being unnecessary for adsorption. Adsorbed enzymes were still catalytically active. Importantly, Frommherz demonstrated transfer of electronic excitation energy from appropriate dyes incorporated into the lipid layer, to the heme chromophores of hemoglobin and catalase; this energy transfer obeyed the distance rules of Förster (30). Because of its obvious pertinence, the approach of Kuhn and associates and related techniques are now being applied to models of biomembranes.*

Ullrich and Kuhn (31) have also utilized compound B to study photoelectric effects in artificial lecithin bilayers (32) and extending the investigations of Tien and Verma (33) and Verma (34). The cyanine dye is adsorbed at one surface of the bilayer at a concentration of about 1 molecule/200 A^2. Upon activation of the fluorophore, every excited molecule, B*, accepts an electron from an electron donor in the electrolyte solution on the dye side within about 10^{-3} sec. An adsorbed dipole B^-E^+, with a lifetime of $\sim 10^{-1}$ sec, results. During its lifetime, electrolyte charges are attracted to both sides of the membrane, charging this as a capacitor. Upon sufficiently long exposure to exciting light, the

* The development of sophisticated new techniques has permitted measurement of the fluorescent characteristics of ANS bound to black lipid films (H. P. Zingsheim and D. A. Haydon, *Biochim. Biophys. Acta*, **298:** 755, 1973). The absorption of the dye could be measured thermodynamically, allowing evaluation of its quantum yield *in situ.* The emission spectra and quantum yield obtained with the films were in accord with data obtained from sonicated lipid dispersions. The data suggest that ANS localizes near the water-film surfaces.

charge difference is dissipated across the membrane resistance. Related studies have been carried out with chlorophyll (35–37).

When a monolayer of chlorophyll is spread at an air-water interface, the angular distribution of fluorescence indicates that the transition dipole of the red absorption band lies at an angle 20° to the water plane, while that of the blue excitation band $\simeq 28°$. Accordingly, the molecules must orient with the planes of their aromatic residues at an angle to the air-water interface, with the apolar phytyl chains oriented into the air. In mixed monolayers of chlorophyll a and copper pheophytin, energy transfer occurs in accordance with Förster's equations.*

The fluorescence properties of flavine- , mono- , and dinucleotides incorporated into 25μ, transparent methyl-cellulose films suggest that the molecules lie in cavities with a lower polarity and greater microscopic viscosity than water.

The fluorescence parameters, Q, λ_{Em}, and efficiency of energy transfer of ANS incorporated into various protein-lipid-water models depend markedly upon phase structure, determined by X-ray, and the nature of the lipid-protein interaction (electrostatic vs. apolar) (38). In the lamellar phase of lipid-water systems, ANS appears to lie as described above. For the studied systems where protein (lysozyme) binds to the lipid by electrostatic interactions, ANS localizes in the lipid or at the protein-lipid interface in all phases. In systems where apolar protein-lipid interactions predominate, ANS appears to localize at the areas of hydrophobic contact. The models studied are very pertinent to the study of biomembranes. Unfortunately, they depend heavily on the fluorescent parameters of ANS, which do not mirror the perturbing effect of the dye.

Lipid dispersions—gangliosides (sialoglycosphingolipids)

The fluorescent probes most widely applied to lipid systems have been aniline naphthylsulfonates, but before their introduction into this area, Albers and Koval (9) examined the fluorescence yield and polarization of the neuroactive bases, *d-tubocurarine* and *berberine*, as well as of the bases rhodamine and acridine orange, complexed to ganglioside micelles in aqueous dispersion. (The micellar weight, $\sim 2.10^5$, suggests about 133

* The visible absorption spectra of chlorophyll can now be measured on phosphatide bilayers doped with the pigment (R. J. Cherry, K. Hsu, and D. Chapman, *Biochim. Biophys. Acta* **267:** 512, 1972). On the assumption that the transition moments of the main red and blue absorption bands lie parallel to the plane of the chlorophyll porphyrin, and are mutually perpendicular, the porphyrin moiety appears tilted at about 48° to the bilayer plane.

Related photodichroic methods have been employed in studying the rotational diffusion of rhodopsin-digitonin micelles (L. Strakee, *Biophys. J.* **11:** 728, 1971). It appears that in this system transient photodichroism arises solely from rotational diffusion of pigment moieties.

lipid molecules per micelle.) All of the bases increase fluorescence when complexed to the acidic gangliosides, and correlations with solvent effects suggest that the bases enter into regions of low dielectric in the ganglioside micelles.

Enhancement of ANS fluorescence by sonicated dispersions of gangliosides is negligible in 0.01 M Tris HCL, pH 7.4 (39). Ca^{++} has little effect, but $1–5 \times 10^{-3}$ M butacaine increases ANS fluorescence about 50 fold. The fluorescence yield also rises sharply below pH 3; this is also true for phosphatide dispersions and erythrocyte membranes.

The blue-shifted, enhanced fluorescence emitted by the bound ANS indicates that the dye molecules lie in weakly polar regions of ganglioside micelles. However, the effects of butacaine and pH show that electro-

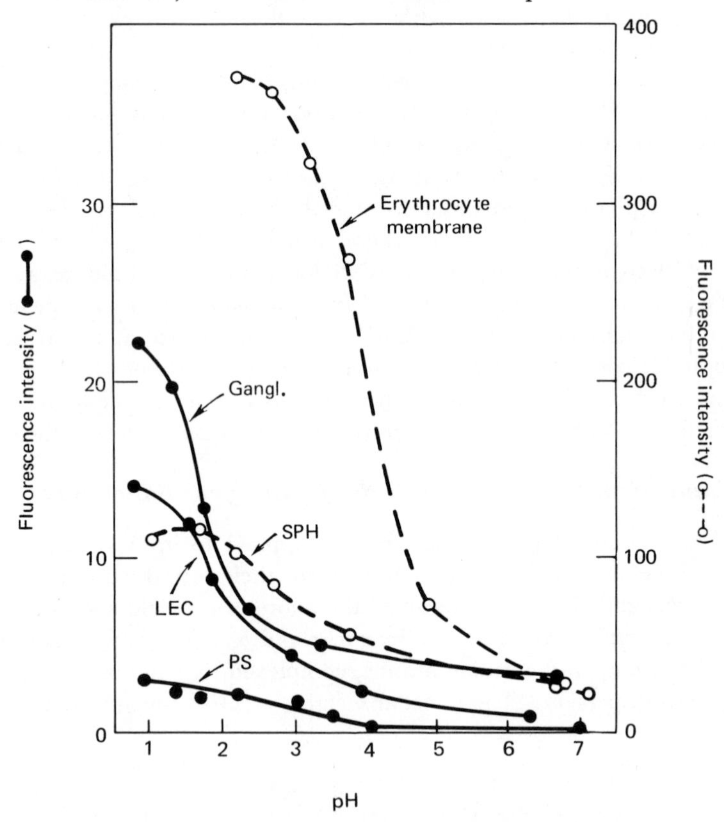

Fig. 7.3 Effect of pH on ANS binding to erythrocyte membrane (0.119 mg protein/ml) ganglioside (GANGL), sphingomyelin (SPH), egg lecithin (LEC), and phosphatidylserine (PS). The lipids were at a concentration of 0.05 mg/ml except for ganglioside, which was present at 0.1 mg/ml because of its higher molecular weight. ANS concentration was 8.5 μM. [From Feinstein et al. (39). Courtesy of North Holland Publishing Co.]

static repulsions between ANS anions and the anionic sialate head groups of the lipid can hinder dye penetration into these regions. The pH dependence of the enhancement of ANS fluorescence by phosphatides and erythrocyte membranes is clearly distinct from the ganglioside case; we shall return to this.

Phospholipids

In the study just cited, Feinstein et al. (39) also compared the fluorescence enhancement of ANS by sphingomyelin, lecithin, and phosphatidyl serine, all in aqueous dispersion, with the effect of gangliosides. The acidic phospholipid, phosphatidylserine, behaves very much like the gangliosides. Sphingomyelin, however, strongly enhances fluorescence even at neutral pH, with only moderately further increases by Ca^{++} and butacaine. With all of the lipids, the enhancement of ANS fluorescence is markedly augmented at low pH (Fig. 7.3).

The enhancement of ANS fluorescence by phosphatides was also examined by Vanderkooi and Martonosi (13), using 5×10^{-5} M ANS, and 1–10 mg sonicated lipid per milliliter. All lipids tested shifted the emission maximum from 510 nm to 470 nm. Fluorescence enhancement followed the sequence lysolecithin > lecithin > phosphatidylethanolamine > phosphatidylserine. In all cases, enhancement was greater at higher ionic strengths and/or in the presence of multivalent cations. Titrations to low pH suggest that protonation of lipid phosphates facilitates penetration of ANS into hydrophobic regions of the phosphatide micelles. Interestingly, the pH-fluorescence relationships of the various phosphatide dispersions differed significantly from that of muscle microsomes; these have a sharp inflection at about pH 3.3.*

The excitation spectra of ANS-lysolecithin complexes change with the micellar transition of this phosphatide (2). Moreover, below the critical micelle concentration, the *emission* maximum of the ANS-lysolecithin complex was at 500 nm, whereas it lay at 485 nm with micellar lysolecithin and at 475 nm with micellar lecithin.

* Alteration of the surface charge of phosphatide liposomes by pH manipulation, introduction of charged components, and variation of divalent cation concentration influences ANS binding (M. T. Flanagan and T. R. Hesketh, *Biochim. Biophys. Acta* **298:** 535, 1973). The more negative the surface charge, the less the ANS binding. In contrast, binding of the neutral dye N-phenyl-l-napthylamine is not influenced by charge manipulations, but is inhibited by addition of cholesterol, suggesting a competition between this type of probe and the steroid. The effect of the divalent cations appears to depend upon their binding of phosphate residues. ANS binding to sphingomyelin liposomes is temperature dependent and suggests a broad phase transition above 32°C. The temperature dependence is similar to that observed during the crystalline to liquid-crystalline transition of dipalmitoyllecithin. With rising temperature the ANS fluorescence increases as the lipid enters the transition region and drops once the hydrocarbon chains are in the liquid-crystalline state (H. Träuble, *Naturwissenschaften* **58:** 277, 1971).

Detergents

About 1 ANS molecule binds per micelle of neutral (Triton X-100) or cationic (cetyltrimethylammonium bromide) detergent, independent of ionic strength (40). The same study claims that enhancement of ANS fluorescence by *anionic* micelles occurs only at high ionic strength, but this is contrary to the data given in Fig. 5 of (40).

The fluorescence of TNS is not significantly altered by 0.006 mg/ml sodium dodecyl sulfate (41), but this detergent level is below the critical micelle concentration. Hasselbach and Heimberg (42) found the enhancement ratios for ANS by aqueous dispersions of lecithin:cetyltrimethylammonium bromide:oleylamine to be 1:1.76:2.70. They suggest that ANS binding and fluorescence enhancement would be most favorable at a positively charged site in an apolar environment.

In contrast to ANS, the cationic ethidium bromide fluoresces more efficiently in the presence of anionic than neutral detergent micelles (16, 43). With sodium dodecyl sulfate, the greatest rise in fluorescence with detergent concentration occurs near the critical micelle concentration.

Charged probes such as ANS, TNS, and ethidium bromide can bind to micellar dispersions of gangliosides (39), various phospholipids (e.g., 2, 13, 39, 44), and diverse detergents (e.g., 16, 40, 42, 43), provided this is not countered by electrostatic repulsions. This matter is discussed in some details in (44), which shows that with suitable control of ionic strength and composition, ANS can act as a versatile monitor of ion binding, its intrinsic fluorescence properties being independent of ionic environment.

Orientation studies on lipid dispersions

Most experiments on the interactions of fluorophores with "model membranes" utilize aqueous dispersions of phospho-, glyco-, and other lipids in predominantly micellar arrangements, but often of indeterminate phase composition. One can expect that the phase of a lipid influences its interaction with and the fluorescent properties of a fluorophore; moreover, adsorption of the dye probably alters lipid packing. These matters have not been extensively explored, and studies on micellar systems are, therefore, less interpretable than the experiments of Kuhn and associates.*

* The fluorescene polarization of the fluorophore perylene can be used to monitor thermotropic lipid phase transitions in vesicles of dipalmitoyllecithin and dipalmitoylphosphatidylglycerol (D. Papahadjopoulos, K. Jacobson, S. Nir, and T. Isac, *Biochim, Biophys. Acta* **311:** 330, 1973). In both cases fluorescence polarization diminishes with increasing temperature, with an inflection centering near 38°C. However, fluorescence intensity increases abruptly near this temperature. It is proposed that below the transition temperature, the fluorophores cannot be accommodated in the hydrocarbon regions of the lipid vesicles, whereas they can enter these domains above the transition temperature.

A different approach to the study of fluorescent probes in oriented lipid systems utilizes large, water-filled spherules, bounded almost completely by a bilayer of oxidized cholesterol (45). The following fluorophores were incorporated at a ratio of 1 molecule: 1000 cholesterols.

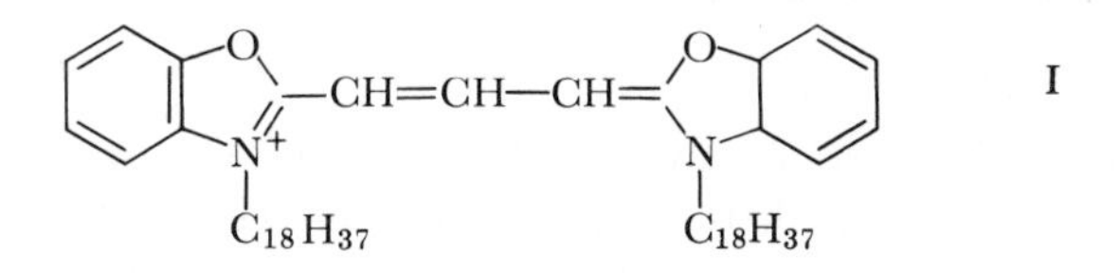

I

Label I is the same as A in the experiments of Kuhn et al. (28).

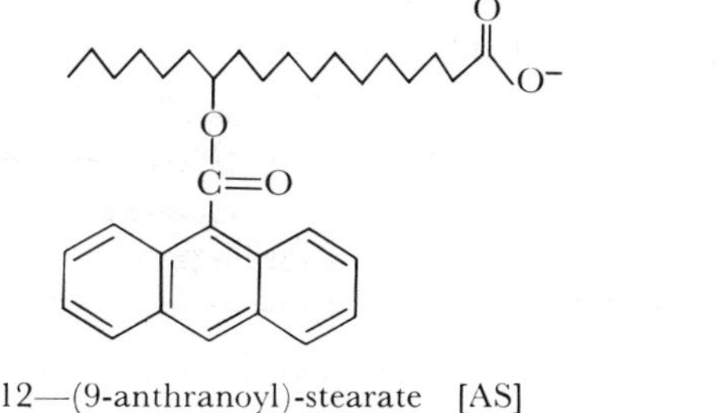

II

12—(9-anthranoyl)-stearate [AS]

CH_3 CH_3

III

p-bis [2-(4-methyl-5-phenyloxazolyl)]-benzene

The orientations and mobilities of the probes were determined by polarization and nanosecond fluorescence measurements, respectively.

I was found to be highly oriented, most likely with the chromophore at the lipid-water interface. The chromophores of II and III probably lie in the apolar core of the lipid shell, and are less aligned. Also, III is oriented perpendicular to I and II. All probes were notably mobile

---

Correlated studies on $^{22}Na^{+}$ self-diffusion through dipalmitoylphosphatidylglycerol vesicles show an increase between 30 and 38°C followed by a drop between 38 and 48°C. In phosphatidylcholine vesicles the reversal occurs at 42°C. The diffusion reversal near the perylene-indicated transition temperature can be explained by the possible existence of highly permeant boundary regions between liquid and solid domains during phase transition. Notably vesicles containing cholesterol (1/1 molar ratio) give no indication of lipid phase transitions by either perylene fluorescence polarization of $^{22}Na^{+}$ self-diffusion.

The fluorescence polarization of dansyl phosphatidylethanolamine and 9 methyl-anthracene can be used to monitor order-disorder transitions in phospholipid vesicles and in liposomes (J.-F. Faucon and C. Lussan, *Biochim. Biophys. Acta*, **307**: 459, 1973). Thermotropic phase transitions appear more sharply defined in liposomes than in small vesicles, perhaps because of the latter's high radii of curvature. This paper also presents technical improvements for the study of fluorescence polarization in turbid suspensions.

parallel to the plane of the lipid shell. A major problem with these elegant experiments is that, unlike in Kuhn's approach, the structure of the lipid is unknown, once the probes are incorporated.

Polypeptides and Proteins

Poly-L-lysine

In an important study, Lynn and Fasman (41) show that synthetic poly-L-lysine (mol wt 35,000) enhances the fluorescence of TNS 150–250 fold, with typical spectral shifts, *when in the β-structure*. Fluorescence enhancement is slight when the polymer is α-helical or unordered. Significantly, the fluorescent polarization p is greater with the β-structured polymer ($p = 0.38$) than with the α-helical ($p = 0.206$) and unordered ($p = 0.165$) species. Presumably, the hydrophobic interactions that stabilize the β-structure (46) also create apolar pockets accessible to TNS.

Bovine serum albumin (BSA)

The ANS complexes of BSA have been studied in detail (47–50) and yield information pertinent to the use of this probe with membranes. At neutral pH, 1 BSA molecule binds 5 molecules of ANS, which emit at 465 nm with a high Q ($Q = 0.75$); all the data point to a hydrophobic association.

Because ANS per se is *symmetric*, but becomes *asymmetric* when bound to BSA, its circular dichroism changes with the degree of saturation of the protein, allowing Anderson (50) to distinguish two components

Table 7.4 Polarization of Fluorescence of Absorbates of 1-Anilinonaphthalene-8-Sulfonate on BSA[a]

Polarization	ANS Adsorbed (mole/mole BSA)
0.303	0.068
0.302	0.083
0.291	0.163
0.283	0.324
0.276	0.437
0.258	0.840
0.240	1.69
0.227	2.55
0.220	3.39
0.218	5.0

[a] From Weber and Young (48).

within the five binding sites. Expansion of BSA at pH 2 generates 40 or more new binding sites with corresponding fluorescence enhancement. ANS circular dichroism at this pH indicates increasing homogeneity of binding sites as $\bar{n}$, the number of occupied sites, increases from 2 to 20.

The favorable overlap integral of the absorption and emission spectra of ANS results in energy transfer *among* ANS molecules bound to the same BSA molecule (48). This is manifested by a decrease in polarization, with increasing $\bar{n}$ (Table 7.4). Pertinently, the rotational relaxation time is 105 nsec when $\bar{n} = 1$, and 128 nsec when $\bar{n} = 5$. This migration of the excited state requires 2 or more ANS molecules to be absorbed to the same BSA molecule, with their emission oscillators neither mutually parallel nor perpendicular and with the distance between them less than $\frac{3}{2} \times R_0$ (where R_0 is the characteristic distance for the transfer of excitation energy).*

Apomyoglobin and apohemoglobin

In an equally pertinent study, Stryer (10) finds that apomyoglobin (AMb) and apohemoglobin (AHb) bind 1 molecule of ANS per equivalent of protein, yielding emission maxima of 454 nm and 457 nm, respectively, and *Q* values of 0.98 and 0.92. The tryptophan fluorescence of both proteins is completely quenched upon ANS binding, due to energy transfer. Addition of heme fully displaces bound ANS. The binding of ANS and its electronic structure are unaffected by a 6-fold viscosity increase produced by sucrose addition. However, the polarization changes from 0.187 (ANS-AMb) and 0.248 (ANS-AHb) to a limiting value (at 380 nm) of 0.396 for both proteins.

2,1-ANS, 5,2-ANS, and 1,4-ANS show emission maxima of 419, 430, 426 nm, respectively, to be compared with values on BSA of 423, 446, and 422 nm. Various other proteins (pepsin, chymotrypsinogen, α-chymotrypsin, bovine gamma globulin, lysozyme, α-lactoglobulin, ovalbumin, trypsin, and ribonuclease) yield less than 3% the fluorescence of ANS-AMb under identical conditions; this may be due to low binding and/or poor quantum yields.

ANS bound to AMb and AHb exhibits more extreme spectral shifts and higher *Q* values than found with other proteins or solvents as nonpolar, *n*-octanol, and dioxane (dielectric constants 10.3 and 2.2, respec-

* Partial chemical modification of BSA with acetic anhydride, formaldehyde, and glyoxal gives insight into the character of the apolar ANS binding sites of the protein (A. Jonas and G. Weber, *Biochemistry* **10:** 1335, 1971). The procedures do not alter the native tertiary structure of BSA. However formaldehyde- and glyoxal-treated BSA with 30% and 80% derivatized arginine residues, respectively, bind ANS about 100 fold less tightly than acetylated and native BSA. The data suggest that arginine residues lie at or near the strong apolar anion binding sites of BSA.

tively); these complexes might, therefore, serve as useful reference standards for the fluorescence characteristics of ANS in highly hydrophobic regions, such as are thought to exist in membranes.

Cytochrome c

The fluorescence enhancement of ANS bound to cytochrome c is not influenced by the redox state of this protein, *unless* it is complexed ionically with certain phosphatides (51). Thus, the 1:4 complex of oxidized cytochrome c and cardiolipin enhances ANS fluorescence 10–20 fold and shifts the emission maximum to 470 nm. Reduction of the cytochrome effects a further 25% increase in fluorescence and shifts the emission to 465 nm. The authors suggest (a) that the lipid allows the ANS to interact with the protein and to record functional changes in hydrophobicity and/or (b) that the dye mirrors structural transitions of the complex.

Membranes (Binding Studies)

Erythrocyte membranes

Freedman and Radda (52) find ANS to bind to intact ghosts with an extremely rapid initial step, followed by a slow phase, complete in about 5 min. Sonication eliminates the slow phase and effects greater fluorescence enhancement. The data, interpreted to reveal both "superficial" and "deep" ANS binding sites, may also mean that the two faces of the membrane react differently.

The number of binding sites under various conditions is evaluated

Table 7.5 ANS Binding by Erythrocyte Membranes[a]

Sample[b]	Enhancement	ANS (μ moles/ g protein)	K_{diss} (μM)	p
Intact stroma	170	23	41	0.21
Stroma, pH 3.1	740	44	9	0.21
Sonicated stroma	23	82	10	0.24
Membrane protein	27	89	19	0.20
Membrane protein, pH 3	800	—	—	—
Stroma, 2.7 M NaCl, pH 6.5	—	—	—	0.23
Stroma (after sialidase)	203	39	17	—

[a] From Brocklehurst et al. (17) and Freedman and Radda (52).

[b] Unless stated otherwise, the samples were in isotonic sodium phosphate buffer at pH 7.4 or in isotonic sodium chloride adjusted to the stated pH.

by a double reciprocal plot of fluorescence against membrane concentration, extrapolated to infinite binding (Table 7.5). Enhancement of fluorescence rises with (H^+) beginning below pH 7 and is still not maximal even at pH 1. The authors erroneously state that the fluorescence of ANS bound to BSA is independent of pH in this range (but cf. 48–50). They conclude that ANS binding reflects its solubility in nonpolar membrane "phases" and that it signals structural rearrangements therein, but the data have many alternative interpretations. The pH effects are attributed to repulsion due to sialic acid, but the effects of sialidase (17) do not support this.

Rubalcava et al. (40) determine the effects of ionic conditions on the binding of ANS by erythrocyte "ghosts" and some detergents. In 20 milliosmolar Tris buffer, pH 7.4, the ghosts bind 15 μmoles/g. Addition of 300 milliosmolar NaCl raises this to 45.5, whereas 3 milliosmolar $CaCl_2$ raises it to 41. The corresponding statistical dissociation constants are 43, 34.5, and 27 μmoles, respectively. The enhancements of ANS binding by these salts is viewed as resulting from decreased electrostatic repulsions between ANS and anionic phosphatides at high ionic strength. This emphasis on the role of phosphatides is not well supported by the data; further, the authors fail to consider ionic influences on ANS binding by proteins.

Feinstein et al. (39) compare the interaction of ANS with erythrocyte "ghosts," as well as various micellar lipid dispersions (pH 7.4), and also study the effect of $CaCl_2$ and butacaine on the binding. Untreated membranes bind 30 μmoles of ANS per gram of protein with an apparent dissociation constant of 30 μM. Ca^{2++} and butacaine (1–4 mM) increases fluorescence intensity similarly, doubling it at 4 mM. This effect is due to increased ANS binding, not a change in Q. Thus, at 2 mM butacaine, about 110 μmoles ANS bind per gram of protein. The fluorescent enhancement due to erythrocyte membranes notably exceeds that caused by various phospholipids [even sphingomyelin, the major phospholipid of bovine erythrocytes (60%)] (Fig. 7.3). Also, the fluorescence of membrane-bound ANS rises dramatically below pH 5; with the lipids, the effects of [H^+] are less marked and occur only at very low pH. The authors assume that butacaine and Ca^{++} shield anionic charges on phosphate or sialate anions. They extrapolate this contention to erythrocyte membranes, but fail to consider the obvious role of membrane proteins.

In a study of energy transfer from membrane fluorophores to ANS, Wallach et al. (2) find human erythrocytes to bind 50 μmoles ANS/g protein in 0.1 M phosphate (pH 7.4). At low pH, fluorescence increases markedly, presumably due to increased ANS binding and analogous to the increment in ANS fluorescence during the acid expansion of BSA.

Erythrocyte membranes in 20 milliosmolar Tris HCl, pH 7.4, bind 15 μmoles ethidium bromide per gram (16) with a 17-fold enhancement of fluorescence over that of ethidium in water; the apparent $K_{diss} = 63$ μM. Interestingly, enhancement by ghosts is twice that obtained with micellar sodium lauryl sulfate, and the number of binding sites is ten times greater than with the detergent, although the affinities are very similar (apparent K_{diss} for SDS = 5 μM).

Erythrocyte membrane perturbation

Treatment of erythrocyte membranes with *phospholipase A* reduces the number of ANS binding sites and changes the binding constants (Table 7.6) (53).

Table 7.6 Effect of Phospholipase A on ANS Binding by Erythrocyte Ghosts

Conditions	Binding Sites (moles ANS/mg protein $\times 10^{-8}$)	Binding Constant $M^{-1} \times 10^5$	Free Energy (kcal/mole)
Erythrocyte ghosts	5.3 ± 0.32	1.64 ± 0.17	-7.0 ± 0.73
Erythrocyte ghosts after phospholipase A treatment			
without bovine serum albumin	3.8 ± 0.33	0.76 ± 0.11	-6.5 ± 0.94
with bovine serum albumin[a]	7.5 ± 0.42	2.48 ± 0.30	-7.2 ± 0.87

[a] Measurements with ^{125}I serum albumin show it contributes ~2% of the protein in the washed ghosts, which would account for ~20% of its effects on ANS fluorescence (53).

Note: Data are mean ± SD.

Polarization values p were higher than noted by others and in accordance with values reported for electroplax membranes. p was independent of solvent viscosity and did not change even after most of the phospholipid had been enzymatically cleaved. These data suggest that the probe is located deep within the membrane, most likely in association with membrane proteins. "Inside-out" membrane vesicles bound ANS more rapidly than normally oriented membranes, attesting to membrane asymmetry demonstrated by other methods. This difference was abolished by phospholipase A, perhaps by blocking slowly reacting sites.

ANS can effectively monitor perturbations of lipid-protein and lipid-lipid interactions induced in erythrocyte ghosts by external agents (54). For example, exposure of erythrocytes to *sublytic* levels of benzyl alcohol (BeOH), or many other hemolysins, stabilizes their membranes against osmotic hemolysis (55). However, when a critical BeOH concentration is reached, lysis occurs rapidly, with dramatic changes in BeOH PMR, as noted further on.

When the experiments are repeated on *erythrocyte ghosts in the presence of ANS*, one finds

(a) a slight decrease of fluorescence in the prelytic stage (0–80 mM, BeOH),
(b) a *marked rise in fluorescence near the critical lytic concentration*, and
(c) a decrease in fluorescence at very high BeOH levels. The pattern follows that reported by BeOH PMR (q.v., paramagnetic probes).

Isolated membrane lipids *do not yield a biphasic curve*; instead, fluorescence diminishes slowly with increasing [BeOH]. Isolated membrane protein produces far greater fluorescence enhancement, which is stable throughout the prelytic phase, but drops off thereafter. When membrane lipids and proteins are separated in 2-chloroethanol (56) and allowed to recombine in water, a membrane-like recombinate forms. However, this reacts with ANS essentially as the sum of its separate components; BeOH-PMR data are concordant. The same is true for membrane recombinates formed from SDS solutions. These data strongly suggest that membrane proteins are perturbed at the critical lytic concentration and that they are not properly reassembled in the recombinates (57). The work would indubitably have been strengthened by energy-transfer measurements, but nevertheless adds important data, confirming results on *electroplax* membranes (58), as well as PMR studies on erythrocyte ghosts. The data also closely parallel results obtained with *Acholeplasma laidlawii* (see below).

Excitable Tissues

Nerves

Tasaki and associates (59–62) have developed elegant techniques to monitor minute changes in the fluorescence, turbidity, and birefringence of nerves, during the nerve impulse. They use nerves from the walking legs of lobsters and spider crabs and the fins of squid (Fig. 7.4). Although these nerves are unmyelinated, they are not "unsatellited" (q.v., discussion under *spin labels*). The total signals observed, thus, contain contributions from axonal membranes, satellite membranes, and, in unperfused axons, axonal mitochondria. In terms of membrane mass, the axonal membrane contribution is small.

Tasaki et al. (59) apply ANS both to the exteriors of nerve trunks and to the interiors of the squid giant axons. When introduced internally, it does not penetrate the axon membrane, suggesting that extracellularly

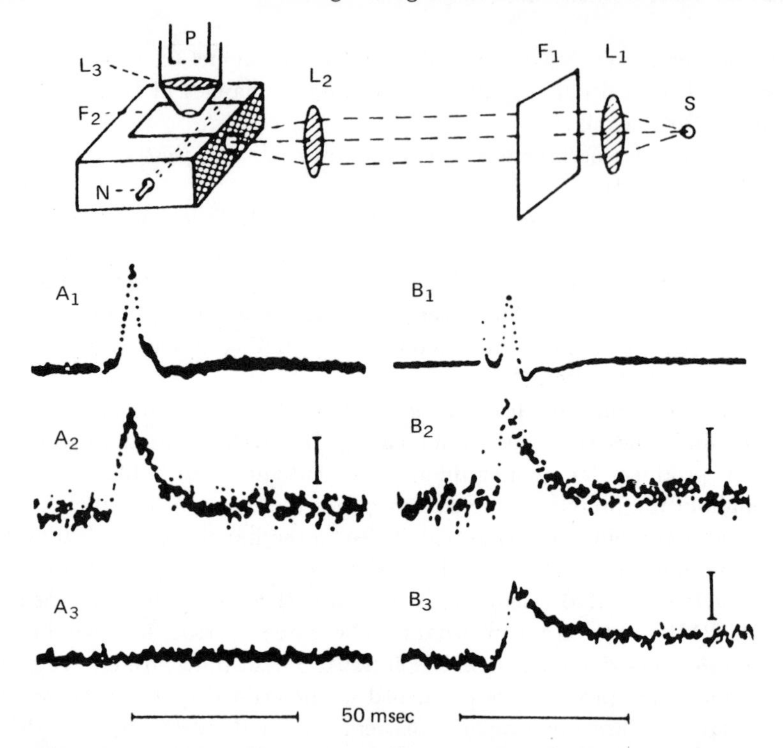

Fig. 7.4 Demonstration of fluorescence changes associated with nerve excitation. *Top:* Schematic diagram of the optical arrangement used. S represents the light source; L_1, L_2, L_3 the lenses; N the nerve; P the photomultiplier; F_1 the ultraviolet transmitting filter; and F_2 the ultraviolet absorbing filter. (Electrodes for stimulating the nerve and for recording action potentials are not shown.) *Bottom:* Action potential (A_1 and B_1) and optical signals (A_2 and B_2) observed with nerves from spider crabs. (Right- and left-hand columns were from two different nerves.) A CAT computer was used to record both action potential and optical signals. The amplitude of the extracellularly recorded action potential was approximately 2 mv. The vertical bars in the second row represent (left) 10^{-4} and (right) 2×10^{-5} times increase in light intensity. Record A_3 was taken with filter F_2 moved to a position between L_2 and N. Record B_3 represents the time course of change in light scattering at 90° observed in the stained nerve; the vertical bar represents a 4×10^{-6} times increase in light intensity. The temperature was 19°C. [From Tasaki et al. (59). Courtesy National Academy of Sciences.]

applied ANS would remain outside the axoplasm. Since the dye is applied *externally* in the other studies, it is assumed that the fluorescence changes during excitation do not involve the axoplasm.

The nerves from lobsters, crabs, and squid, when exposed to ANS and irradiated at 365 nm, emit typical ANS fluorescence. After electrical stimulation, the fluorescence increases slightly, but significantly (5 ×

10^{-5}–4×10^{-4} fold) upon arrival of the nerve impulse at the site of optical irradiation. Light scattering and birefringence also change, as reported by others (13). The authors are very cautious in terms of structural interpretation.

Tasaki et al. (60) also study fluorescence changes during nerve conduction stained by intracellular perfusion with acridine orange (AO). As in the case of ANS, the crab nerves exhibit a small increase of fluorescence upon excitation, maximally about 2–5×10^{-4} times the intensity of the background fluorescence. The squid giant axon yielded a similar pattern with a fluorescence increase of about 2.5×10^{-5} times the intensity at rest. The small signal is considered to indicate that the excitable membrane is only a minute portion of the total structures stained, and fluorescence change in the active nerve regions is thought to be several orders of magnitude greater than those registered. Since the fluorescence of AO is strongly influenced by changes in its microenvironment (9), it appears reasonable to conclude that the fluorescence changes mirror important structural rearrangements in nerve macromolecules during conduction.

Tasaki et al. (61) report very marked extensions in methodology and now include measurements of fluorescence polarization. Some recent data are summarized in Table 7.7. Here, various dyes were applied internally in the case of the squid giant axons, and externally to the squid fin nerves and crab nerves. Acridine orange, acridine yellow, ANS, auramine O, 1-dimethylaminonaphthalene-5-sulfonylchloride, fluorescein isothiocyanate, lysergic acid diethylamide (LSD), pyronine B, pyronine Y, rhodamine B, rhodamine G, rose bengal, and 2-toluildinyl naphthalene sulfonate all yield transient changes in emitted fluorescence upon electrical stimulation. The signals are again small but significant (5×10^{-4} to 5×10^{-5} times background fluorescence). Some signals are positive, some negative. Interestingly, application of LSD gives a signal only when applied to the outside of the squid axon.

In crab nerves stained with pyronin B, the fluorescence polarization decreases during the action potential, perhaps due to lowered membrane viscosity. Fluorescein isothiocyanate, covalently bound, is highly polarized, and other fluorochromes, such as acridine orange, show very small polarization in the resting state; none change polarization during the action potential. The fact that both positive and negative signals are obtained in nerves stained with ANS or pyronin B excludes the possibility that fluorescence changes are due to variations in the potential field within the membrane.

Tasaki et al. (64), continuously improving their already sophisticated methods, have derived significant new information covering structural rearrangements associated with conduction in perfused squid

Table 7.7 Compounds Used to Demonstrate Fluorescence Changes in Nerve[a]

Compound	Excitation Wavelength (nm)	Fluorescence Change During Action Potential		
		Giant Squid Axon	Fin Nerve	Crab Nerve
Acridine orange	465 (20)	+ + +	+ + +	+ + +
Acridine yellow	450 (20)	0	+	+ +
ANS	365 (10)	− − −	+ +	+ + +
Auramine O	450 (5)	0	0	+ +
DNS	365 (10)	0	0	+
FIT	500 (20)	+ (ext)	+ +	+ +
LSD	365 (10)	0	+	− −
Pyronin B	550 (5)	+ +	− − −	− − −
Pyronin Y	550 (5)	0	− −	− −
Rhodamine B	550 (5)	− −	− − −	− − −
Rhodamine G	500 (20)	0		+ +
Rose bengal	550 (5)	0		+ +
TNS	365 (20)	−		−

[a] From Tasaki et al. (61).

Symbols used: ANS, 8-anilinonaphthalene-1-sulfonate; DNS, 1-dimethylaminonaphthalene-5-sulfonyl chloride; FIT, fluorescein isothiocyanate; LSD, lysergic acid diethylamide; TNS, 2-p-toluidinylnaphthalene-sulfonate. Half bandwidths of filters used are given in parentheses. Strong, medium, and weak optical signals are shown by difference in number of + (increase) and − (decrease) symbols; the absence of a measurable signal is shown by 0.

giant axons, using TNS as probe. They show by polarization methods that *those probe molecules that cause the fluorescence change during nerve conduction are oriented firmly with their absorption and emission dipoles parallel* to the long axis of the axon. This finding alone effectively localizes these probe molecules to the *axon* membrane. These data and others (65) indicate that the TNS perturbed by action potentials lies in an apolar, highly ordered structure, a situation quite different from the experiences with model lipid bilayers already described (or with the randomly oriented TNS in axoplasm).

An additional, most important finding is that traces of tetrodotoxin, applied externally, rapidly and profoundly affect the fluorescent signal, in a manner suggesting a highly cooperative response in an ordered lattice.

Electric organ of electrophorus electricus

Kasai et al. (66) determine the amount of ANS bound to these membranes by mathematical extrapolations of the change of fluorescence intensity with increasing dye concentration, from the differences in absorbance between bound and free dye and by centrifugal techniques. Each method yields the same number of binding sites, namely 60 μmoles/g of protein, with an apparent K_{diss} of 4.4×10^{-4} M. A plot of 1/bound ANS vs. 1/free ANS follows a straight line over a large concentration range, indicating that the binding sites are relatively homogeneous. The quantum yield (~0.75) is high. Between 30 and 50°C both the number of ANS binding sites and their affinity for ANS increase. The addition of 10^{-3} M calcium increases the affinity of ANS for the membrane by 5, but does not change the number of sites. Polarization data are discussed below.

Muscle

In frog striated muscle, the fluorescence of bound pyronin B decreases transiently by a small but significant increment (~0.01%) at the onset of an action potential (67). This, as well as associated birefringence and turbidity changes, suggests a macromolecular rearrangement in the muscle membrane, possibly analogous to what is observed in nerves. The effect can clearly not be localized specifically to the plasma membrane.

Acholeplasma laidlawii

Metcalfe et al. (54) describe informative comparisons between native *Acholeplasma laidlawii* plasma membranes and membrane fragments reconstituted from solution in 10 mM sodium dodecyl sulfate by dialysis against Mg^{2+}-containing buffers (68). The reconstituted membranes exhibit much lower ANS fluorescence intensities than untreated controls. Moreover, the normal and reconstituted membranes differ in the responses of their ANS fluorescence to increasing concentrations of the membrane perturbant (BeOH). The former exhibit a constant ANS fluorescence between 0 and 50 mM BeOH, a sharp increase between 50 and 150 mM and a decrease at higher alcohol levels. This biphasic response is also mirrored in the BeOH PMR of erythrocyte membranes (71) and *Acholeplasma laidlawii* membranes (54, 69), as well as BeOH binding by erythrocyte ghosts (57). *Reconstituted membranes, however, lack this biphasic response.* Since concurrent X-ray diffraction data indicate that a lipid bilayer structure reforms during reconstitution (54, 69), the ANS probe experiments, as well as PMR studies (54, 69), are taken to indicate that *membrane proteins are incorrectly "reassembled" in the reconstituted membranes,* the lipid regions being relatively unaffected by the protein state. However, it would appear that the membrane proteins are not only improperly "reassembled," but also at least partly denatured, possibly scrambled and lacking the architectural specificity for native interaction with membrane lipids. Moreover, the low angle X-ray diffraction studies used here seem inadequate to distinguish between a native bilayer array and that noted in the reconstituted membranes.

Mitochondria

With the recognition that certain fluorescent dyes can signal small and localized changes in environmental polarity and/or viscosity came their use in the search for possible multimolecular or macromolecular rearrangements during electron transport and ATP generation in mitochondria.

The initiation of electron transport by the addition of 5 μm oxygen in the presence of succinate and 0.1 mM ANS has a half time of 0.5 msec and is associated with a much slower ANS fluorescence increase ($t_{1/2} = 2$ sec). The process is reversed upon oxygen expenditure, with a $t_{1/2} \sim 7$ sec, much slower than the $t_{1/2}$ of cytochrome a reduction (a measure of the responsiveness of the electron transport system). These data and the effects of uncouplers indicate that the alterations monitored by ANS are slower than the chemical kinetics of electron transport. The changes of ANS fluorescence appear due to alterations of the mitochondrial membrane as a consequence of energy "conservation" and not to alterations in the number of characteristics of dye binding sites.

The changes in ANS fluorescence and nonspecific light scattering of fragmented beef heart mitochondrial membranes occur almost simultaneously with energy coupling (70), but can be separated as follows. *First*, when electron transport and energy coupling are activated in the presence of permeant anions, such as nitrate, at low concentrations, ANS fluorescence rises, but no change in nonspecific light scattering occurs. *Secondly*, activation of electron transport without energy coupling produces significant changes in nonspecific light scattering, which remain unaltered when coupling is activated by the addition of oligomycin; in contrast, ANS fluorescence changes drastically under such conditions. Moreover, whereas the enhancement of ANS fluorescence is lowered by addition of uncouplers, these do not significantly change nonspecific light scattering. *Finally*, activation of electron transport in *uncoupled* particles changes light scattering, but has little influence on ANS fluorescence. Other simultaneous measurements of 90° light scattering and ANS fluorescence enhancement during energized oscillations of mitochondrial volume (71) suggest that possible changes in membrane structure, reflected by alterations of ANS fluorescence, precede energy-linked ion transport.

We have repeatedly stressed the ambiguities inherent in the use of fluorescent probes; however, their use in the study of mitochondrial function raises additional uncertainties. These are suggested by the controversial report of Azzi (72), who examined the interaction of ANS and the basic dye auramine O with intact and sonically derived submitochondrial fragments in various energy states, measuring both fluorescence changes and binding (by centrifugal separation). Activation of coupled electron transport under the conditions employed (nonsaturation with dyes) increased the binding of auramine O by 50% in whole mitochondria, while decreasing ANS binding by 18%. No fluorescence changes were obtained with an uncharged derivative of ANS, 8-anilino-1-naphthalene-sulfonamide. The opposite dye binding of intact and fragmented mitochondria, in their energized and nonenergized states, suggests some electrostatic interaction between the dyes and the membranes and an asymmetrical charge distribution across the mitochondrial membrane, changing when the membrane becomes energized.

The early observations on the interactions of fluorescent probes with mitochondrial membranes have been confirmed and amplified in recent publications (17, 73, 74). Datta and Penefsky (73) found that the addition of ANS, TNS, and MNS (*N*-methyl-2-anilino-6-naphthalene sulfonate) to fragments of beef heart mitochondrial membranes yielded fluorescence enhancements of 50- to 150-fold, plus the familiar blue shift of the emission maxima. 40–55 μmoles of dye bind per milligram of protein with dissociation constants of $2.5–4.5 \times 10^{-5}$ M. Respiratory substrates plus O_2 produce further fluorescence enhancement; this correlates with the respiratory rate, doubling the intrinsic fluorescence augmentation as a maximum, however, without appearance of new binding sites. ATP addition acts similarly. The metabolically dependent fluorescence enhancement is diminished or blocked by respiratory inhibitors and certain uncouplers of oxidative phosphorylation. The data also indicate that mitochondrial *ATPase* may be an important, metabolically varying site of dye binding; thus, purified mitochondrial ATPase binds 6 μmoles of TNS per mole in the presence of ATP and 20 μmoles per mole without it.

Brocklehurst et al. (17) observe similar fluorescence enhancements and spectral shifts when ANS and MNS are added to resting beef heart mitochondrial fragments and obtain a 2.5- and 2.3-fold increase in fluorescence, respectively, when the particles are "energized" by the addition of succinate plus O_2. The second fluorescence enhancement is attributed to an increased quantum yield of the bound probes, together with an increased affinity of the "energized" membranes for the dyes (Table 7.8). More recent data (75) support this view. They also indicate that lipid-depleted submitochondrial particles bind ANS as do those in an "energized" state.

Clearly, dyes such as ANS, TNS, and MNS can function as sensitive indicators of energy coupling in mitochondrial membranes and can be used to monitor the kinetics of structural changes occurring during mito-

Table 7.8 Interaction of ANS and MNS with "Resting" and "Energized" Submitochondrial Particles[a]

Membrane State	Probe	Limiting Enhancement	K_{diss} (μM)	No. Sites (μmoles/g)
"resting"	ANS	104	35	80
"energized"	ANS	251	20	80
"resting"	MNS	20	96	260
"energized"	MNS	45	16	260

[a] From Brocklehurst et al. (17).

chondrial metabolism. We suspect that these fluorescent probes report, at least in part, energy-linked conformational rearrangements in membrane *proteins* because (a) electron transport is associated with an increase in the amount of antiparallel β-structure in mitochondrial protein (see discussion in the section on infrared), and (b) β-structured poly-L-lysine yields a far greater enhancement of TNS fluorescence than its helical or "unordered" conformational isomers (41). However, it is not clear from present studies how fluorescent probes can be used to yield specific information as to the structural rearrangements that accompany mitochondrial energy metabolism particularly since changes in probe binding occur also.

Chance (76) presents a reasonable working hypothesis for this complex matter. He suggests that the energy-dependent membrane domain lies at a structured water interface between a protein and a lipid layer. Membrane "energization" arising from electron transport involves alterations of protein structure, with associated changes in field charge, which reduce the volume of structured water and induce lateral and transmembrane ion movements, including charged probe molecules themselves.

Sarcoplasmic reticulum

ANS binding to muscle microsomes effects a large fluorescence enhancement of the dye and a shift of the emission maximum to 470 nm (42, 77). Binding values obtained by measuring the amount of bound ANS in membranes sedimented by centrifugation or separated by millipore filtration are maximally 20 μmoles protein in 0.01 M imidazole (pH 7.0), and 50 μmoles/g with 10^{-3} M Mg^{2+}. The ANS binding by micellar lecithin is also increased markedly by Mg^{++}. It is generally assumed that increasing [Mg^{++}], i.e., raised ionic strength, allows greater ANS binding by both membrane and phospholipid dispersions.

ANS fluorescence increases in both the natural membranes and lipid dispersions below pH 7, but the microsomes behave more like the erythrocyte membranes than the lipids. They (and phosphatidylethanolamine) exhibit peak fluorescence near pH 3, but with lecithin and phosphatidylserine maximal fluorescence occurs at lower pH; this may indeed be due to titrations of the ionic groups. At pH 3, which represents peak fluorescence intensity, the fluorescence intensity of the microsomes is approximately 3600/mg/ml, compared with 2600 for lecithin, 280 for phosphatidylethanolamine, and 300 for phosphatidylserine. Clearly, the binding characteristics of the membranes are very different from those of the phospholipids; at neutral pH the phospholipid micelles bind ANS much more strongly than the microsomal membranes, but at low pH the fluorescence enhancement is much greater with the microsomes than with the lipid micelles.

The interaction of ANS with sarcoplasmic vesicles indicates a not unexpected heterogeneity of binding sites (77). Moreover, there is considerable transfer of excitation energy from membrane protein tryptophan to ANS, and polarization studies show that the dye is less mobile in the membranes than in phosphatide dispersions and is largely unaffected by medium viscosity. The data suggest a significant involvement of membrane protein in interaction of ANS and sarcoplasmic vesicles, but allow no more specific conclusions.

Hasselbach and Heimberg (42) also argue that ANS binds to the proteins of sarcoplasmic membranes. Digestion of these particles with phospholipase A and extracting the resulting lysophosphatides and fatty acids into defatted serum albumin (78) remove the major phosphatides (and 80% of the total lipid) from the membranes (without use of denaturing solvents), but do not significantly influence the fluorescence of previously added ANS. It also seems that oleic acid and sodium dodecyl sulfate can displace ANS from its membrane binding sites (42), which could well be a protein amino group in a hydrophobic environment. One difficulty in these experiments is that the albumin may not have been completely removed, or may bind to the membranes, as in the case of erythrocyte ghosts (53). Since albumin strongly enhances ANS fluorescence, this could lead to an overestimate of the contribution of protein to ANS fluorescence (78).

Vanderkooi and Martonosi (77), cognizant of the previously discussed effects of cations, particularly divalent cations, upon ANS binding to sarcoplasmic membranes, examine the fluorescence of ANS bound to the latter during energy-dependent Ca^{++} transport. Concordant with other data, Ca^{++} accumulation in the sarcoplasmic vesicles at low ionic strength enhances ANS binding, hence fluorescence. Leaky vesicles do not show this effect, and Ca^{++} complexers abolish it. The use of ANS in this system, thus, provides a sensitive means of measuring Ca^{++} fluxes in this and possibly other membrane systems.

Microsomal membranes

Brain microsomal membranes, indubitably a heterogeneous membrane mixture, bind ANS with concurrent characteristic fluorescence changes (44). The degree of binding depends upon cationic environment and allows evaluation of the relative affinity of the membranes for monovalent cations, which follows the order $Cs^+ > Rb^+ > K^+ > Na^+ > Li^+$. Divalent cations and some basic drugs are more strongly and specifically bound.

A related study (79) shows that the smooth microsomal membranes of rat liver have a higher affinity for ANS and lesser affinity for ethidium bromide than rough microsomal membranes. This appears to reflect primarily these particles' surface charge, since neutralization of the net negative surface charge by cations increases ANS binding. However, under identical binding conditions, ANS fluoresces more strongly when associated with smooth microsomal membranes, suggesting some difference in binding site polarity and/or viscosity.*

* The negatively charged fluorescent probe ANS and the apolar fluorescent probe N-phenyl-l-naphthylamine have been used to evaluate lipid phase transitions in *Escherichia coli* membranes and in aqueous dispersions of their extracted lipids (H. Träuble and P. Overath, *Biochim. Biophys. Acta,* **307:** 491, 1973; P. Overath and H. Träuble, *Biochemistry,* **12:** 2625, 1973). At low concentrations ($< 10^{-5}$ M) N-phenyl-l-naphthylamine effectively monitors the state of lipid hydrocarbon regions. At higher levels it acts as a structural perturbant and decreases the lipid transition temperature. A comparison of the limiting fluorescence changes at the phase transition obtained with membranes and their lipids suggests that 80% of the membrane lipid participates in the phase transition. ANS fluorescence data are taken to indicate that 42% of the lipid is accessible from the aqueous phase, the other being "covered" by protein. (These estimates do not allow for the association of the two fluorophores with the protein of the native membranes.) The authors suggest that "integral" proteins are embedded in a lipid matrix, which has "surface" proteins attached to the lipid headgroups. The likelihood that a single membrane protein may interact with both the hydrocarbon moieties and polar head groups of the lipids is not explored. The authors argue that approximately 600 lipid molecules surround each "integral" protein, of which about 130 are closely associated with the protein, forming a lipid shell in which apolar interactions between hydrocarbon chains are perturbed by the protein. The hypothesis resembles that earlier proposed by one of us (Chapter 9) and also suggested by Griffith and co-workers on the basis of the interaction of cytochrome oxidase with lipid spin labels (Chapter 8).

N-phenyl-l-napththylamine has been used to evaluate the effects of colicin on *Escherichia coli* membranes (W. A. Cramer, S. K. Phillips, and T. W. Keenan, *Biochemistry* **12:** 1177, 1973). A fluorescence increment produced by colicin is taken to indicate that the action of colicin involves its diffusion, or that of induced substances, through hydrocarbon regions of the membrane. The probe is assumed to report from lipid regions only, but there are no data excluding possible interactions with membrane protein.

Polarization

Electric organ of electrophorus electricus

Kasai et al. (66) find that the polarization p of ANS bound to the membranes of the electric organ is 0.31 at 20° (i.e., *the same* as BSA-bound ANS under similar conditions). However, p does not vary with the number of molecules bound (maximally 6×10^{-5} moles/g) and unlike the polarization of ANS bound to BSA to soluble proteins from *electrophorus electricus* and to apomyoglobin, does not depend upon solvent viscosity. The ANS appears strongly immobilized by a component sequestered deeply within the membrane phase.

The polarization of 5-dimethyl-amino-1-naphthalene sulfonyl chloride (DANS), coupled covalently to the proteins of *electrophorus* membrane fragments, as that of ANS, is high (0.240 vs. 0.255 for DNS-BSA) and is not normally sensitive to changes in solvent viscosity (58). However, experiments separating the membrane lipids and proteins and subsequently recombining them, as well as the effects of Triton X-100, indicate that a high p, insensitive to solvent viscosity, requires normal *association of lipids and proteins into an organized membrane structure.* Disruption of membrane organization lowers p and makes it responsive to solvent viscosity.

The DANS is bound to protein only, and there is no competition between the binding of ANS and DANS. Apparently, the ANS is primarily protein bound because (a) the p observed (0.31) is greater than in lipid dispersions (0.15); (b) transfer of electronic excitation energy between membrane protein chromophores and ANS is highly efficient.

These experiments have been extended by nanosecond fluorescence polarization spectroscopy (80), which also indicates strong immobilization of membrane proteins binding both ANS and DANS. The study suggests the possibility, often stressed in the past, that both rigid and fluid domains may occur in the same membrane and that one might reasonably expect heterogeneity of molecular motions in *biomembranes.*

Erythrocyte ghosts

Freedman and Radda (52), exciting bound ANS at 380 nm and measuring at 480 nm, observed a lower polarization than in albumin and attribute this to localization of ANS in a "mobile phase." However, albumin binds about 7×10^{-5} moles ANS/g, similar to their values of 2 to 9×10^{-5} moles/g, and at ANS saturation the polarization values for albumin are in the order of 0.22, due to migration of the excited state (48). These data, thus, do not justify conclusions as to the mobility of bound ANS. In any event, the ANS could be packed in clusters where migration of the excited state would prevent interpretation of polarization data. Moreover, under somewhat different conditions (4.5 μmoles ANS/mg protein and more extensive cell purification), Weidekamm et al. (53) find a p of 0.31 which is unchanged by phospholipase A action and changes of solvent viscosity. They reason as in (58) that the ANS is bound by protein sites sequestered in the membrane interior.

Mitochondria

Azzi et al. (81) found $p = 0.194$ when beef heart mitochondria contain 3 μmoles ANS/g protein; p drops to 0.159 when the binding is 15 μmoles/g. This concentration dependence is attributed to intermolecular energy transfer. Brocklehurst et al. (17) reported p values of 0.22 and 0.18 for unenergized and energized particles, respectively. Since the ANS binding sites have not yet been specified, the meaning of these observations is unclear.

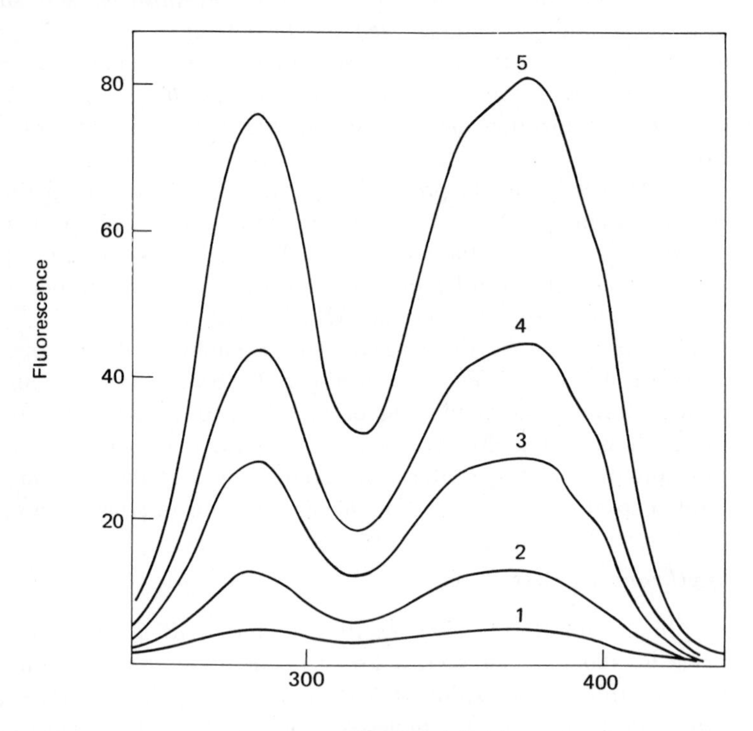

Fig. 7.5(a) Excitation spectrum of ANS in various ethanol-water mixtures. (1) 38% ethanol; (2) 57% ethanol; (3) 77% ethanol; (4) 87% ethanol; (5) 96% ethanol. ANS concentration 20 μM; emission at 480 nm; excitation bandwidth 2 nm. Spectra not corrected.

Energy Transfer and Quenching Studies

Erythrocyte membranes

Measurements of the binding and altered spectral properties of fluorescent probes may not give precise information about membrane structure, but evaluation of the transfer of excitation energy from membrane protein fluorophores to inserted fluorescent probes could signal the localization of probes relative to tryptophan and/or tyrosine-bearing regions of membrane proteins (2). Naphthalene sulfonates are suitable for such studies because of a favorable overlap integral between their long-wavelength excitation band and the fluorescence emission of tryptophan (10), and Wallach et al. (2) show that there is indeed transfer of excitation energy from some of the membrane tryptophans to membrane-bound ANS; this is apparent from both the excitation spectra of free and membrane-bound ANS [Fig. 7.5(a), (b)] and the quenching of tryptophan fluorescence by binding of ANS [Fig. 7.5(c)]. Compatible with studies on albumin, ANS binding and the degree of energy transfer are much

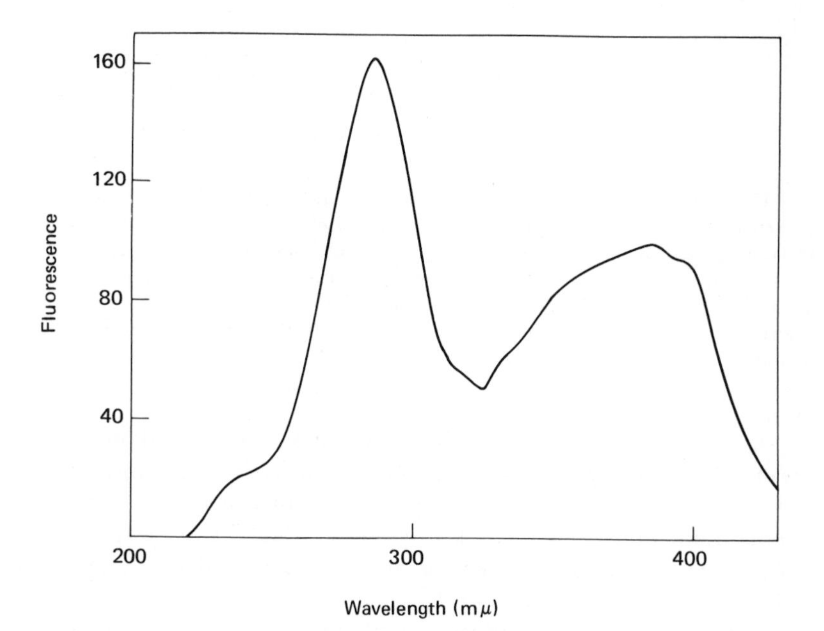

Fig. 7.5(b) Excitation spectrum of erythrocyte membranes in the presence of 12 μM ANS. Protein concentration 42 μg/ml; emission 480 nm; excitation bandwidth 2 nm; emission bandwidth 4 nm. ANS emission upon excitation at 280 nm is markedly increased compared with the dye alone. This is due to transfer of excitation energy from tryptophan to ANS.

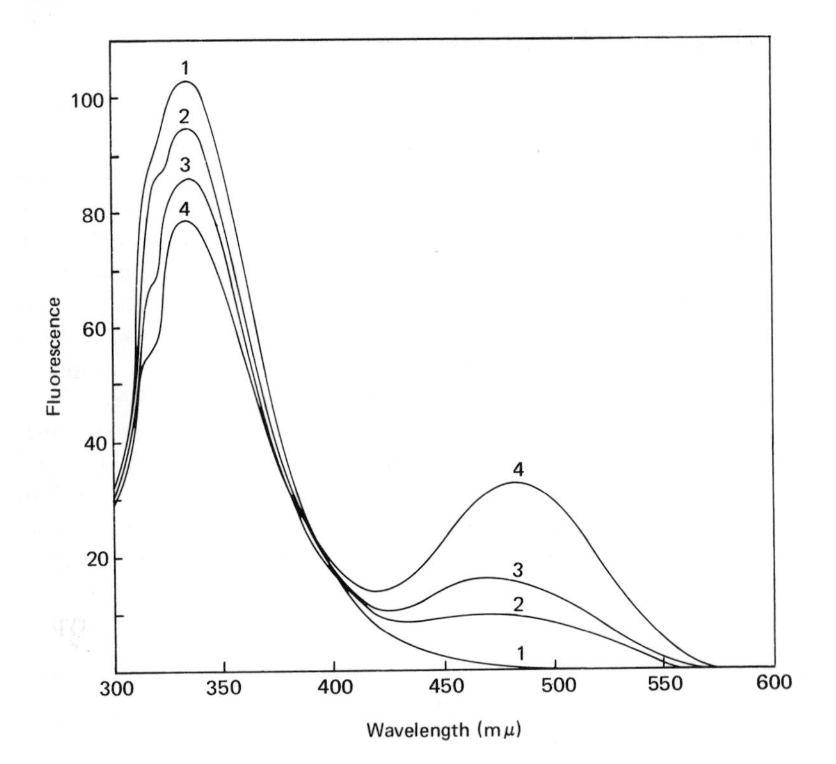

Fig. 7.5(c) Emission spectra of erythrocyte membranes alone (1), and in the presence of 20 μM ANS (2), 40 μM ANS (3), and 80 μM ANS (4). Protein concentration 9 μg/ml. Excitation bandwidth 3 nm; emission bandwidth 4 nm; excitation at 286 nm. Spectra corrected. The shoulders near 320 nm are due to Raman scattering by water. The decrease of TRP fluorescence and increase of ANS fluorescence with excitation near 280 nm indicates energy transfer. [From Wallach et al. (2). Courtesy Elsevier Publishing Co.]

greater at pH 3 than at neutrality. These workers could not obtain accurate quantitative estimates of transfer efficiency. However, Brocklehurst et al. (17) report very high transfer efficiencies for erythrocyte stroma in the presence of high ANS levels, particularly at low pH (Table 7.9). Extensive energy transfer has also been reported for sarcoplasmic reticulum coupled with ANS (77).

Not all fluorescent probes potentially able to accept excitation energy from membrane tryptophans actually do so. Thus, the polyene antibiotic filipin binds strongly to erythrocyte "ghosts," presumably forming complexes with membrane cholesterol, but is not excited by tryptophan emission and does not cause quenching of tryptophan fluorescence (82), despite a favorable overlap integral of tryptophan emission

Table 7.9 Energy-Transfer Parameters for Probe-Membrane Complexes

Membrane	(Probe)$_{total}$ (μM)	Transfer Efficiency $(T)^b$
Stroma (0.07 mg/ml) isoosmotic phosphate buffer, pH 7.4	4 ANS	0.06
	10	0.15
	20	0.28
	30	0.36
	40	0.40
Stroma (0.07 mg/ml) (0.15 M NaCl adjusted to pH 3.1)	4 ANS	0.22
	10	0.51
	20	0.68
	30	0.75
	40	0.79
Stroma (0.07 mg/ml) (M NaCl, pH 7.0)	4 ANS	0.03
	10	0.15
	20	0.26
	30	0.29
	40	0.33

[a] From Brocklehurst et al. (17).
[b] $T = 1 \cdot Q_T/Q_0$; where Q_T is the quantum yield of the donor in the presence of transfer, and Q_0 is that without transfer.

and filipin excitation. The efficient energy transfer found with membrane-bound ANS suggests that this localizes in the membrane protein and/or near lipid-protein interfaces. Brocklehurst et al. (17) argue for an additional type of localization—in pockets of "structured water."

Mitochondria and chloroplasts

Brocklehurst et al. (17) observe efficient energy transfer also with electron transport particles isolated from ox heart mitochondria and ANS and MNS. They found no change in energy transfer upon initiation of electron transport by the addition of succinate under aerobic conditions.

Kraayenhof (83) has employed a different approach to this problem. He stresses the limitations of binding studies in the following energized state of mitochondria and suggests that for the quantitative determination of the "energized state" a fluorescent probe should have sufficiently high affinity for the system to dissipate the energy-utilizing reaction; i.e., the fluorochrome should be an uncoupler.

He has studied the relationship between energy generation in spinach chloroplasts and the fluorescence quenching of the uncoupler *atabrine*, "energizing" the membranes (a) by photosynthesis, (b) by ATP addition, and (c) by moving the chloroplasts from low to high pH. All methods of producing an energized state in the chloroplast quench atabrine fluorescence, probably reflecting combination of the uncoupler with structures participating in the energy-conservation mechanism, or a change of the environment in which the uncoupler is situated. Since uncouplers are closely linked to the energized state, fluorescent changes of such compounds should be useful tools in monitoring such a state.

Excitable membranes from the electric organ of electrophorus electricus

Kasai et al. (58) report energy transfer of high efficiency between membrane tryptophans and membrane-bound ANS. They reason that the ANS-binding sites of these membranes are primarily at lipid-protein interfaces and that these areas of contact are structurally sensitive to temperature.

Epilogue

The use of fluorescent probes has yielded a vast body of information. Unfortunately, nearly all data can be interpreted in more than one way. The great strengths of the technique lie in its sensitivity and its applicability to dynamic situations. Its greatest weakness derives from the fact that the electronic properties of fluorophores do not relate unambiguously to individual atomic and molecular positions and/or interactions. There is also a constant danger that fluorescent probes locate selectively and also perturb their environment. We suspect that these techniques will realize their potential only when joined to other approaches, such as X-ray diffraction, nuclear magnetic resonance, infrared and laser Raman spectroscopy, as well as biochemical methods.

References

1. Wallach, D. F. H., and Zahler, P. H. *Proc. Natl. Acad. Sci.* (U.S.) **56:** 1552, 1966.
2. Wallach, D. F. H., Sellin, D., Weidekamm, E., and Fischer, H. *Biochim. Biophys. Acta* **203:** 67, 1970.
3. Sonenberg, M. S. *Proc. Natl. Acad. Sci.* (U.S.) **68:** 1051, 1971.
4. Sonenberg, M. S. *Biochem. Biophys. Res. Commun.* **36:** 450, 1969.
5. Harkins, W. D. *The Physical Chemistry of Surface Films.* New York: Reinhold, 1952, p. 299.
6. Corrin, M. L., and Harkins, W. D. *J. Am. Chem. Soc.* **69:** 679, 1947.
7. Träuble, H. In *Symposium on Passive Permeability of Cell Membranes*, Rotterdam, 1971 (in press).

8. Lesslauer, W. L., Cain, J. E., and Blasie, J. K. *Proc. Natl. Acad. Sci.* (U.S.) **69:** 1499, 1972.
9. Albers, R. W., and Koval, G. J. *Biochim. Biophys. Acta* **60:** 359, 1968.
10. Stryer, L. *J. Mol. Biol.* **13:** 482, 1965.
11. Stryer, L. *J. Am. Chem. Soc.* **88:** 5708, 1966.
12. Oster, G., and Nishijima, Y. *J. Am. Chem. Soc.* **78:** 158, 1956.
13. Vanderkooi, J., and Martonosi, A. *Arch. Biochem. Biophys.* **133:** 153, 1969.
14. McClure, W. O., and Edelman, G. M. *Biochemistry* **5:** 1908, 1966.
15. Lepecq, J. B., and Paoletti, C. *J. Mol. Biol.* **27:** 104, 1967.
16. Gitler, C., Rubalcava, B., and Caswell, A. *Biochim. Biophys. Acta* **193:** 479, 1969.
17. Brocklehurst, J. R., Freedman, R. B., Hancock, D. J., and Radda, G. K. *Biochem. J.* **116:** 721, 1970.
18. Waggoner, A. S., and Stryer, L. *Proc. Natl. Acad. Sci.* (U.S.) **67:** 5A, 1970.
19. Förster, T., and Kasper, K. *Z. elektrochim.* **59:** 976, 1955.
20. Förster, T. *Angew. Chem.* **81:** 364, 1969.
21. O'Sullivan, M., and Testa, A. C. *J. Am. Chem. Soc.* **90:** 6245, 1968.
22. Deering, R. A., and Setlow, R. B. *Biochim. Biophys. Acta* **68:** 526, 1963.
23. Eisinger, J., Gueron, M., Shulman, G., and Yamane, T. *Proc. Natl. Acad. Sci.* (U.S.) **55:** 1015, 1966.
24. Lamola, A. A., and Eisinger, J. *Proc. Natl. Acad. Sci.* (U.S.) **59:** 46, 1968.
25. Keleti, T. *FEBS Letters* **7:** 280, 1970.
26. Feinstein, M. B., and Felsenfeld, H. *Proc. Natl. Acad. Sci.* (U.S.) **68:** 2037, 1971.
27. Seliskar, C. J., and Brand, L. *Science* **171:** 799, 1971.
28. Büecher, H., Drexhage, K. H., Fleck, M., Kuhn, H., Möbius, D., Schäfer, F. P., Sondermann, J., Sperling, W., Tillman, P., and Weigand, J. *Molecular Crystals* **2:** 199, 1967.
29. Blodgett, K. L. *J. Am. Chem. Soc.* **57:** 1007, 1935.
30. Förster, T. *Ann. Physik.* **2:** 55, 1948.
31. Ullrich, H. -M., and Kuhn, H. *Biochim. Biophys. Acta* **266:** 584, 1972.
32. Mueller, P., and Rudin, D. O. *Nature* **217:** 713, 1968.
33. Tien, H. T., and Verma, S. P. *Nature* **227:** 1232, 1970.
34. Verma, S. P. *Biophysik* **7:** 288, 1971.
35. Tweet, A. G., Gaines, G., Jr., and Bellamy, W. D. *J. Chem. Phys.* **40:** 2596, 1964.
36. Gaines, G., Jr., Tweet, A. G., and Bellamy, W. D. *J. Chem. Phys.* **42:** 2193, 1965.
37. Penzer, G., and Radda, G. K. *Nature* (Lond.) **213:** 251, 1967.
38. Gulik-Krzywicki, T., Schechter, E., Iwatsubo, M., Rank, J. L., and Luzatti, V. *Biochim. Biophys. Acta* **219:** 1, 1970.
39. Feinstein, M. B., Spero, L., and Felsenfield, H. *FEBS Letters* **6:** 245, 1970.
40. Rubalcava, B., Martinez de Munoz, D., and Gitler, C. *Biochemistry* **8:** 2742, 1969.
41. Lynn, J., and Fasman, G. *Biochem. Biophys. Res. Commun.* **33:** 327, 1968.
42. Hasselbach, C. W., and Heimberg, W. K. *J. Memb. Biol.* **2:** 341, 1970.
43. Caswell, A., and Gitler, C. *Federation Proc.* **29:** 605, 1970.
44. Gomperts, B., Lantelme, F., and Stock, R. *J. Membrane Biol.* **3:** 241, 1970.

45. Yguerabide, J., and Stryer, L. *Proc. Natl. Acad. Sci.* (U.S.) **68:** 1217, 1971.
46. Davidson, B., and Fasman, G. *Biochemistry* **6:** 1616, 1967.
47. Weber, G., and Laurence, D. J. R. *Biochem. J.* **56:** 31, 1954.
48. Weber, G., and Young, L. *J. Biol. Chem.* **239:** 1415, 1964.
49. Anderson, S. R., and Weber, G. *Biochemistry* **8:** 371, 1969.
50. Anderson, S. R. *Biochemistry* **8:** 4838, 1969.
51. Azzi, A., Fleischer, S., and Chance, B. *Biochem. Biophys. Res. Commun.* **36:** 322, 1969.
52. Freedman, R. B., and Radda, G. K. *FEBS Letters* **3:** 150, 1969.
53. Weidekamm, E., Wallach, D. F. H., and Fischer, H. *Biochim. Biophys. Acta* **241:** 770, 1971.
54. Metcalfe, S., Metcalfe, J. C., and Engelman, D. M., *Biochim. Biophys. Acta* **241:** 422, 1971.
55. Seeman, P. *Int. Rev. Neurobiol.* **9:** 145, 1966.
56. Wallach, D. F. H., and Zahler, H. P. *Proc. Natl. Acad. Sci.* (U.S.) **56:** 1552, 1966.
57. Colley, C. M. C., Metcalfe, S. M., Turner, B., and Burgen, A. S. V. *Biochim. Biophys. Acta* **233:** 720, 1971.
58. Kasai, M., Podleski, R., and Changeux, J. P. *FEBS Letters* **7:** 13, 1970.
59. Tasaki, I., Watanabe, A., Sandlin, R., and Carnay, L. *Proc. Natl. Acad. Sci.* (U.S.) **61:** 883, 1968.
60. Tasaki, I., Carnay, L., and Sandlin, R. *Science* **163:** 683, 1969.
61. Tasaki, I., Carnay, L., and Watanabe, A. *Proc. Natl. Acad. Sci.* (U.S.) **64:** 1362, 1969.
62. Tasaki, I., Barry, W., and Carnay, L. In *Physical Principles of Biological Membranes*, F. Snell, J. Wolken, G. J. Iverson, and J. Lam, eds. New York: Gordon and Breach, 1970, p. 17.
63. Cohen, L. D., Keynes, R. D., and Hille, B. *Nature* (Lond.) **218:** 438, 1968.
64. Watanabe, A., and Hallett, M. *J. Memb. Biol.* **8:** 109, 1972.
65. Tasaki, I., Watanabe, A., and Hallett, M. *Proc. Natl. Acad. Sci.* (U.S.) **68:** 938, 1971.
66. Kasai, M., Changeux, J. P., and Monnerie, L. *Biochem. Biophys. Res. Commun.* **36:** 420, 1969.
67. Carnay, L. D., and Barry, W. H. *Science* **165:** 608, 1970.
68. Razin, S. *Biochim. Biophys. Acta* **265:** 241, 1972.
69. Metcalfe, S., and Engelman, D. M. *Biochim. Biophys. Acta* **241:** 412, 1971.
70. Chance, B., and Lee, C. *FEBS Letters* **4:** 181, 1969.
71. Packer, L., Donovan, M. P., and Wrigglesworth, J. M. *Biochem. Biophys. Res. Commun.* **35:** 832, 1969.
72. Azzi, A. *Biochem. Biophys. Res. Commun.* **37:** 254, 1969.
73. Datta, A., and Penefsky, H. S. *J. Biol. Chem.* **245:** 1537, 1970.
74. Chance, B. *Proc. Natl. Acad. Sci.* (U.S.) **67:** 560, 1970.
75. Harris, R. A. *Arch. Biochem. Biophys.* **147:** 436, 1971.
76. Chance, B. *Proc. Natl. Acad. Sci.* **67:** 560, 1971.
77. Vanderkooi, J. M., and Martonosi, A. *Arch. Biochem. Biophys.* **144:** 87, 1971.
78. Vanderkooi, J. M., and Martonosi, A. *Arch. Biochem. Biophys.* **144:** 99, 1971.

79. Dallner, G., and Azzi, A. *Biochim. Biophys. Acta* **255:** 589, 1972.
80. Wahl, P., Kasai, M., and Changeux, J. P. *Eur. J. Biochem.* **18:** 332, 1971.
81. Azzi, A., Chance, B., Radda, G. K., and Lee, C. P. *Proc. Natl. Acad. Sci.* (U.S.) **62:** 612, 1969.
82. Price, H. D., and Wallach, D. F. H. (unpublished observations).
83. Kraayenhof, R. *FEBS Letters* **6:** 161, 1970.

CHAPTER 8

SPIN-LABEL PROBES

Theory

General

Spin-labeling studies employ nitroxides with the following general structure: (see also Fig. 8.1).

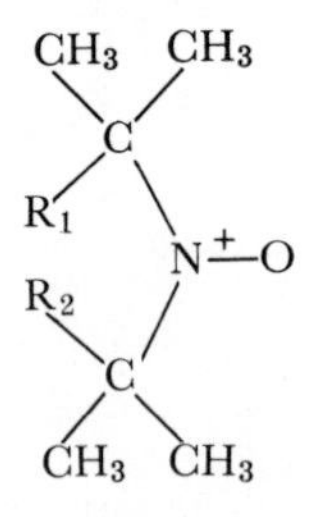

The paramagnetic properties of these compounds derive from an unpaired electron localized primarily on the nitrogen. While relatively stable, nitroxides can be inactivated, i.e., their free electron spin canceled by a number of mechanisms, e.g., H-donation by some other molecule producing an OH bond. Vitamin C is believed to react in this fashion and is often used for this purpose experimentally (see below). Spin labels are also deactivated rapidly by many cells and also membranous organelles, perhaps by related mechanisms.

The magnetic moment $\mathcal{N}$ precesses in a constant magnetic field, H, at a frequency ω, given by

$$\omega = \frac{2\pi \mathcal{N} H}{h} \tag{8.1}$$

where h is Planck's constant. Rewriting this equation for paramagnetic substances in terms of the Bohr magneton β and the g value,

$$\omega = \frac{2\pi g \beta H}{h} \tag{8.2}$$

ESR spectra are usually obtained at a fixed frequency, varying the magnetic field *H*. To avoid interference due to broad microwave absorption by various paramagnetic molecules and electric dipoles, spectra are usually recorded as first derivatives (i.e., dI/dH vs. H, where I is in absorption units and H in gauss).

The magnetic field experienced by the unpaired electron is perturbed by the nuclear magnetic moment of ^{14}N. This, having a spin-quantum

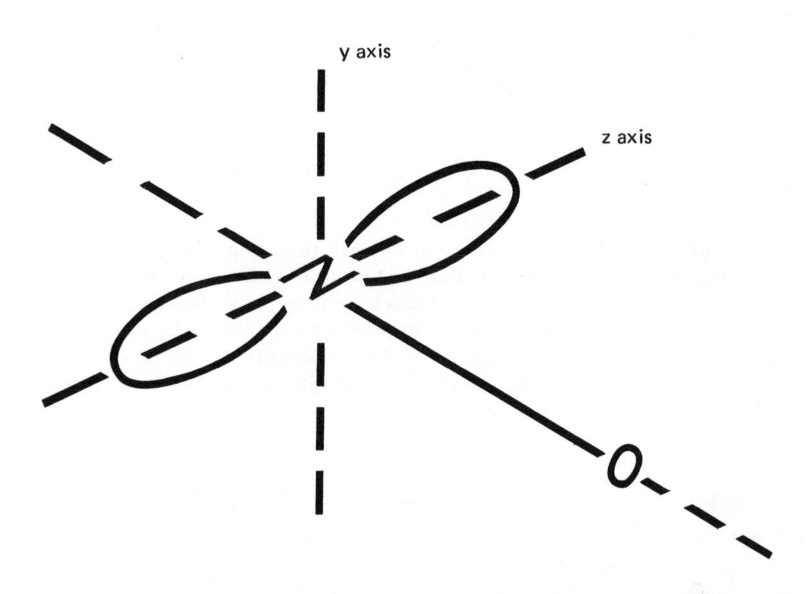

Fig. 8.1 Electronic structure of the nitroxide radical. The paramagnetic properties of these compounds derive from an unpaired electron localized 80–90% in a $2p\pi$ atomic orbital on the nitrogen. The z axis is parallel to the $2p\pi$ orbital; the x axis is parallel to the N—O bond; the y axis is perpendicular to both x and z axes. The spin-orbital interaction is much smaller than interatomic Coulombic forces. The angular momentum of the unpaired electron is, thus, essentially negligible, and the magnetic properties of nitroxides, therefore, arise predominantly from the free electron spin.

number of 1, allows only three orientations—parallel, antiparallel, and perpendicular—of the nuclear magnetic moment with respect to the laboratory field. These three perturbing fields split the hypothetical single electron resonance band into the three equally spaced components (Fig. 8.2) (1). Each ESR spectrum is characterized by two important parameters: (a) the g_0-value = center of spectrum; (b) a_0, the isotropic coupling constant (splitting) = distance between adjacent lines. Different values of g_0 signify shifts of the entire three-line spectrum to higher or

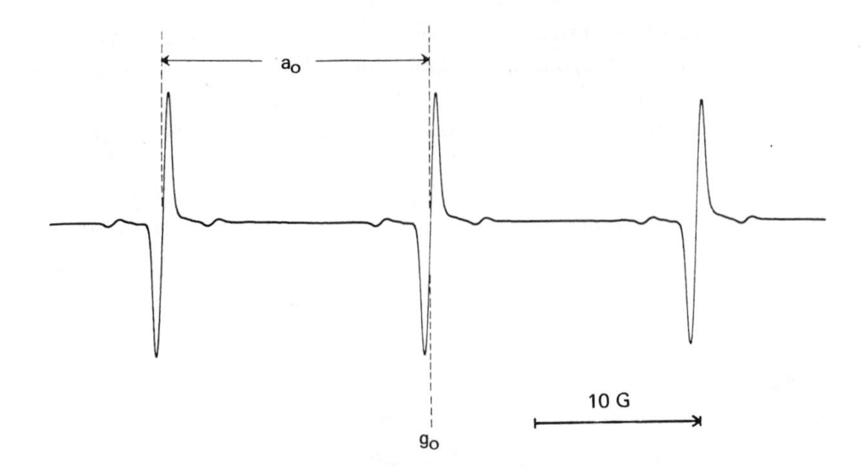

Fig. 8.2 Typical ESR spectrum of a freely tumbling nitroxide. The hypothetical single-electron resonance band is split into the three equally spaced components. Each ESR spectrum is characterized by (a) the g_0 value = center of spectrum; (b) a_0, the isotropic coupling constant = distance between adjacent lines. Different values of g_0 indicate shifts of the entire three-line spectrum to higher or lower fields, without necessary changes in a_0. Conversely, a_0 may vary without any change in g_0. See text for details.

lower fields, without necessary changes in a_0. Conversely, a_0 may vary without shift in g_0.

Anisotropy and Environmental Effects

The term "anisotropic motion" broadly implies that a certain molecule does not orient itself with equal probability in all directions. An important example is rapid rotation about one axis, without any other motion. Both a_0 and g_0 are anisotropic in that they vary with the orientation of the probe in the magnetic field. It is this anisotropy of a_0 and g_0 that is responsible for the effects of the rotational motion of the label on its ESR spectrum and that allows detection of the orientation of the label.

Role of medium polarity

Both g_0 and a_0 are solvent dependent, because the interaction between the solvent molecules and the N—O bond changes the spin density at the N-nucleus. Typically, a_0 is 1–2 G smaller and g_0 somewhat larger (+0.0005) in apolar solvents than in water. Sometimes, if regions of two distinct polarities are accessible to a rapidly tumbling nitroxide, the proportion of probes in each can be measured.

Role of medium viscosity

Anisotropic effects average out to nearly zero when the nitroxide radicals are in rapid motion tumbling rates of 10^{11}/sec. As the probe mobility is decreased, e.g., by progressively raising solvent viscosity, the spectrum also changes, showing first a broadening of the high-field resonance band (tumbling rate 10^9/sec = weak immobilization), then asymmetric broadening of all bands (tumbling rate 10^8/sec = intermediate immobilization), and finally when the radicals are strongly immobilized (tumbling rates $<10^8$/sec), appearance of the "rigid-glass" spectrum, with high- and low-field bands moved to higher and lower fields, respectively. The last spectrum exhibits every possible orientation relative to the applied magnetic field. Intermediate spectra arise from independent changes of a_0 and g_0 with viscosity (Fig. 8.3).

Orientation

As shown in model studies, the ESR spectra of immobilized nitroxide radicals allow detection of the orientation of the probes with respect to the laboratory magnetic field. The application of this technique to membranes will be discussed below.

Other Potentially Useful Effects

Nitroxide-nitroxide interactions

The magnetic properties of nitroxides can be altered by nearby molecular O_2, paramagnetic ions, other free radicals, and other nitroxides, When nitroxides are present in concentrations 10^{-3} M, interactions among individual electron spins can produce marked broadening of the ESR lines. The resulting spectra may be complex, but when the probes tumble very rapidly, the three-line spectrum broadens symmetrically. This phenomenon has potential applicability to the study of membranes; thus, information concerning *lipid-protein interactions* might be obtained by spin-labeling-selected lipid and protein sites.

Nitroxide effects on fluorescence and NMR

Nitroxide radicals profoundly affect fluorescent molecules by paramagnetic quenching. For this, the nitroxide must act as energy acceptor for the excited singlet state of the fluorophore (2). This offers an opportunity to locate the label relative to intrinsic or extrinsic membrane fluorophores. The large magnetic moment of unpaired electrons also broadens the resonance of protons and other nuclei lying near spin labels or paramagnetic ions. [Green et al. (*J. Chem. Phys.* **58**: 2690, 1973) present an important, timely discussion of fluorescence quenching by nitroxides.]

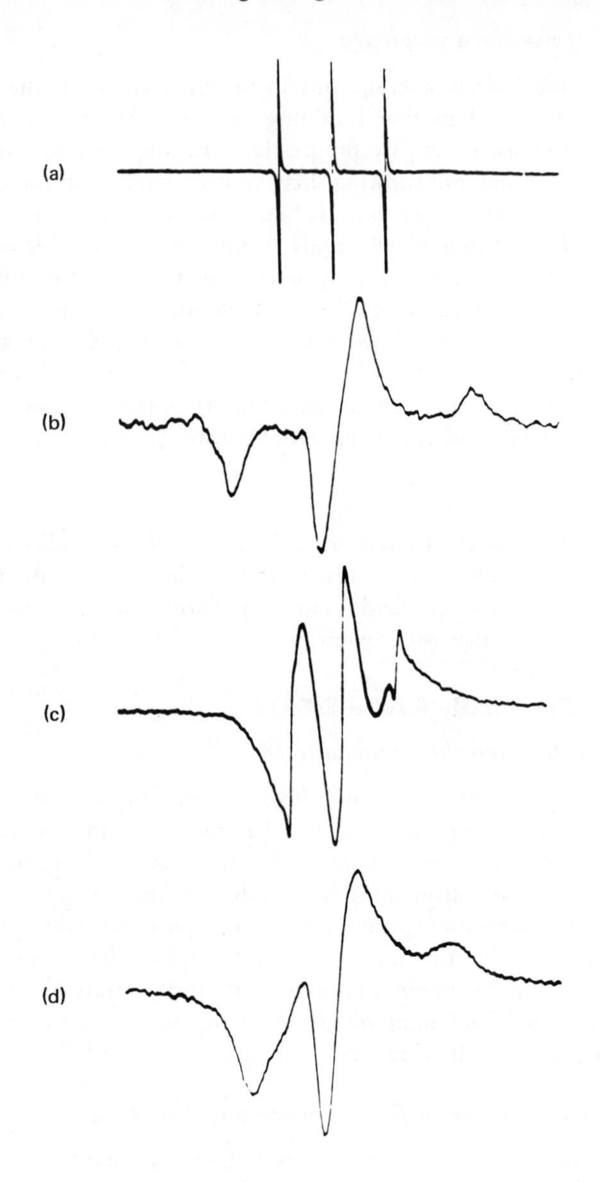

Fig. 8.3 An example of viscosity effects on ESR spectra. The figure shows the ESR spectra of (a) 10^{-4} M nitroxide IV in water at room temperature, (b) 10^{-4} M nitroxide IV in ethylene glycol at −65°C, (c) spin-labeled *Neurospora* mitochondria at room temperature, and (d) spin-labeled *Neurospora* mitochondria at −15°C. The distance between the outermost lines of spectrum (a) is 31 Gauss. All four spectra have the same horizontal scale; however, each has a different vertical scale. [From Keith et al. (46) through the courtesy of the authors and the National Academy of Sciences.]

Model Systems

Detergents

Waggoner et al. (3) have used nitroxides I and II, where the paramagnetic label is attached to aromatic residues, in their studies of micellar

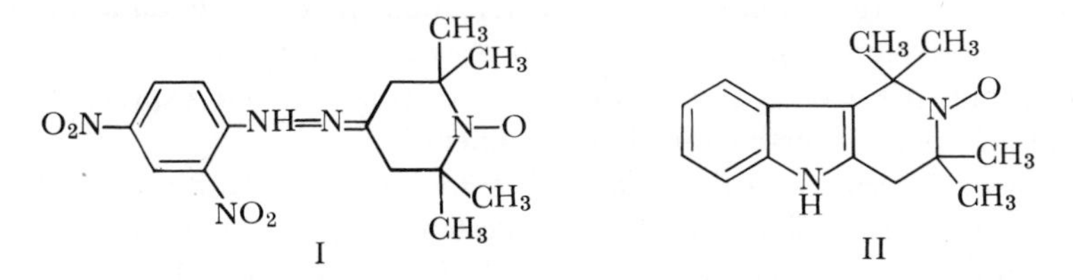

sodium dodecyl sulfate (SDS). In water, the ESR spectra of I and II consist of three lines of nearly equal width, a pattern typical of rapidly tumbling nitroxides. However, in 5% SDS, the spectra broaden asymmetrically, due to a decreased tumbling rate, manifested quantitatively by an increase in the rotational correlation time τ_C. Well below the critical micelle concentration (CMC) of SDS (0.23 percent), the values of τ_C are typical of water. Above the CMC concentration, τ_C is much greater, but independent of detergent concentration. The value of τ_C observed, 6×10^{-6}, deviates from the 10^{-8} expected if the probes were bound to rigid SDS micelles of typical radius 22 A, indicating that this association model is unlikely.

The magnitudes of the coupling constants reflect the polarity of the nitroxide environment, as do the optical absorption maxima of the aromatic residues in I and II. Accordingly, Waggoner et al. (3) compared the paramagnetic and optical properties of these probes in water, dodecane, and 5% SDS (Table 8.1). The data indicate that in SDS both

Table 8.1 Solubility, Hyperfine Coupling Constants, and Optical Absorption Maxima for Nitroxide I[a]

Solvent	Solubility (M)	Hyperfine Coupling Constant (Gauss)	Optical Absorption Maxima (A)
Water	5.0×10^{-6}	16.16	3690
Dodecane	2.4×10^{-4}	14.30	3440
5% NaDS	1.1×10^{-3}	15.72	3640

[a] From Waggoner et al. (3).

Note: All values for 23°C; relative error for coupling constants is ±0.05 gauss; absorption values are accurate to ±10 A. In case of SDS, 99% of the probe is bound to detergent micelles.

ends of the probe sense a similar environment—about 20% that of dodecane and 80% that of water. This excludes orientation of the molecules with the aromatic portion within the micelles and the paramagnetic section externally, or solution of the entire probe in the hydrocarbon core. These two models and the previous one—binding of the probe to the surface of a rigid micelle—are clearly inadequate. The data rather suggest a dynamic association of the probes with SDS micelles, in which the probe preserves "a random spatial orientation and experiences a relatively polar, time-averaged environment."

More recently, Dodd et al. (4) demonstrated linear relationships between solvent polarity and the hyperfine coupling constants of III, 2,2,6,6-tetramethyl-peperidine-1-oxyl (TEMPO) and IV, 4,4-dimethyl-oxazolidine-3-oxyl derivative of methyl 12-ketosterate.

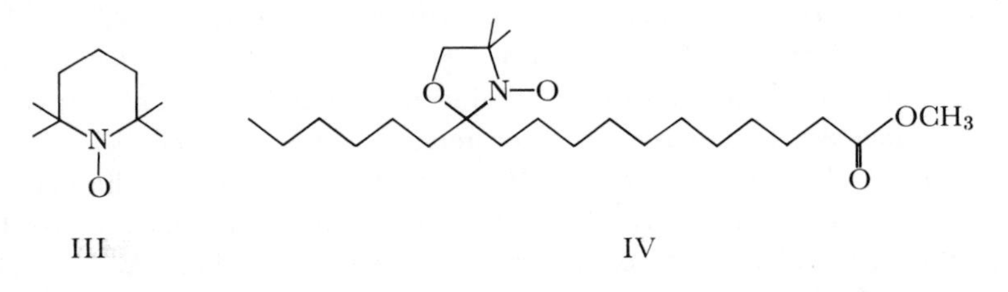

According to the calibration curve of IV, the nitroxide portion of this "lipid probe" lies, as expected, in the alkyl-chain region of lecithin and lysolecithin micelles. Importantly, however, the probe senses a water environment in the case of micellar SDS. Apparently, the 18-carbon molecule does not fit easily into a 12-carbon micelle and folds about carbon 12 so that the nitroxide protrudes into water. The data illustrate the difficulty in localizing the reporter group, even in simple systems, as well as possible perturbations by the probe, and call for caution in the interpretation of spin-labeling experiments on biological membranes.

Phosphatide Multilayers and Dispersions

Localization and orientation

Libertini et al. (5) incorporated nitroxide-labeled stearic acid, V, and a nitroxide-labeled steroid (cholestane), VI, into lecithin films

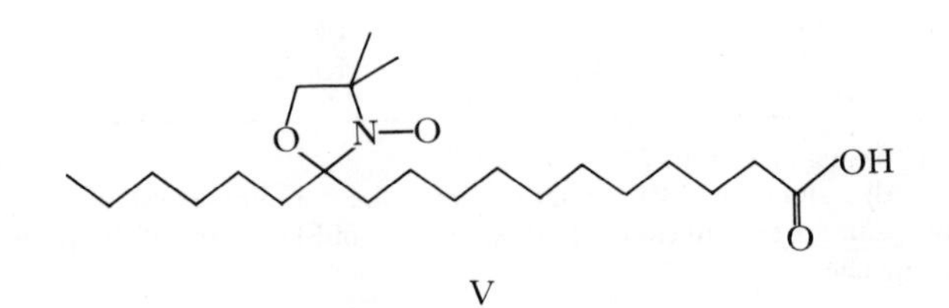

V

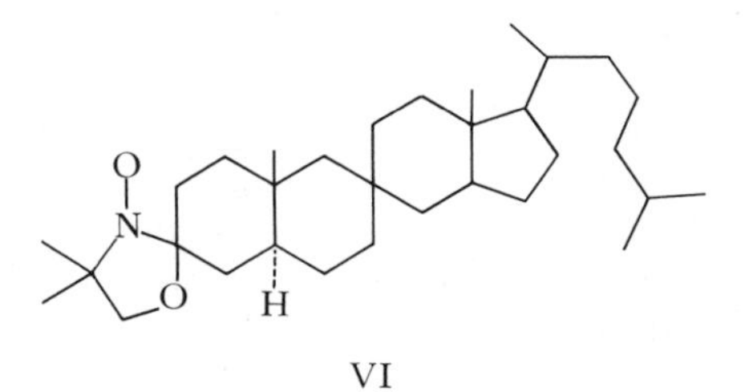

prepared by drying aqueous lipid dispersions on glass slides, assuming the lipid to form an oriented multilayer. Unfortunately, precise orientation of the lipid was not ensured by depositing it with the Blodgett technique (6)

Nevertheless, as the films are rotated in the magnetic field, their spectra show a smooth change in the value of a', the anisotropic coupling constant, corresponding to the distance in gauss between the central line and the low field. For the nitroxide-labeled stearic acid, V, a' is greatest with the film plane perpendicular to the magnetic field and minimal when the plane is parallel [Fig. 8.4 spectra (a) and (b)]. The film orientation for the maximum splitting is the opposite for nitroxide-labeled cholestane, VI, and the maximum value of a' is here found with the film plane parallel to the magnetic field [Fig. 8.4 (c) and (d)].

The largest values of a' occur when the magnetic field is parallel to the nitroxide Z axis (parallel to the nitrogen $2p\pi$ orbital), i.e., the Z axis of V tends to align at right angles to the film plane and that of VI parallel to the plane. Since the Z axis of the stearic acid nitroxide is parallel to

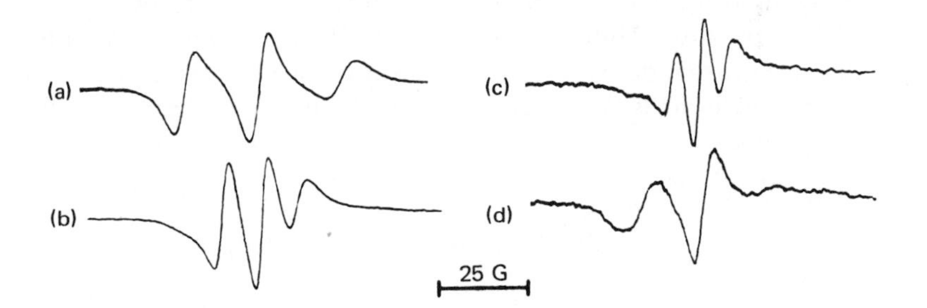

Fig. 8.4 Room-temperature ESR spectra of nitroxide-labeled lipids in lecithin films, showing the changes in spectra that result when the lecithin films are rotated in the magnetic field. Spectra (a) and (b) are of nitroxide V with the magnetic field perpendicular and parallel, respectively, to the plane of the glass slide supporting the lecithin film. Spectra (c) and (d) are of nitroxide VI with the magnetic field perpendicular and parallel, respectively, to the plane of the glass slide. The mole ratio of the nitroxides to lecithin is 5×10^{-3}. (From Libertini et al. (5) by courtesy of the authors and the National Academy of Sciences.)

the hydrocarbon chain and that of the cholestane derivative perpendicular to the long axis of the molecule, the data indicate that both molecules orient preferentially with their long axes perpendicular to the film plane. However, comparison of the present data with those obtained from single crystals indicates that the nitroxides are not *all* aligned perpendicular to the film plane. This may be due to defective multilayers (7) and/or a broad distributional orientation about a direction normal to the film plane. More recently, Jost et al. (8) extended the study of lecithin multilayers, using nitroxide radicals bonded to the 5, 7, 12, and 16 positions of stearate and employing 2-oxyl propane, oriented in single crystals of 2,2,4,4-tetramethyl-cyclobutane-dione, as reference for the ESR tensor. All spin labels oriented with their long axes preferentially perpendicular to the lecithin film, but the molecular motion increased (a) as the relative humidity was raised, and (b) the further the label is from the carboxyl end. The humidity effects are maximal when the label is near the —COO— residue, where it is anchored into the lecithin layer.

TEMPO, III, bound to phosphatide micelles and various membranes, yields a three-line ESR spectrum, representing the sum of signals arising from (a) label in aqueous solution and (b) label in hydrophobic regions of low viscosity (9). The spectrum can be duplicated by recording the ESR of a dual sample: one, a solution of the probe in water and the other, a solution of the probe in dodecane. The band shapes point to rapid tumbling in the hydrophobic phase ($\tau_C = 10^{-9} - 10^{-11}$ sec); either the "binding region" is naturally fluid or TEMPO makes it so. TEMPO inserted into micelles of SDS or sodium deoxycholate yields spectra that are more an average of hydrophobic and aqueous environments.

More recently, Hubbell and McConnell (10) studied the ESR spectra of nitroxide derivatives of 12-ketostearic acid, 5-ketostearic acid, 5-ketotricosanoic acids, $V_{m,n}$, codispersed with purified soybean phospha-

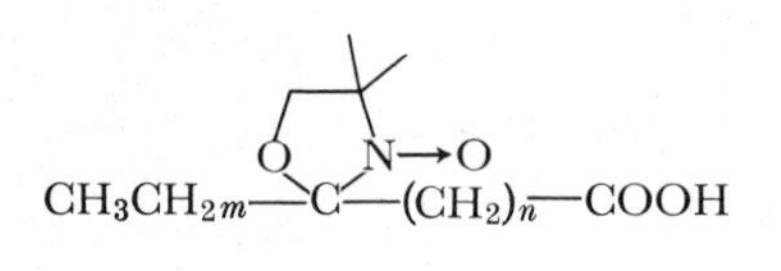

$V_{m,n}$

tides in 0.05 M Tris HCl, pH 8. These substances are useful because (a) the nuclear hyperfine tensor of the nitroxide group is axially symmetric; (b) the unique principal hyperfine axis of the ^{14}N nuclear hyperfine tensor is parallel to the hydrocarbon chain, and (c) the values of *m* and *n* can be adjusted to place the label closer or further from the hydrophilic end of the molecule.

The ESR spectrum of $V_{17,3}$ in a sonicated dispersion of phosphatides is unusually sharp, even though the motion of the label is evidently severely restricted. This indicates rapid rotational diffusion of the labels about the hydrocarbon axis of the molecule. The labeled fatty acids appear packed into a bilayer array of phosphatides and, since the mobility of the nitroxide diminishes the closer it is to the polar head group, the bilayer is considered more rigid at the aqueous interface than in the hydrophobic domain. However, packing near the reporter group must differ from that in a normal bilayer (Chap. 3).

Spin-labeled steroids, such as VII, incorporated into phosphatidyl-

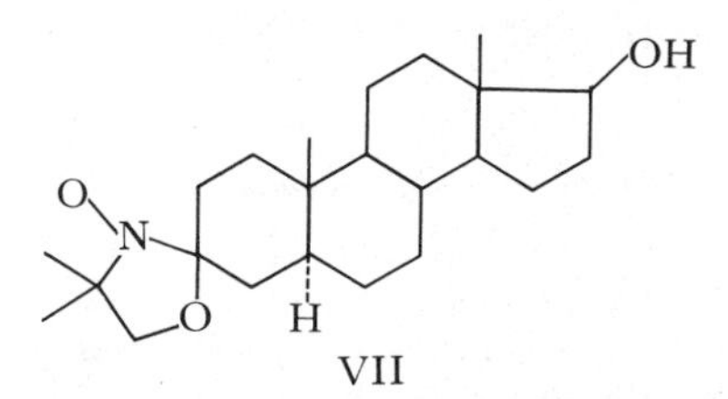

VII

serine liposomes yield ESR spectra indicating high mobility of the nitroxide group ($\tau_C = 10^{-7} - 10^{-8}$ sec). Since the label is rigidly linked to the nucleus, the data indicate that the entire molecule undergoes rapid rotational motion in the hydrophobic regions of the phosphatide micelles. Again, this may reflect native fluidity or a perturbed state.

Whether dissolved in water-glycerol mixtures of varying viscosities or incorporated into phosphatidylserine micelles, these spin labels give no evidence of strong anisotropic motion. Indeed, no tested environmental variables (e.g., ionic strength and composition) produce marked rotational anisotrophy in phospholipid systems. This contrasts to the properties of nerve "membranes" discussed below.

Hsia and Boggs (11) evaluated the effects of cholesterol and pH on the structure of phosphatidylethanolamine multibilayer films, using nitroxide-cholestane as an ESR probe. Lecithin multibilayers were used as controls. They found that the spin label oriented significantly only in hydrated films. In lecithin-cholesterol films, pH had almost no effect on probe alignment, but in phosphatidylethanolamine multibilayers, there is a sharp decrease in orientation of the spin label between pH 8.3 and 10.4, an effect partially countered by cholesterol. The data suggest that well below the pK of the amino group, this residue neutralizes the negative charge of the phosphate group, allowing close packing of the head groups. This effect diminishes as the amino group is titrated, allowing electrostatic repulsion between the phosphate residues. The role of cholesterol appears to be via restriction of hydrocarbon chain mobility.*†

Peterson et al. (12) have used steroid spin labels to study the effects of *aliphatic alcohols* on the structure of phosphatide multibilayers. As noted elsewhere here, the cholestane spin label assumes an orientation with its long axis aligned preferentially perpendicular to the bilayer plane. Various aliphatic alcohols perturb the structural integrity of the bilayers at concentrations correlating well with those inducing anaesthesia in certain biomembranes and inhibiting osmotic lysis of erythrocytes. Benzyl alcohol, promethazine, and chlorpormazine behaved analogously.‡

Lateral and transmembrane diffusion

Phospholipids in a bilayer may diffuse in two directions, namely, *parallel* to the bilayer plane or *perpendicular* to it (by moving from one

* Marsh and Smith (D. Marsh and I. C. P. Smith, *Biochim. Biophys. Acta* **298:** 133, 1973) recently introduced the use of interacting spin-labeled cholestane pairs, to clarify the effect of cholesterol on phosphatide packing in lecithin multibilayers; they monitor lateral molecular separations of the probes through their spin-spin interactions.

They find that, in the case of dipalmitoyl lecithin, an increase in the ratio [cholesterol:lecithin] increases the label separation, presumably by decreasing the packing of the fatty acid chains. In contrast, cholesterol condenses egg yolk lecithin and dioleoyl lecithin. This effect can be attributed in part to closer packing of the fatty acid chains induced by their interaction with the sterol. However, some of the effect seems to involve sequestration of cholesterol into "cavities" within the lipid bilayers.

† The ESR spectra of stearate nitroxides ($V_{m,n}$) incorporated into a lamellar lipid phase (sodium-cardiolipin), a hexagonal cylindrical phase (calcium cardiolipin), as well as ganglioside micelles, lamellar 1-monoglycerides, and hexagonal monoglycerides all yield spectra typical of anisotropic motion (J. M. Boggs and J. C. Hsia, *Proc. Natl. Acad. Sci.* (U.S.), **70:** 1406, 1973). The spectra from micellar and hexagonal phases cannot be distinguished from those from lamellar dispersions. The spin probes exhibit preferred orientation in cylindrical and spherical micelles, with the long probe axes parallel to the lipid hydrocarbon chains. In the hexagonal phase the mobility of the acyl chains increases toward the terminal methyl, as in the case of lamellar phases.

‡ A spin label (nitroxide cholestane) study of the effects of local anaesthetics on ox-brain lipid multibilayers (K. W. Butler, H. Schneider, and I. C. P. Smith, *Arch. Biochem. Biophys.* **154:** 548, 1973) indicates that such agents tend to disrupt the lipid bilayer structure at pHs above their pK. In films containing a low proportion of cholesterol, local anaesthetics engender formation of multilamellar lipid arrays. This effect requires lower anaesthetic levels than the disordering produced in the presence of cholesterol. It appears that a local anaesthetic can either disrupt or stabilize artificial lipid multibilayers, depending upon their cholesterol content and pH, as well as the concentration of the anaesthetic.

Trudell et al. (J. R. Trudell, W. L. Hubbell, and E. N. Cohen, *Biochim. Biophys. Acta* **291:** 321, 328, 1973) show that spin-labeled phosphatidyl choline, incorporated into phosphatidyl choline liposomes, decreases in order upon exposure to inhalation anaesthetics. They show this to be due to a general fluidization of the lipid membranes. They also document reversal of this fluidization effect by 150 to 200 atmospheres of helium or hydrostatic pressure, conditions known to reverse the anaesthetic effects in diverse organisms. They reason that anaesthetic action involves disordering of membrane lipid domains and that this can be blocked by high pressure.

monolayer to the other). These processes have been evaluated using the spin-labeled phosphatidylcholine VIII:

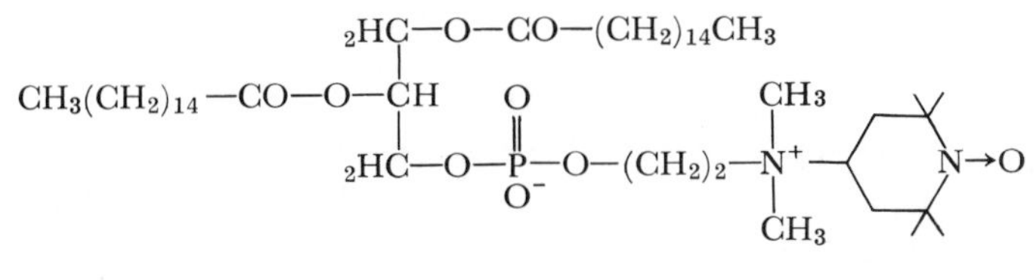

VIII

This material was designed because of its resistance to oxidation and analogies to egg lecithin.

Lateral diffusion was evaluated by PMR (13), taking advantage of the fact that the $—N^+(CH_3)_3$ PMR of sonicated, egg lecithin residues, doped with VIII, is broadened by the spin label to an extent depending on its proportion. This is proved to be unrelated to vesicle collision or fusion (by lysolecithin), as well as spin-label exchange. Rather, the broadening (9.7 Hz at 35°C with a VIII: egg lecithin ratio = 0.01) reflects the rapid lateral diffusion of the spin label (>0.05 μ/sec). It remains to be determined whether this rate is representative of normal lecithins.

To establish transmembrane diffusion ("flip-flop"), Kornberg and McConnell (14) utilize the fact that lecithin bilayer residues are not permeable to ascorbate, which rapidly reduces nitroxides. Hence, in a vesicle doped with VIII, any spin label located on the outside is inactivated by added ascorbate, while internal label is protected. With the vesicle size used, about 65% of the label is on the outside and is immediately inactivated upon ascorbate addition. The resulting distributional assymetry decays with a half time of 6.5 hr at 30°C, corresponding to an inside-to-outside translation at a rate of $<2 \times 10^{-5} \cdot sec^{-1}$ and a reverse movement at a rate of $\sim 10^{-5} \cdot sec^{-1}$.

It is difficult to evaluate these rates in biologic terms, since the "flip-flop" exchange of a labeled lecithin for an unlabeled one, which the authors propose, has multiple alternatives. Also, slow as the observed rate is, that of normal lecithin, which is more polar than VIII, is assuredly slower. It is clearly highly desirable to support these ingenious experiments by large-angle X-ray diffraction, to determine how VIII fits into lecithin bilayers.

Transmembrane potential

In a logical extension of the previous studies, Kornberg et al. (15) synthesized IX,

CO_2^-—CHOH—CHOH—CO—NH—⟨ N→O

tempotartrate

IX

which can permeate into egg lecithin vesicles.

The transmembrane distribution of this label, defined from determinations of *ascorbate-inaccessible* tempotartrate and measurements of intravesicular water, reflects the electrical potential across the H^+ permeable vesicle membrane and is equal to the transmembrane [Cl^-] gradient. It appears that the [Cl^-] gradient produces the [H^+] gradient, i.e., that the egg lecithin membranes are HCl permeable.

Artificial Lipid-Protein or Lipid-Peptide Systems

There exists a large armamentarium of nitroxide derivatives, which can be covalently linked to reactive groups on proteins, e.g., nitroxide isocyanate (16), nitroxide maleimide (17), nitroxide iodoacetamide (18, 19).

Accordingly, Barratt et al. (20) labeled several proteins with a nitroxide maleimide, X, to gain some insight into lipid-protein interactions.

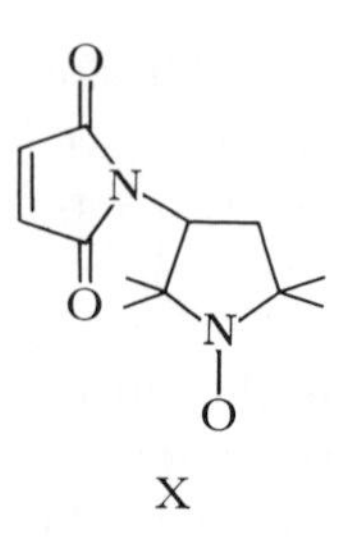

X

Cytochrome c, lysozyme, ribonuclease A, histone, protamine, and poly-L-lysine reacted with X, yielded narrow ESR spectra in aqueous solution. Since none of the proteins possesses reactive SH groups, the spectra are due to X-labeled —NH_2 groups. The spectra, showing the appreciable mobility of the nitroxide groups relative to the protein molecule, indicate that the labeled —NH_2s are on the protein surface.

All of the labeled substances formed water-insoluble complexes upon interactions with aqueous dispersions of phosphatidylserine:lecithin (3:2), and all of these complexes were partially soluble in isooctane. Accordingly, ESR spectra of the labeled substances were recorded from aqueous solutions, aqueous dispersions of the lipid-protein complexes, and isooctane solutions of the same complexes. The data are consistent with the expected ionic combination of phospholipids and basic proteins. However, they also show that mobility of the probe with respect to the protein-lipid complex is much greater in water than in isooctane. The authors suggest very plausibly that this arises from the orientation of the lipid hydrocarbon chains; these, not being directly involved in the lipid-protein interaction, take up a conformation determined by the solvent.

Hong and Hubbell (21) have presented a thought provoking study on phosphatide bilayers containing the visual pigment rhodopsin. This protein is rather insoluble in water, contains an unusual proportion of apolar amino acids, and, according to X-ray diffraction data (22), appears to penetrate at least partially through the photoreceptor membrane.

The authors used XI:

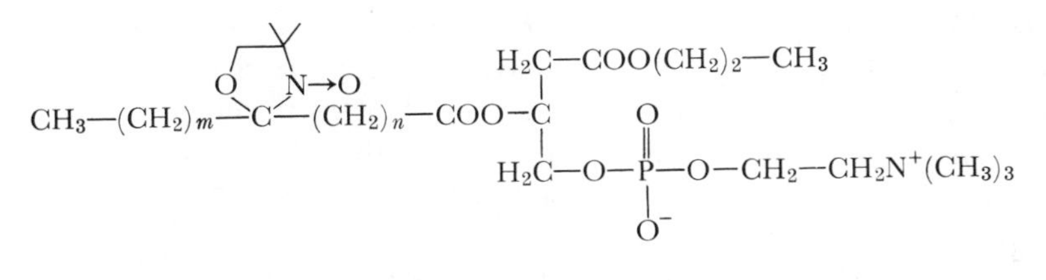

XI

as spin label and also carried out some pertinent experiments by freeze etching–freeze cleaving electron microscopy. Rhodopsin was extracted by dodecyl-trimethyl-ammonium bromide (^{14}C labeled) and combined with egg lecithin containing spin label after removal of the dodecyl-trimethyl-ammonium bromide by dialysis.

The ESR data indicate that rhodopsin restrains the segmental motion of phosphatide hydrocarbon chains in a manner analogous to the role of cholesterol. The further the spin label lies from the ester linkage, the greater the mobility of the spin label. Interestingly, freeze fracture electron micrographs of rhodopsin-phosphatide recombinants show clusters of particles in the fracture plane, which are lacking in pure phosphatide liposomes. In both, the outer (etchable) surface is smooth. All in all, the data are consonant with the view first proposed by Wallach and Zahler (23) that the surfaces of membrane proteins may have a structure allowing close interactions with phosphatide acyl chains. The data do not yet allow conclusions as to the functional state of the incorporated rhodopsin.*

* Jost et al. (P. C. Jost, O. H. Griffith, R. A. Capaldi, and G. Vanderkooi, *Proc. Natl. Acad. Sci.* **70:** 480, 1973) performed an experiment highly pertinent to membrane structure and also supporting the hypothesis earlier proposed by one of us (D. F. H. Wallach and P. H. Zahler, *Proc. Natl. Acad. Sci.* **56:** 1552, 1966; D. F. H. Wallach and A. Gordon, *Fed. Proc.* **27:** 1263, 1968). They examined the interactions of cytochrome oxidase with phosphatides, using 16-nitroxide stearate as probe. At low phospholipid: protein ratios ($\leq$0.19 mg lipid/mg protein) the probe is highly immobilized, but at higher phosphatide proportions an additional signal characteristic of fluid bilayers appears. Analysis of the data suggests a single layer of immobilized lipid ensheathing the protein surrounded by a more fluid lipid bilayer region.

Verma et al. (24) have carried out very closely related studies on egg lecithin multibilayers containing light-sensitive pigments and the spin-labeled steroid 3-spiro [2′-(*N*-oxyl-4′,4′-dimethyloxazolidine)]-cholestane (VI). Importantly, they observe significant changes in the ESR spectra of the spin-labeled multibilayers containing chlorophyll a, retinal or methylene blue, when these are illuminated at appropriate wavelengths. Maximal photoresponses occurred at 450 nm–650 nm with chlorophyll a, 350 nm–375 nm with retinal, and 670 nm with methylene blue. The spectra suggest that chlorophyll a and retinal reside in apolar regions, there inducing ordering effects similar to those of cholesterol, while methylene blue very likely accumulates in both polar and apolar domains of the phosphatide multibilayers. It appears that illumination disorders the microenvironment of the spin label, widening the average angular deviation of its long axes from 90° to the bilayer plane, but still maintaining a net preferential orientation of the molecular axes normal to the bilayer surface.

The interaction of the small amphiphilic polypeptide *melittin* with phosphatides is of considerable interest, since this substance lyses a number of biomembranes (25). This polypeptide from bee venom has a molecular weight of 2848 and the following unusual amino acid sequence:

```
1                5                   10
GLY ILE GLY ALA VAL LEU LYS VAL LEU THR
                                15                 20
    THR GLY LEU PRO ALA LEU ILE SER TRP ILE
                                        25
                   LYS ARG LYS ARG GLN GLN
```

Clearly, the segment from 1 to 20 is quite apolar. However, because the polypeptide is so short, it cannot assume a globular shape with buried hydrophobic residues and, instead, associates into a teramer in aqueous solution (26). Williams and Bell (27) have incorporated 5- and 12-nitroxide stearates into diverse phosphatide liposomes to monitor the effect of melittin on these structures. In this, they employed sonically dispersed egg lecithin and phosphatidylserine as well as isolated *E. coli* membranes and sonically dispersed lipids extracted therefrom.

Although melittin perturbs the permeabilities of liposomes at levels as low as 10^{-6} M (26), no ESR effects were observed below 10^{-4} M. Then the ESR spectra show only slight decreases of spin-label *mobility*, but a large increase of spin-label incorporation in all the membrane systems. The authors suggest that at low concentrations, melittin monomerizes allowing its hydrophobic segments to totally interfere with phosphatide packing and producing permeability changes as a consequence. At higher concentrations, the perturbations are more extensive, allowing greater incorporation of spin labels that do not fit well into a normal, tightly packed phosphatide bilayer.*

These results differ substantially from ESR studies on the interaction of the peptide antibiotics polymyxin-B and gramicidin-S with phosphatide liposomes as reported by 12-nitroxide stearate (28). Neither substance interacts with negatively charged liposomes at pH 7 (cardiolipin, phosphatidylserine). No effect is observed with egg lecithin either. However, 12-nitroxide-stearate incorporated into dipalmitoyl lecithin liposomes is mobilized by polymyxin-B below the phosphatide transition temperature but immobilized by gramicidin-S. The ESR data, together with experiments utilizing differential scanning calorimetry, NMR and IR spectroscopy, suggest that polymyxin-B interacts with both the polar and apolar regions of phosphatide bilayers, while gramicidin-S interacts with the polar domains predominantly.

A very interesting aspect of these studies concerns the disruption of lecithin liposomes by gramicidin-S into polypeptide-phosphatide complexes of mol wt$<$100,000 in which the phosphatide is arrayed as a highly packed bilayer. The authors speculate that membrane proteins could transport membrane lipids from their sites of biosynthesis to their membrane loci in an analogous fashion.

Membranes

Erythrocyte Membranes

Noncovalent insertion of paramagnetic probes

Landsberger et al. (29) have performed a study which is very fundamental to the exploration of erythrocyte membranes by spin-label probes. They found the ESR spectra of intact human erythrocytes and ghosts isolated therefrom, spin-labeled with nitroxide-stearates $V_{12,3}$ and $V_{1,4}$, to be identical as far as the membrane lipids are concerned. However, the labels were transferred to the membranes from a bovine-serum-albumin complex, which has a characteristic spectrum, and rigorous analysis of the data shows that the complex is *not adsorbed to*

* Further information on the membrane actions of mellitin has been obtained using erythrocyte ghosts and their aqueously dispersed lipids, both spin-labeled with exogeneously added 1-nitroxide stearate, 5-nitroxide stearate, or 12-nitroxide stearate. Melittin produces a dose-dependent increase of the orientation of the 5-nitroxide derivative whether added to the membranes or their extracted lipids. In the case of 5-nitroxide stearate, incorporated into intact erythrocytes melittin first increases and then decreases molecular motion (D. Hegner, U. Schummer, and G. H. Schnepel, *Biochim. Biophys. Acta* **291:** 15, 1973). It would appear likely that the high degrees of immobilization observed are analogous to the immobilization effects of cytochrome oxidase noted by others (p. 317).

intact cells, but strongly so to *isolated membranes*. The authors suggest that this arises from an *increased accessibility* of the inner membrane surface to the albumin-spin-label complex.

Although TEMPO does not label erythrocytes and erythrocyte membranes (9), nitroxide derivatives of 12-ketostearate, 5-ketostearate, 5-ketotricosanoate, and 5 α-androstan-3-one 17 β-ol ($V_{m,n}$ and VII) do (10). These probes have a greater avidity for fluid, hydrophobic regions; indeed, they readily bind to BSA, yielding ESR signals indicative of strong immobilization of the probe.

Labeling of the erythrocytes here was by transfer from serum, which had been saturated with the label, the excess of the latter being removed by gel filtration. The authors say the labeling process did not cause hemolysis and did not affect the shape of the erythrocytes. All of the spin-labeled substances are more strongly immobilized in erythrocyte membranes than in phosphatide dispersions, but probe mobility is greater with $V_{5,10}$ than $V_{12,3}$.

In an attempt to determine the orientation of the probes relative to the cells' surfaces, ESR spectra were obtained on cell suspensions passing through a thin, flat aqueous sample chamber at a velocity such that higher flow rates produced no further spectral changes. Under these conditions of approximately laminar flow, the erythrocyte cylinder axes are preferentially aligned perpendicular to the larger flat face of the sample chamber. By proper alignment of the chamber, spectra were obtained with the erythrocyte cylinder axes oriented preferentially perpendicular or parallel to the laboratory magnetic field.

Nitroxide $V_{5,10}$ and the steroid nitroxide exhibit considerable spectral anisotropy, even though the hydrodynamic erythrocyte orientation is not perfect, and most of the membrane of a biconcave erythrocyte is neither perpendicular nor parallel to the cylinder axis of the cell. The sense of the preferred orientation is one in which the unique ($2p\pi$ orbital) axis of the nitroxide (i.e., the long amphiphilic axis of the molecule) is perpendicular to much of the membrane surface. The authors interpret the immobilization and anisotrophy to mean that the hydrophobic regions of erythrocyte membranes are similar to those of lipid bilayers, but are more tightly packed than those of phosphatide dispersions. However, since both types of label are strongly bound and "immobilized" by serum albumin, we see no evidence to exclude immobilization and orientation of the label in erythrocyte membranes by membrane proteins, alone or in conjunction with membrane lipids.

Kroes et al. (30) studied the ESR spectra of a nitroxide-labeled hydrocarbon, fatty acid, and steroid, all bound to erythrocyte membranes. The spectra were characteristic of probes in apolar environments. Ascorbate reduced the membrane-bound probes at a slow rate, also suggesting location of the probes in nonaqueous regions. Immobilization was weak for the hydrocarbon, moderate for the fatty acid, and strong for the steroid. The mobility of the steroid nitroxide could be increased irreversibly by raising temperature. The probe mobilities were similar to those found with aqueous codispersions of lecithin and cholesterol.

More recently, Kroes et al. (31) used 4-nitroxide stearate and 12-nitroxide stearate inserted into erythrocyte membranes to explore the role of cholesterol in the membranes. High- and normal-cholesterol guinea pig erythrocytes were prepared by feeding the animals high- and normal-cholesterol diets for 5–8 weeks before bleeding. The spin labels, added to the washed cells in minute amounts of ethanol, exhibited markedly different spectra in the two-cell types. The spectra of the cholesterol-laden cells report a marked increase in the viscosity of the probe's microenvironment. The nitroxide groups appear to be localized in the membrane core, since they are not subject to reduction by ascorbate, and their spectra were unaffected by treating the cells with proteases.

Cholesterol-rich and control cells yielded the same ESR spectra after labeling of protein —SH groups with nitroxide maleimide. This is compatible with the notion that the membrane proteins are not affected by cholesterol loading. The authors suggest that the generally decreased permeability of high-cholesterol membranes (32) is due to cholesterol-induced, high viscosity of the lipid domains in the membranes.

Chemical coupling of spin labels to membrane proteins

Chapman et al. (33) have initiated spin-labeling studied on the proteins in human erythrocyte membranes. As before (20), they utilized a nitroxide maleimide, which reacts with the —SH and $—NH_2$ groups of proteins.

The ESR spectrum of the labeled membranes consists of (a) three narrow lines arising from freely tumbling nitroxide and (b) three broad bands, due to spin-label probes with correlation times of 10^{-8} sec or more. These signals represent two types of binding site on membrane *proteins*, since all free labels had been removed earlier, only traces of label occurred in membrane lipid, and the amino sugars of membranes, being *N*-acetylated, do not react with X.

Importantly, treatment of the membranes with *N*-ethyl-maleimide before spin labeling almost fully abolishes the broad spectral component, indicating that this arises from labeled SH-groups. Griffith and McConnell (17) found that the broad ESR component of the BSA-X complex could also be removed by prior treatment with *N*-ethyl-maleimide. Some membrane SH-groups are thus accessible to the spin label, but do not allow it free mobility. The narrow ESR component is presumably due principally to spin-labeled lysines.

Membrane Modification and Recombination of Membrane Components

Berger et al. (34) were first to investigate the interaction of sialic-acid-free erythrocyte membrane protein (s.a.f.) (35) and its recombinates with membrane lipids (36). In these experiments, the authors utilized a nitroxide anhydride, XII:

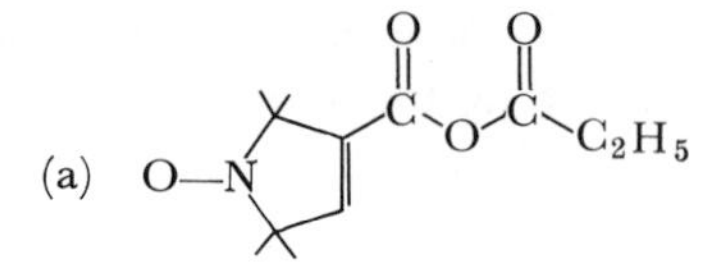

XII

(b) $V_{10,5}$ and

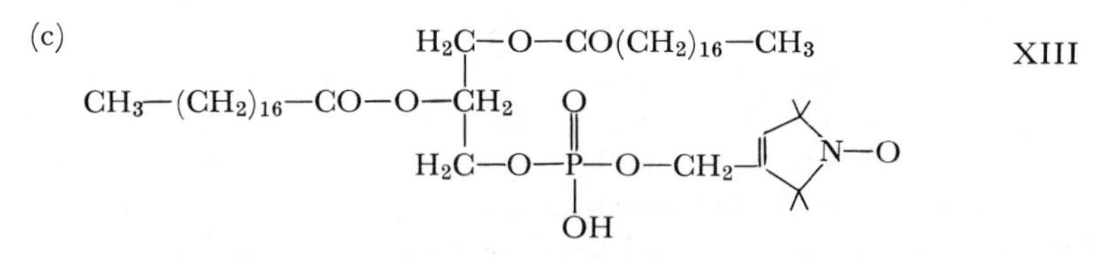

The ESR of spin-labeled s.a.f. exhibits a broad component attributed to reacted —SH, and a well-resolved contribution due to labeled NH_2 and —NH groups. The spectra of labeled protein are not influenced by addition of sonicated membrane lipids at pH 7, but increase the broad component drastically when recombined with lipids at pH$\sim$4, according to Zwaal and van Deenen (36). These data indicate that *recombination with lipid* alters protein *structure.*

Looking at the lipid side, use of $V_{10,5}$ and XIII indicates that these compounds can be extensively immobilized upon binding to the erythrocyte membrane "apoprotein," suggesting extensive architectural changes, which are lipid stabilized. Both apolar and polar lipid-protein interactions appear involved, judging from the noted perturbations of lipid probes with the nitroxides in apolar and polar regions.

The work is extended in Berger et al. (37). Here certain effects of lysolecithin were also measured. Combination of apoprotein with this phosphatide limits the mobility of $V_{10,5}$ more than when phosphatidylserine, phosphatidylcholine, or mixed membrane lipids are used; lysolecithin appears more strongly bound above pH 5. Moreover, lysolecithin appears to interact more strongly with apoprotein than with bovine serum albumin or acid casein. At neutral pH, where the apoprotein and lipids are negatively charged, the interactions are apolar.

At pH 4, the protein bears a positive charge and reacts with the anionic lipid by electrostatic mechanisms also. These studies suffer from a disadvantage, other than inherent in the probe approach, namely, the membrane "apoprotein" prepared by their techniques assuredly represents a dubious yield of heterogeneous, membrane protein aggregates. Moreover, other studies show that "recombination" does not reproduce the native structure.

The problem of lipid-protein interactions in erythrocyte ghosts has also been investigated in Simpkins et al. (38), following the response of the ESR labels

(a) 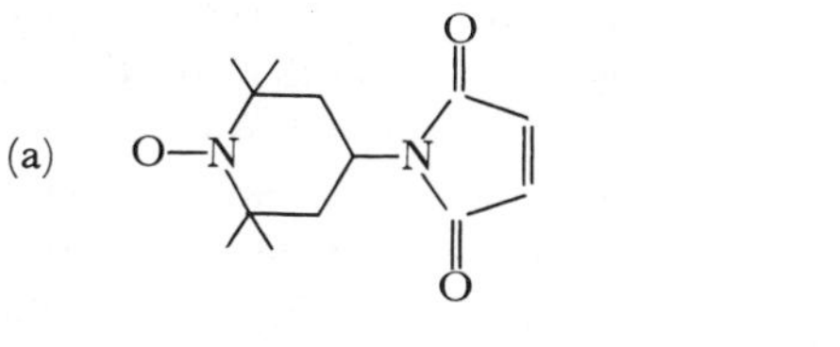

X

(b) O—N NHCCH₂I 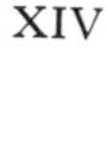 XIV

(c) 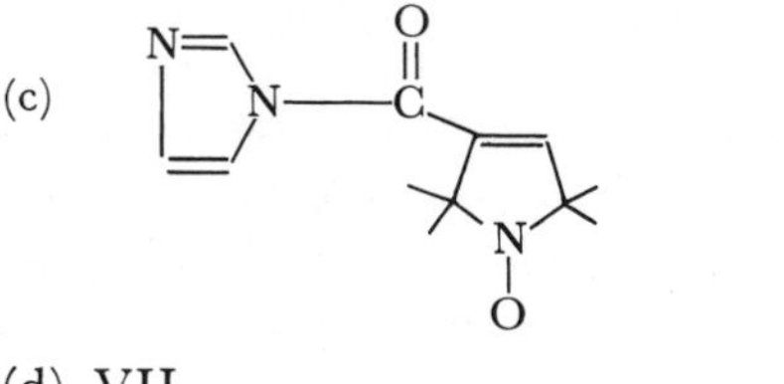

XV

(d) VII
(e) 12-nitroxide stearate (V)
(f) 14-nitroxide stearate (V)

to treatment of the ghosts with phospholipase C. Fat-free serum albumin was used as vehicle for transfer of the labels to the membranes.

The spin labels covalently bound to membrane protein (a)–(c) remain unchanged after phospholipase C treatment. The steroid probe, VII, also remained unchanged. Since several others' data place this compound fairly near the water interface, one must assume that the label is associated primarily with the 30–40% phosphatide resistant to phospholipase C.

In contrast, the stearate probes appear to become much more mobile after membrane treatment with phospholipase C. All in all, the data are taken to show that membranes contain discrete phosphatide regions, which can change state without influencing protein architecture. Comparative data, on mitochondrial membranes, are consistent with this notion.*

The results of Berger et al. (34) and Simpkins et al. (38) are not contradictory (a) if one accepts the view (23) that membranes are topologically heterogeneous lipid-protein arrays and (b) if one appreciates the quantitative uncertainties of ESR.

Metcalfe et al. (39) have correlated the erythrocyte membrane effects with those observed in lecithin vesicles using probe III as reporters and cholesterol and BeOH as "perturbants." Both BeOH and cholesterol narrowed the ESR bands. However, unlike the effects observed with BeOH in both cases, this reagent had no effect on the ESR spectra in the presence of cholesterol. This is attributed to lesser steric interactions between the tempoyl group of the probe and vicinal $—N^+(CH_3)_3$ groups than between these themselves. Cholesterol primarily increases the spacing of the native lecithins, allowing greater mobility of the nitroxide. At the same time, cholesterol diminishes the chain motion of the lecithins drastically. The authors believe that the two effects balance in the presence of benzyl alcohol, so that this produces no detectable effect. This finding clearly indicates further caution in using spin labels to monitor membrane behavior.

Metcalfe's data lead to an important generalization, namely, that even "lipoidal" spin labels bind to both protein and lipid components of membranes and, in both cases, yield similar responses to perturbing agents, because of *the common nature of the chemical forces involved in the interactions with either component. Hence attempts to determine the localization of small molecules in the membrane must generally rest on quantitative rather than qualitative differences in the binding to membrane lipids and protein.*†‡

* Morin et al. (F. Morin, S. Tay, and H. Simpkins, *Biochem. J.* **129:** 781, 1972) report major differences between plasma membrane, smooth- and rough-endoplasmic reticulum, using spin-labeling techniques. Thus the spectra obtained using 5-nitroxide stearate suggest that the lipids in rough endoplasmic reticulum are more mobile than in the smooth. In contrast, 7-nitroxide stearate clearly distinguishes between plasma membrane and smooth endoplasmic reticulum; this is attributed to the greater cholesterol level in plasma membrane and its "condensing" effect on phosphatides. Finally, the ESR signals from *N*-ethyl-maleimide labeled plasma membrane and endoplasmic reticulum indicate a greater proportion of mobile —SH groups on the former.

† Spin-labeling of SH groups on erythrocyte ghost proteins with nitroxide maleimide allows one to monitor perturbations of protein structure produced by various membrane solubilizers (F. H. Kirkpatrick and H. E. Sandberg, *Biochim. Biophys. Acta*, **298:** 209, 1973). Anionic detergents, guanidine HCl, and KSCN produce changes in the ESR spectrum consistent with alterations of protein structure. These correlate with solubilization. The threshold for structural change in the case of sodium dodecyl sulfate coincides with the critical micelle concentration. Triton X-100 apparently causes no structural change accessible to the spin label employed.

‡ The fluorescence of ANS bound to erythrocyte ghosts is enhanced by 2-(N, N-diethylamino) ethyl p-alkoxy benzoates (local anaesthetics) (D. D. Koblin, S. A.

"Excitable" Membranes

Hubbel and McConnell (10) have employed the spin-label TEMPO, III (5×10^{-4} M, in physiologic media) in their ESR studies of membrane structure. They found no probe accumulation by human erythrocytes, erythrocyte "ghosts," glycerinated rabbit skeletal-muscle fibers, *Neurospora* mitochondria, and bovine serum albumin. That TEMPO did not label the hydrophobic pockets of BSA appears curious, since these are readily accessible to ANS, a much larger molecule and, in the case of defatted BSA, to nitroxide-labeled steroids (40); perhaps the BSA employed was already saturated with fatty acids.

However, when added to concentrated human serum, rabbit vagus, frog muscle fibers, rabbit sarcoplasmic reticulum, and, to a lesser extent lobster walking-leg nerves, TEMPO yields an ESR spectrum similar to that obtained with phospholipid vesicles. Binding by vagus nerve is abolished by prior extraction with ethanol-ether, but not when ether alone is used. There is no evidence for spin interaction, excluding formation of lipid TEMPO aggregates.

The authors suggest that TEMPO concentrates in poorly viscous, lipidlike, hydrophobic regions of certain "excitable membranes," possibly the apolar cores of phosphatide bilayers, but there is little to substantiate the argument that the binding is a particular feature of *excitable membranes*; thus, one cannot specify the anatomic location of the probe in the case of intact muscle fibers, whose surface coat is by no means a simple membrane (41). Since isolated mitochondria (although from *Neurospora*) do not bind TEMPO, but sarcoplasmic vesicles do, its location in the muscle could be in the sarcoplasmic reticulum. Probe localization in nerves is equally uncertain, since both vagus and amphibian nerves, although "unmyelinated," are enveloped by satellite cells, with their own extensive and specialized membrane systems (cf. 42, 43) (Fig. 8.5). The axons are completely covered by a layer of satellite (Schwann) cells, intimately apposed to the axon surface and containing extensive, elaborate membrane systems, which form a complex array of channels running from the extracellular space proper to the extraaxonal space. In addition, both satellite cells and axons (unless extensively perfused) contain numerous mitochondria, which in the case of axons are intimately associated with the axonal membrane (44). From the data

---

Kaufman, and H. H. Wang, *Biochem. Biophys. R. Commun.* **53:** 1077, 1973). However spin-labeled analogs of the anaesthetics produce strong paramagnetic quenching of ANS fluorescence. Since much of the ghost-bound ANS is protein associated (Chap. 7) and paramagnetic quenching involves interaction distances of 4–6 Å, the data suggest interaction of anaesthetics with membrane proteins, consistent with data in (66–68). Interaction with membrane lipids is not excluded.

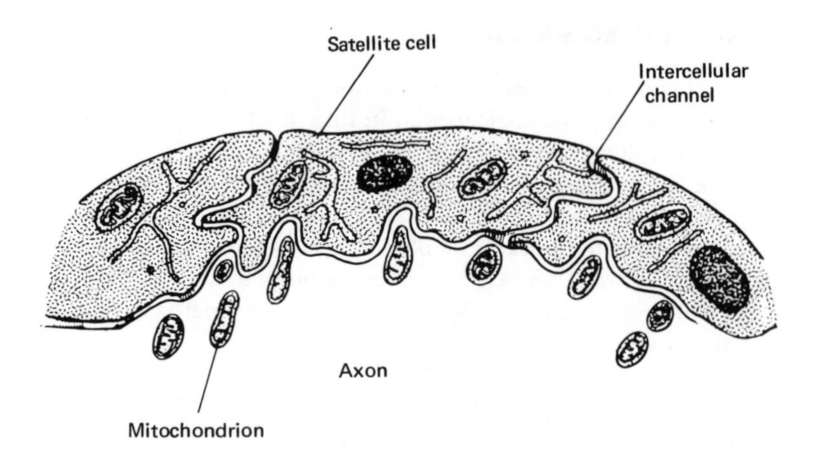

Fig. 8.5 Schematic of an unmyelinated invertebrate giant axon. The axon contains numerous mitochondria aligned roughly perpendicular to the average axon membrane plane; they are concentrated near the axonal periphery. The axon is surrounded by a layer of "satellite" cells, each containing many membranous organelles and special membrane elaborations. The ESR signal of a spin-label probe located in such a system cannot be unambiguously attributed to the axonal membrane. [Drawn from electron micrographs in Villegas and Villegas (43) and Geren and Schmitt (44).]

available, TEMPO might be localized in any one or more of the membrane systems in nerves, or possibly in highly structured, water-filled narrow membrane channels. We shall return to this matter.

In an extension of their work with TEMPO, Hubbell and McConnell (10) studied the orientation and motion of nitroxide-labeled fatty acids, $V_{m,n}$ and a spin-labeled steroid, VII, bound by the walking-leg nerves of *Homerus americanus*, in minced nerve fibers where the specimen is isotropic and shows no preferred orientation of the label. However, when *intact* axons were aligned with their axes perpendicular and parallel to the laboratory magnetic field, the various spectral bands change markedly with orientation. The data show that the preferred orientation of the $2p\pi$ nitroxide orbital is perpendicular to the axon. Also, the longer the hydrophobic region of the amphiphilic spin label and the closer the nitroxide lies to the polar end of the molecule, the more strongly is the label immobilized. In the case of the spin-labeled steroid, the anisotropy is small, but again the preferred orientation of the long amphiphilic axis is perpendicular to the nerve surface.

The authors conclude that the data strongly suggest a preferred orientation of the labels with the long hydrocarbon chains extended perpendicular to the membrane surface. They also argue that their results show a substantial fraction of the axonal membrane to be lipid bilayer. This conclusion appears unjustified by the data, particularly since the authors' elegant physical methods do not circumvent the anatomical complexities pointed out above; the data might fit with a bilayer array in the axonal membrane, but could also reflect orientation of satellite cell membrane and/or axonal mitochondria.

Hubbell and McConnell (40) also examined the ESR spectra of spin-labeled steroids taken up by walking-leg nerves of lobsters. In these studies, the probe was transferred to the nerve from a 5% solution of fatty-acid-poor BSA. Importantly, the ESR spectra of steroid spin labels in serum albumin show strong immobilization, a picture that is quite distinct from that of the probe bound to nerves.

The nerve-bound nitroxide gives τ_C values of $4 \times 10^{-7}-10^{-8}$ sec and a rather anisotropic motion, particularly when the axis of rapid motion is perpendicular to the $2p\pi$ orbital axis. This is attributed to (a) low resistance to rotation about the long axis of the molecule, (b) localization of the steroid nucleus in the oil phase, i.e., intercalated between phosphatidyl alkyl chains, with its hydroxyl group in water, (c) anchoring of the steroid hydroxyl in the polar region of phosphatide bilayers in nerve membranes.

Like most nitroxides, VII is quickly reduced by ascorbate in aqueous solution at room temperature and pH 7.5. However, when the nitroxide was bound to nerves, reduction was slow, whether the nerves were intact or homogenized. It appears that the label lies in a protected region, but it is not clear what this means anatomically. Interestingly, label XVI (dodecyl dimethyl tempoyl ammonium),

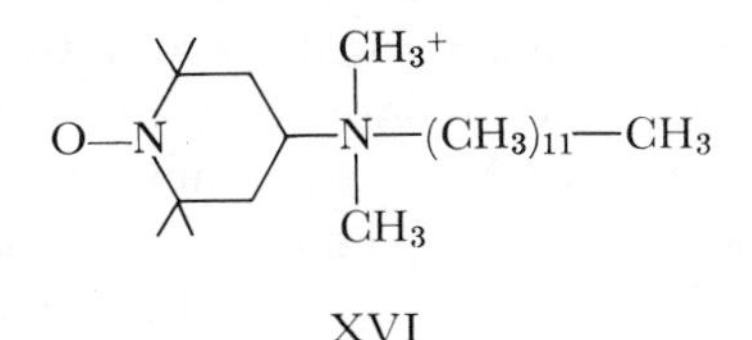

XVI

whose paramagnetic group is expected to lie in an aqueous environment, is reduced rapidly upon addition of ascorbate to labeled nerves. It would be of interest to extend these studies to BSA labeled with XVI and steroid nitroxides.

The effect of pH on the ESR spectrum of the labeled membranes is informative; taking the intensity of the low-field line as a measure of the amount of freely tumbling label, this appears minimal at pH 3.5–4.5 and maximal below pH 2 and above pH 10. The pH of minimum intensity corresponds to the overall isoelectric region of the protein. If the narrow ESR component arises from the charge sites of the membrane protein, interactions at these sites due to charge equalization in the isoelectric region would inhibit probe tumbling and produce a loss of narrow line intensity.

Mitochondria

We have commented that the spin-label TEMPO does not label mitochondria (9). However, Koltover et al. (45) were able to introduce the spin label, XVII,

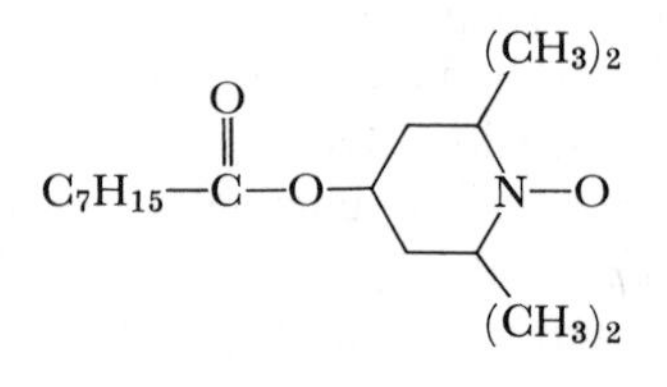

XVII

into electron transport particles from beef heart mitochondria. The resulting band shapes and correlation times indicate that the spectra represent superpositions of a weakly anisotropic signal, due to probes located in regions with large effective free volumes, and a strongly anisotropic one, arising from probes highly immobilized in other membrane regions. Reduction of the respiratory chain by substrate addition reversibly increased anisotrophy (τ_C red. $= 20 \times 10^{-10}$ vs. τ_C ox. $= 4 \times 10^{-10}$). The data are taken to demonstrate conformational transitions in mitochondrial membranes concomitant with electron transport.

Keith et al. (46) have pioneered in an elegant and promising approach, introducing spin label into the mitochondrial lipids of *Neurospora crassa* biosynthetically. The questions asked were: (1) How can one incorporate a nitroxide into the lipid portion of the membrane in a meaningful way? (2) Is the nitroxide moiety sufficiently stable to remain paramagnetic in a living system? (3) Will the system survive in the presence of the nitroxide-free radicals?

They added IV, 12-nitroxide methyl stearate, together with stearic-acid-1-^{14}C (as control) to *Neurospora* cultures and harvested the organisms after 72 hr growth. The nitroxide did not affect growth, even though present at a concentration of 5×10^{-4} M. Mitochondria were isolated from the organisms for ESR spectroscopy and also for lipid analysis. The lipids were separated into phospholipids, free fatty acids, and neutral lipids; nitroxide-labeled lipid migrated chromatographically as a neutral lipid. In addition, the phospholipid fraction was saponified and the fatty acids methylated and analysed by gas chromatography. Nitroxide concentration was determined by ESR spectroscopy.

After 72 hr growth, less than 5% of the original nitroxide and ^{14}C remained in the culture medium. The ^{14}C appeared in both saturated and unsaturated C_{16} and C_{18} fatty acids; in contrast, *the spin label could not be located in any unsaturated fatty acids* or in saturated acids *other than stearic acid*. Assuming an average phospholipid molecular weight of 1000, the measured spin-label concentration in mitochondrial phospholipids corresponds to about 1–2% of the fat chains having spin label.

The ESR spectra of mitochondria labeled with IV in comparison with those of a reference nitroxide, XVIII,

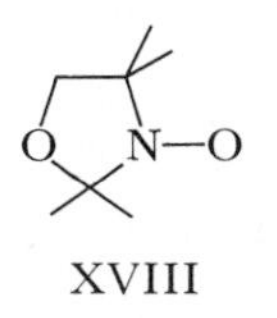

XVIII

represent the summed signals of freely tumbling nitroxides and highly immobilized ones. When the mitochondrial phospholipids are extracted and dispersed in water, they yield spectra rather similar to those of whole mitochondria.

This work clearly indicates an important experimental direction. The spin label is not merely adsorbed physically unto cell surfaces or dissolved into some hydrophobic region. Rather, some of the probe has been enzymatically coupled with glycerol derivatives to form phospholipid-molecules. Although the data do not yield the proportion of label remaining in the cells after 72 hr, the 1–2% labeled phosphatide fatty acids are by no means a trace amount. Interestingly, the hydrocarbon chain of the recovered probes appears largely unchanged, in contrast to the ^{14}C-stearate administered simultaneously. The authors do not attempt analysis of membrane structure on the basis of their ESR data.

Raison et al. (47) employed spin label XIX

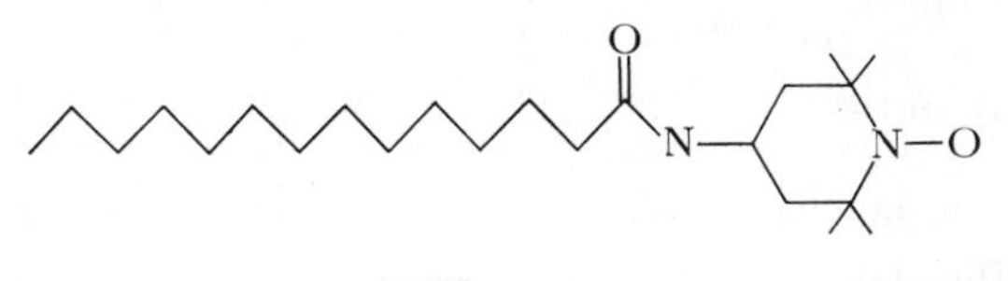

XIX

to study temperature-induced changes in the mitochondria of sweet potato tubers, potato tubers, rainbow trout livers, and rat livers. The probes, dissolved in ethanol, were added to the mitochondrial suspensions. The probes indicate a temperature-sensitive transition (12–23°C) in the mitochondria of chilling-sensitive plant or animal mitochondria. This was not found in mitochondria from the poikilothermic trout and chill-insensitive plants. The data further show that the enzymes participating in electron transport in rat liver and sweet potato mitochondria depend upon lipid state. Implicit in this finding is the possibility that probes of different structure than native lipids might perturb membrane enzyme function.

Hsia et al. (76) recently introduced a new approach to the study of mitochondrial function, synthesizing a metabolically active nitroxide derivative of 2,4-dinitrophenol (DNP), XX,

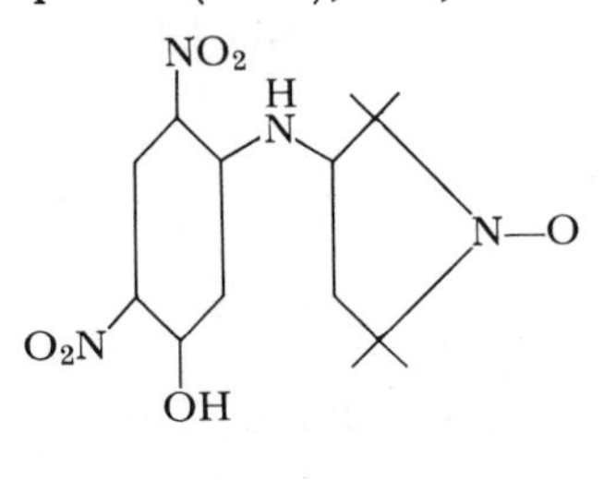

XX

According to their somewhat sparse data, this probe compound uncouples mitochondrial oxidation from phosphorylation as well as the parent compound. Moreover, the nitroxide moiety per se does not uncouple. A disadvantage of the reagent is that it becomes rapidly reduced by the mitochondrial electron transport system.

A subsequent ESR study by the same authors (77) using the DNP nitroxide claims to provide proof for phosphatide bilayer in the inner mitochondrial membranes of mammalian mitochondria. Unfortunately, the authors pay less than usual heed to the cautions listed in Chapter 3, apply analytical techniques that have not been shown to apply to this probe, and neglect some pertinent experiments. Thus, the ESR spectra of the probe in intact mitochondrial membranes exhibit band broadening attributed to protein binding, but this is dismissed because lipid-depleted mitochondria yield different spectra; but such solvent-treated mitochondria are not functional. There is no comment on the possible reduction of the probe in functional membranes and, amazingly, no study of the probe's behavior in various mitochondrial respiratory states. We hope that spin-labeled DNP will be utilized in a more rigorous manner in the future.

Acholeplasma Laidlawii

Keith and associates (48) have extended the biosynthetic approach by incorporating $V_{5,10}$, a nitroxide derivative of 12-ketostearic acid, into the lipids of *Acholeplasma laidlawii*, grown on unsupplemented media and media containing added fatty acids. The lipids were separated into "polar and nonpolar" fractions. The nitroxide migrated exclusively with the "polar" lipids, in contrast to the experience reported for *Neurospora crassa*, where the biosynthetic derivatives of $V_{5,10}$ methyl ester always chromatographed as a neutral lipid. The authors unfortunately do not specify the composition of their "polar" fraction nor the proportion of spin-labeled phosphatide fatty acids. The ESR signal from aqueous suspensions of total lipids was identical to that of an aqueous suspension of "polar" lipids.

The ESR spectra of stearate-enriched *Acholeplasma* membranes and their extracted lipids are similar and demonstrate stronger probe immobilization than the membranes and lipids of oleate-enriched cells. All of the spectra are somewhat broad, indicating restricted mobility of some of the nitroxide radicals, but this effect is much smaller than that seen with BSA-bound $V_{5,10}$, which at 30°C is severely immobilized, but is somewhat freed up by protein unfolding at 60°C.

Importantly, the ESR spectra reflect the fatty acid composition of the membranes and "polar" lipids. Moreover, the measured correlation times (τ_C) decreased monotonically with increasing temperature, giving no evidence of thermal protein denaturation—or lipid phase changes for that matter—even at 95°C.

The authors argue that their data "attest that the lipids of *Acholeplasma laidlawii* are in a bilayer." This conclusion appears premature, however, since there are no data indicating that the behavior of the labeled lipids is representative of other membrane lipids. It is clear, nevertheless, that a well-controlled and quantitative extension of the present approach could be highly informative.

Rottem et al. (49) labeled *Acholeplasma laidlawii* membranes with various nitroxide stearates ($V_{m,n}$) by exchange from spin-labeled serum albumin. They find a steep temperature dependence of the probe in membranes, a variation of the probe mobility with membrane lipid composition and increasing molecular motion of the radical the further it lies from the carboxyl terminal. Despite the fact that the probe's motion, in general, resembled that of aqueous dispersions of *Acholeplasma* lipids and is unaffected by gluturaldehyde fixation, the authors suggest that membrane proteins do exert some influence upon the incorporated nitroxide stearates. This is (a) because the motion is distinctly less in native membranes than in their extracted lipids and (b) because the rotational freedom of the probe in membranes reaggregated from SDS increases as the lipid:protein ratio of the recombinates is increased. The last finding is not easily interpreted in view of the fact discussed elsewhere

(p. 340) that the proteins in reaggregated *Acholeplasma* membranes are either improperly assembled and/or denatured.*

Influenza Virus Membranes

Stearate nitroxides ($V_{m,n}$) and androstane nitroxide (VII) exchange from labeled, defatted serum albumin to influenza virus particles (50). The resulting ESR spectra indicate free and bound probe. The resonances from the latter resemble those from labeled erythrocyte ghosts in the case of the three nitroxides studied ($V_{12,3}$, $V_{1,14}$, and VII), except that they appear *less* mobile in the virus particles.

The glycoproteins of the viral envelope do not influence the ESR spectra, and these are thought to reflect solely the lipid structure of the viral envelope. Moreover, this is considered to be a phospholipid bilayer.

Apart from the general ambiguities of the probe approach, this monolithic interpretation appears premature, since there is no evidence that the labels are solely lipid associated. Binding to protein other than the glycoprotein is not excluded, which is surprising since defatted serum *albumin* was used as a vehicle for virus labeling and since this protein is known to strongly immobilize the probes used in this study (9). As noted before, further ESR studies of these probes' association with serum albumin (and other proteins) are essential.

Sarcoplasmic Reticulum

Hubbell and McConnell (9) introduced the spin-label TEMPO into the sarcoplasmic vesicles of rabbit muscle as did Davis and Inesi (51) and Robinson et al. (52). All authors agree that the label localizes in low viscosity regions, but Robinson et al. (52) argue that TEMPO binding by these membranes is more extensive than previously appreciated. They find that isolated sarcoplasmic lipids bind 73% as much as the intact membranes and that butanol-liberated proteins (53) bind very little TEMPO. However, such butanol isolates cannot be assumed to possess the same binding properties of the native proteins. Indeed, not knowing the specific interactions between protein and lipid in the native membrane, one cannot assume *a priori* that the sum of the binding properties of separated protein and lipid isolates represents that of the intact membrane.

* When viable *Acholeplasma laidlawii* are incubated with sonically produced vesicles of spin-labeled phosphatidyl choline (XI) the artificial- and bio-membranes apparently fuse (C. W. M. Grant, and H. M. McConnell, *Proc. Natl. Acad. Sci.* (U.S.), **70:** 1238, 1973). The vesicular form of XI yields a single, broad ESR band, because of spin-spin interaction. Upon fusion, XI gradually diffuses into unlabeled lipids; this is reflected in the gradual appearance of a typical, three-line nitroxide spectrum. There is also progressive reduction in the intensity of the spectrum as well as appearance of some free spin-labeled fatty acid, apparently through phospholipase A action.

Moreover, TEMPO may distort the structure of native membranes differently from that of lipid and protein isolates.

Landgraf and Inesi (54) went somewhat further, using spin probes covalently bound to sarcoplasmic membrane proteins to detect possible conformational changes within the membranes during ATP-dependent calcium accumulation. They were able to label the membranes with nitroxide iodoacetamide (XIV) under conditions that did not impair calcium uptake and ATPase activity.

The resulting ESR spectra show a weakly immobilized component, *W*, and a more tightly immobilized one, *T*, indicating at least two different binding sites (Fig. 8.6). In resting membranes the *W*/*T* amplitude ratio was 2.65/2.75. Importantly, the ratio shifted to 3.00/3.17 in the presence of ATP. A similar effect was produced by ADP and, to a lesser extent, by ITP, but not by pyrophosphate. The shift was independent of the presence of Ca^{++} or Mg^{++}. A similar change in *W*/*T* ratio occurs as the pH is raised from 8 to 10. Solubilization of the membranes almost fully abolishes the *T* component.

The membranes could also be labeled with nitroxide isothiocyanate, reacting primarily with amino groups, but the ESR spectra of such labeled membranes did not change upon addition of ATP.

The lesser immobilization of the iodoacetamide bound to site *T* after addition (and presumably hydrolysis) of ATP, but independently of Ca^{++} and Mg^{++}, suggests formation of a membrane-ATP complex involving alteration of the environment of some SH-groups. ADP, an ATPase inhibitor, also binds, and ITP can act as a substitute for ATP. The authors reason that the conformational change must be localized, since it affects only the spin label reacted with SH and not those probes reacted with amino groups.*

The presumptive orientation of lipid hydrocarbon chains in sarcoplasmic membranes has been further studied by Eletr and Inesi (55), using nitroxide stearates, V, with the spin label 4, 9, and 12 carbons from the carboxyl. The membranes were packed centrifugally into stacks of flattened disks (56), oriented perpendicularly or parallel to the laboratory magnetic field. The ESR spectra indicate that the long axes of the stearates are oriented preferentially perpendicular to the membrane

* Spin-labeled substrates for membrane enzymes can be used to gain insight into lipid protein interactions (A. Stier and E. Sackmann, *Biochim. Biophys. Acta*, **311:** 400, 1973). The authors determined how temperature influences the reduction rate of a lipophilic nitroxide (nitroxide stearate) and a water soluble one (TEMPO-phosphate ester) by microsomal cytochrome P450-cytochrome P450-reductase. Analysis of the data suggests a mosaic structure for the membrane, with the reductase system enclosed in a "rather rigid phospholipid halo." This is in a quasicrystalline state below 32°C and undergoes a crystalline-liquid crystalline transition at 32°C. In contrast, the bulk of the membrane lipid appears to be in a rather fluid state.

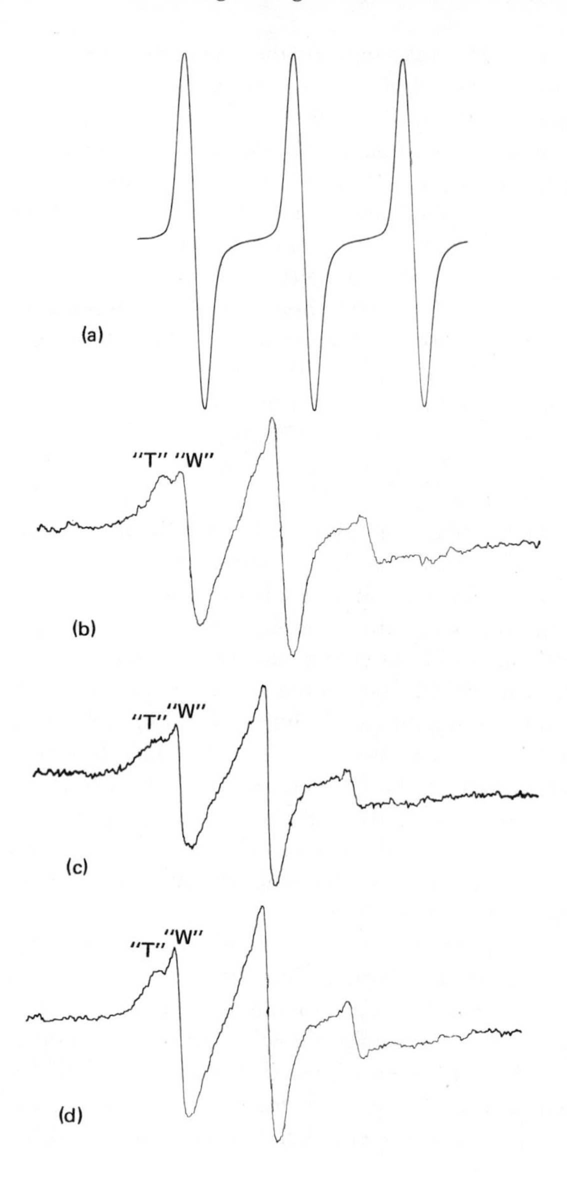

Fig. 8.6 EPR spectra of nitroxide iodoacetamide in water (a) and of fragmented sarcoplasmic reticulum labeled with nitroxide iodoacetamide (b), (c), and (d). Before initiating the EPR scan, the labeled membrane suspension was diluted to 7.5 mg/ml in (b) 16 mM Tris-maleate, pH 6.8, 80 mM KCl, and 0.1 mM EDTA; (c) as in (b) plus 3 mM ATP; (d) 100 mM bicarbonate buffer, pH 10. The EPR spectra were obtained with 1.0 G mod. ampl., 70 mw power, 1 sec time constant, 100 G/8 min field scan rate. [From Landgraf and Inesi (54) by courtesy of the authors and Academic Press, Inc.]

surfaces. Unfortunately, the orientation techniques available yield only a diffuse, qualitative picture, which does not allow unambiguous structural information.

We have earlier described the use of the spin-labeled phosphatides for measurement of the diffusion rates of phosphatides in the planes of artificial phospholipid bilayers. Related studies have now been reported for sarcoplasmic vesicles by Scandella et al. (57).

For labeling of the membranes, the 60–300 nm vesicles were equilibrated with sonicated dispersions of VIII, and the uncombined label separated by sucrose density gradients. Spin-labeled sarcoplasmic vesicles retained their ability to accumulate Ca^{++}.

At temperatures below 20°C, dipolar broadening of the ESR spectra suggests that VIII, or complexes with high VIII content, segregate as a separate phase. At temperatures >40°C, the label's bandwidth becomes principally dependent on spin exchange and reflects the collision frequency of the spin-labeled molecules. The spin-exchange data yield a diffusion constant for the label in sarcoplasmic vesicles of 6×10^{-8} cm²/sec. This is close to the value of $\sim 10^{-7}$ cm/sec² for sonicated, total, sarcoplasmic lipids or 4:1 egg lecithin:cholesterol liposomes. The value should also be compared with the apparent diffusion constant of 2×10^{-6} cm²/sec of certain mammalian surface antigens (58).

The data, taken literally, are consonant with an extremely high lateral mobility of phosphatides in these biomembranes. However, many uncertainties remain. Thus, the mobility in the biomembrane may be quite localized; the label contains a bulky nitroxide in its hydrophobic portion, which is likely to perturb hydrophobic interactions and facilitate diffusion, and measurements at 40°C are likely to reflect the thermotropism of many proteins including membrane proteins (59–61).

Gas-Vacuole Membranes

Jones et al. (62) have spin-labeled the lipid-free vacuolar membranes of blue-green algae, using a nitroxide maleimide, X, and also *N*-(1-oxyl-2,2,5,5-tetramethyl-pyrrolidinyl)-ethyl anhydride.

An aqueous suspension of intact gas vacuoles has a milky appearance, but clears upon application of hydrostatic pressure; concomitantly, the ESR spectrum of anhydride-labeled membranes becomes considerably more symmetric. The data indicate a rearrangement of the protein architecture under mechanical stress, so that the paramagnetic probe becomes less restricted in mobility relative to the protein surface.

Kidney Plasma Membranes

Several authors, particularly Henry and Keith (63), express concern that some spin-label probes might accumulate as aggregates of impurity within apolar membrane domains. In that case, changes in the ESR spectra of the probes due to some membrane alteration might actually reflect a perturbation of the aggregate or pool rather than a reorientation of the natural analogues of the spin-label probes. This mechanism is distinct from any possible molecular perturbations induced by single-probe molecules, as well as the likelihood that many spin-label probes do not adequately represent their "parent" molecules.

In their early studies, McConnell and associates point out that formation of spin-label "lenses" is improbable at the spin-label concentrations used in various membrane systems. Barnett and Grisham (64) have looked at this matter specifically for the case of kidney plasma membrane fragments. They employed the methyl esters of 5- and 12-nitroxide stearate, as well as a spin-labeled steroid in their experiments. Importantly, at probe concentrations within the membranes that do not interfere with cation-sensitive ATPase, spin exchange occurs rather freely, with a second-order rate constant of 10^7 M^{-1} sec^{-1}. The data suggest that the probes are "fairly" uniformly distributed in the membranes, without significant pooling, and that the viscosity for lateral diffusion is 10^2–10^3 times that of water.

Role of Unsaturation in Membranes

Until recently, the type and degree of unsaturation of membrane-phospholipid hydrocarbon chains were not investigated by "probe techniques." However, this matter has now been approached by use of a yeast mutant, deficient in fatty acid desaturase (65).

These organisms were enriched with stearolic acid and octadecanoic acids containing a *cis*-double bond 6, 9, or 11 carbons from the ester linkage, by presenting these fatty acids in the medium. The cells, whose predominant membrane system is the plasma membrane, were then studied with spin-labeled stearates with nitroxides at 4, 6, 9, and 12 carbons from the carboxyl terminal, transferred to the yeast cell from small amounts of ethanol.

The authors explain their data, assuming that their probe is incorporated into a phosphatide bilayer. If this assumption is correct, this work clearly illustrates the importance of degree and location of unsaturation. As expected, *cis*-unsaturated bonds disorder the packing of a lipid bilayer. With a single unsaturated site, the *position* of the *cis*-unsaturation circumscribes the ordered domain of bilayer phosphatides to between the ester linkage and the double bond. Beyond the double bond, the hydrocarbon chains become disordered. In systems with multiple double bonds at various sites along the fatty acid chain (i.e., a typical animal cell membrane case), the situation may become very complex. If the unsaturated fatty acids are distributed at random, their impact would average out. However, we suspect that the protein composition of a given membrane domain influences the fatty acid composition in this region and that a realistic picture of complex animal cell membranes can only derive from the study of their functional domains. We doubt that this goal is within the range of present spin-labeling and/or fluorescent probe techniques, but welcome realization of an important biological problem.*

Proton Paramagnetic Probes

This method was first developed to monitor the interactions of anesthetic agents such as benzyl alcohol (BeOH), neopentyl alcohol, and xylocaine with erythrocyte and model membranes. All have distinctive PMR bands, whose widths and intensities differ in the free and membrane-bound states, allowing deductions about the viscosity of the probe environment, the number of binding sites, and the nature of the molecular events occurring during certain membrane perturbations (66).

These compounds, like most membrane-active agents, *protect*, rather than disrupt, erythrocyte membranes when present *below their critical lytic concentration* (63) (Fig. 8.7); the critical lytic concentrations vary widely.

Metcalfe and associates (66) have explored the phenomenon in some detail, using the aromatic protons of BeOH to report on perturbations produced by this agent and/or by other membrane-active substances.

* The spin label TEMPO reports two characteristic temperatures in fatty acid auxotrophs of *Escherichia coli* grown on single essential fatty acids (C. D. Linden, K. L. Wright, H. M. McConnell, and C. F. Fox, *Proc. Natl. Acad. Sci.* (U.S.) **70:** 2271, 1973). These correlate with marked slope changes in the Arrhenius plots for beta-glucoside and beta-galactoside transport. The higher temperature signals the initial phase in the formation of solid lipid patches during a thermotropic lateral phase separation. The lower temperature represents the end point of the process, where all lipids have solidified. At the higher temperature one expects a sharp increase in the isothermal lateral compressibility of membrane lipids. It is here, where organisms grown on elaidic acid exhibit a nearly twofold increase in beta-glucoside transport with a $< 1°$ temperature change (38.6°C–38.1°C). The authors suggest that large packing fluctuations at high, lateral, isothermal compressibility may favor insertion of transmembrane transport proteins. Other possibilities suggested are an increase in the number of "buried," functional transport proteins, as well as facilitation of the "expansion-contraction" cycles of buried transport proteins.

Freeze-etching electron microscopy on *Tetrahymena* membranes during thermotropic phase transitions (F. Wunderlich, V. Speth, W. Batz, and H. Kleinig, *Biochim. Biophys. Acta*, **298:** 39, 1973) which show lateral segregation of membrane intercalated particles during the transition, suggest another possibility—facilitation of an essential interaction between transport entities.

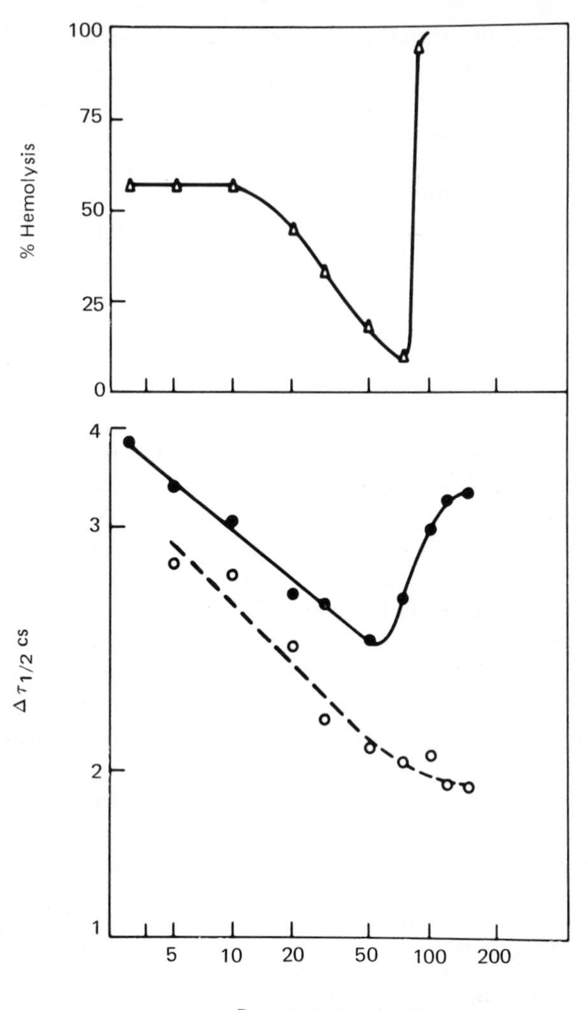

Fig. 8.7 An illustration of the use of benzyl alcohol (BeOH) as a PMR probe. At low concentrations, this and many other membrane-lytic reagents actually protect against osmotic lysis of erythrocytes (top panel). Above, a critical BeOH concentration lysis occurs abruptly. This can be correlated with the changing BeOH PMR signals at differing concentrations (lower panel). During the stabilization phase, the bandwidth (ordinate) of the PMR spectrum decreases progressively in both erythrocyte ghosts (————) and their lipids (– – – –). This suggests partition of BeOH from the water phase into apolar regions. However, at the critical lytic concentration, the PMR signal reverses abruptly in the membranes (not lipids). This suggests a sudden increase of BeOH binding to membrane protein. [From Wallach (75); redrawn from Metcalfe (71). Courtesy Springer-Verlag, Heidelberg.]

The aromatic resonance band of membrane-bound BeOH is wider (i.e., the relaxation rate, $1/T_2$, is greater) and of lesser amplitude than that of the alcohol in water, indicating restricted mobility of the probe. With increasing concentrations of BeOH, the resonance band narrows progressively and reversibly until the critical lytic concentration is reached (Fig. 8.7) (67); there, the band broadens abruptly and irreversibly. Independent studies by Colley et al. (68) indicate that new *protein* binding sites for BeOH appear at the critical lytic concentration. These workers found that at 25°C, the partition of BeOH erythrocytes/water diminishes somewhat up to 80 mM BeOH in the bulk phase. At higher concentrations, BeOH uptake by the membranes increases sharply; below 80 mM binding is reversible, and the partition coefficient into the membrane is slightly less than for the separated lipid and much less than for the separated protein. Above 80 mM there is a progressive irreversible change in membrane state associated with markedly increased BeOH binding at an interface with essential protein-lipid and possibly protein-protein interactions. The maximum binding is very high (1 molecule BeOH per molecule of lipid and 1 per 2–3 amino acid residues). The authors attribute the protein's high BeOH to unusual amino acid sequence. As pointed out by Wallach (69), the phenomenon suggests that membrane proteins constitute a cooperative system (70). Clearly, the biphasic response of the intact membrane as signaled by BeOH PMR is not simply the summed contribution of its component parts; however, aqueous dispersions of membrane lipids can simulate the prelytic phase. The data indicate that, in the prelytic range, the alcohol molecules partition into an increasingly fluid environment with rising BeOH concentration. It appears that at the lytic concentration, the protein of the membrane is perturbed, exposing binding sites previously inaccessible. It is possible that in the intact membrane, the lipid masks protein sites (which become exposed when lipid dissociates from protein at high BeOH levels) and/or that the protein changes conformation.

Neopentyl alcohol does not exhibit a concentration dependence of relaxation rates at low levels (71). Xylocaine, an organic base, can be displaced from the membrane by Ca^{++}, but this is not true for BeOH; apparently, the binding of BeOH is determined primarily by hydrophobic forces.

Importantly, the characteristic PMR pattern found in going from prelytic to lytic levels of BeOH is also obtained when BeOH is used at prelytic levels—as a "reporter" molecule—to monitor changes wrought by other membrane-active substances. This approach is of obvious importance alone and to supplement information obtained by other spectroscopic methods.

The use of BeOH as a PMR probe has recently been extended by

Metcalfe et al. (72) to compare the structures of "native" plasma membranes from erythrocytes and *Acholeplasma laidlawii* with corresponding membrane recombinates, derived by dissolving the membranes in sodium dodecyl sulfate and removing most of the detergent by dialysis against Mg^{2+}-containing buffers (73). The reaggregates proved closely similar to the native membranes by transmission electron microscopy, equilibration in sucrose density gradients and composition. Moreover, low-angle X-ray diffraction showed that both native and reconstituted membranes contain appreciable proportions of lipid in a bilayer array.

What is most interesting in this study and parallel fluorescence probe experiments discussed earlier is the fact that *both* membrane types exhibit a biphasic response to increasing concentrations of BeOH (Fig. 9.7) (72, 68), the second phase (at high BeOH levels) corresponding to altered architecture of membrane-proteins and/or lipid-protein interactions. This suggests a common structural principle for both membrane types despite their differences of composition (23, 74). Importantly, the reconstituted membranes lack this biphasic response, i.e., renaturation of membrane proteins and/or correct reassembly of membrane components fails to occur under the conditions employed. Thus, while reconstitution may appear successful according to some common criteria, these are clearly inadequate to describe the details and dynamics of membrane structure.

Epilogue

We suspect that no one physical method now available combines the versatility, sensitivity, and informational content of the spin-labeling approach and feel that its large, rapid impact upon membrane research is well justified. Indeed, we anticipate that spin labeling, *applied with the necessary biological* and *biochemical controls*, can become a tool of unparalleled value in studies of a host of important biomedical membrane domains; here we include topics such as the interactions of cells with (a) other cells—normal and malignant, (b) parasites, (c) viruses, (d) hormones (and other regulatory agents, as well as immunologic recognition phenomena, membrane changes consequent to neoplastic transformation, a.s.f.). The relatively high expense of ESR spectrometers is more than outweighed by their versatility, informational yield, and established computer compatibility, as well as the wide variety of spin labels commercially available and/or readily synthesized.

However, writing at a time when the first wave of spin-label technology has just passed, we feel forced to temper our enthusiasm and express our criticism here (and Chap. 3) of what has been done so far.

Quite understandably, the first thrust of the spin-label approach sought to solve the problem of *membrane structure* in general. Yet, no matter how elegant the experimentation, spin labels cannot be taken to properly represent their native analogues; indeed, here lies the techniques' Achilles' heel. Indubitably, spin-label probes can contribute invaluable information to the elucidation of the absolute structure of a given membrane domain, *but only in conjunction with other discriminating methods.* However, we see no proved competitor for the spin-label approach as a structural-analytical tool of great versatility and major dynamic capacity.

In promoting the spin-label approach, we do not wish to denigrate other paramagnetic probe techniques, but their applicability, sensitivity, and versatility appear more limited.

References

1. Griffith, D. H., and Waggoner, A. S. *Acc. Chem. Res.* **20:** 17, 1969.
2. Budachenko, A. L., Khloplyankina, M. S., and Dobryanov, S. N. *Opt. Spektrosk.* (USSR) **22:** 554, 1967.
3. Waggoner, A. S., Griffith, O. H., and Christensen, C. R. *Proc. Natl. Acad. Sci.* (U.S.) **57:** 1198, 1967.
4. Dodd, G. H., Barratt, M. D., and Rayner, L. *FEBS Letters* **8:** 286, 1970.
5. Libertini, L. J., Waggoner, A. S., Jost, G. C., and Griffith, O. H. *Proc. Nat. Acad. Sci.* (U.S.) **64:** 13, 1969.
6. Blodgett, K. L. *J. Am. Chem. Soc.* **57:** 1007, 1935.
7. Levine, Y. K., and Wilkins, M. H. F. *Nature New Biology* **230:** 69, 1971.
8. Jost, P. J., Libertini, L. J., Herbert, V. C., and Griffith, O. H. *J. Mol. Biol.* **59:** 77, 1971.
9. Hubbell, W. C., and McConnell, H. M. *Proc. Natl. Acad. Sci.* (U.S.) **61:** 12, 1968.
10. Hubbell, W. C., and McConnell, H. M. *Proc. Natl. Acad. Sci.* (U.S.) **64:** 20, 1969.
11. Hsia, J. C., and Boggs, J. M. *Biochim. Biophys. Acta* **266:** 18, 1972.
12. Paterson, S. J., Butler, K. W., Huang, P., Labelle, J., Smith, I. C. P., and Schneider, H. *Biochim. Biophys. Acta* **266:** 597, 1972.
13. Kornberg, R. D., and McConnell, H. M. *Proc. Natl. Acad. Sci.* (U.S.) **68:** 2564, 1971.
14. Kornberg, R. D., and McConnell, H. M. *Biochemistry* **10:** 1111, 1971.
15. Kornberg, R. D., McName, M. G., and McConnell, H. M. *Proc. Natl. Acad. Sci.* (U.S.) **69:** 1508, 1972.
16. Stone, T. J., Buckman, T., Nordio, B. L., and McConnell, H. M. *Proc. Natl. Acad. Sci.* (U.S.) **54:** 1010, 1965.
17. Griffith, O. H., and McConnell, H. M. *Proc. Natl. Acad. Sci.* (U.S.) **55:** 9, 1966.
18. Ogawa, F., and McConnell, H. M. *Proc. Natl. Acad. Sci.* (U.S.) **58:** 19, 1967.
19. McConnell, H. M., and Hamilton, C. C. *Proc. Natl. Acad. Sci.* (U.S.) **60:** 776, 1968.

20. Barratt, N. D., Green, D. K., and Chapman, D. *Biochim. Biophys. Acta* **152:**
21. Hong, K. H., and Hubbell, W. L. H. *Proc. Natl. Acad. Sci.* (U.S.) **69:** 2617,
22. Blasie, J. K. *Biophys. J.* **12:** 191, 1972.
23. Wallach, D. F. H., and Zahler, P. H. *Proc. Natl. Acad. Sci.* (U.S.) **56:** 1552, 1966.
24. Verma, S. P., Schneider, H., and Smith, I. C. P. *FEBS Letters* **25:** 197, 1972.
25. Habermann, E., and Jentsch, H. *Hoppe-Seyler's Z. Physiol. Chem.* **348:** 37, 1967.
26. Sessa, G., Freer, J. H., Colaccio, G., and Weissman, G. *J. Biol. Chem.* **244:** 3575, 1969.
27. Williams, J. C., and Bell, R. M. *Biochim. Biophys. Acta* **288:** 255, 1972.
28. Pache, W., Chapman, D., and Hillaby, R., *Biochim. Biophys. Acta* **255:** 358, 1972.
29. Landsberger, F. R., Paxton, J., and Lenard, J. *Biochim. Biophys. Acta* **266:** 1, 1972.
30. Kroes, J., Ostwald, R., and Keith, A. *Federation Proc.* **29:** 366 Abs., 1970.
31. Kroes, J., Ostwald, R., and Keith, A. *Biochim. Biophys. Acta* **274:** 71, 1972.
32. Kroes, J., and Ostwald, R. *Biochim. Biophys. Acta* **249:** 647, 1971.
33. Chapman, D., Barratt, M. D., and Kamat, V. B. *Biochim. Biophys. Acta* **173:** 154, 1969.
34. Berger, K. U. B., Barratt, M. D., and Kamat, V. B. *Biochem. Biophys. Res. Comm.* **40:** 1273, 1970.
35. Maddy, A. H. *Biochim. Biophys. Acta* **117:** 193, 1966.
36. Zwaal, R. F. A., and Van Deenen, L. L. M. *Chem. Phys. Lipids* **4:** 29, 1970.
37. Berger, K. V., Barratt, M. D., and Kamat, V. B. *Chem. Phys. Lipids* **6:** 351, 1971.
38. Simpkins, H., Panko, E., and Tay, S. *J. Memb. Biol.* **5**: 334, 1971.
39. Hubbell, W. L., Metcalfe, J. C., Metcalfe, S. M., and McConnell, H. *Biochim. Biophys. Acta* **219:** 415, 1970.
40. Hubbell, W. L., and McConnell, H. M. *Proc. Natl. Acad. Sci.* (U.S.) **63:** 16, 1969.
41. Wallach, D. F. H. In *Specificity of Cell Surfaces*, B. Davis and L. Warren, eds. Englewood Cliffs, N.J.: Prentice-Hall, 1967, p. 129.
42. Schmitt, F. O., and Geschwind, N. *Progr. Biophys. Biophys. Chem.* **8:** 166, 1957.
43. Villegas, G. M., and Villegas, R. *J. Ultrastruct. Res.* **8:** 197, 1963.
44. Geren, B. B., and Schmitt, F. O. *Proc. Natl. Acad. Sci.* (U.S.) **40:** 863, 1954.
45. Koltover, V. K., Goldfield, M. G., Hendel, L. Y., and Rozantzev, E. G. *Biochim. Biophys. Res. Comm.* **32:** 421, 1968.
46. Keith, A. D., Waggoner, A. S., and Griffith, W. H. *Proc. Natl. Acad. Sci.* (U.S.) **61:** 891, 1968.
47. Raison, J. K. R., Lyons, J. M., Melhorn, R. J., and Keith, A. D. *J. Biol. Chem.* **246:** 4036, 1971.
48. Tourtellotte, M. E., Branton, D., and Keith, A. D. *Proc. Natl. Acad. Sci.* (U.S.) **66:** 909, 1970.
49. Rottem, S. R., Hubbell, W. L., Hayflick, L., and McConnell, H. *Biochim. Biophys. Acta* **219:** 104, 1970.
50. Landsberger, F. R., Lenard, J., Paxton, J., and Compans, R. *Proc. Natl. Acad. Sci.* (U.S.) **68:** 2579, 1971.
51. Davis, D. G. D., and Inesi, G. I. *Biochim. Biophys. Acta* **241:** 1, 1971.
52. Robinson, J. D. R., Birdsall, N. J. M., Lee, A. G., and Metcalfe, J. C. *Biochemistry* **11:** 2903, 1972.

53. Maddy, A. H. *Biochim. Biophys. Acta* **88:** 448, 1964.
54. Landgraf, W. C., and Inesi, G. *Arch. Biochem. Biophys.* **130:** 111, 1968.
55. Eletr, S., and Inesi, G. *Biochim. Biophys. Acta* **282:** 174, 1972.
56. Inesi, G., and Asai, I. *Arch. Biochem. Biophys.* **126:** 469, 1968.
57. Scandella, C. J. S., Deveaux, P., and McConnell, H. M. *Proc. Natl. Acad. Sci.* (U.S.) **69:** 2056, 1972.
58. Frye, L. D. F., and Edidin, M. E. *J. Cell. Sci.* **7:** 319, 1970.
59. Blazyk, J. F., and Steim, J. M. *Biochim. Biophys. Acta* **266:** 737, 1972.
60. Schechter, E., Gulik-Krzywiki, T., and Kabak, H. R. *Biochim. Biophys. Acta* **274:** 466, 1972.
61. Lin, P. S., Wallach, D. F. H., and Tsai, S. *Proc. Natl. Acad. Sci.* (U.S.) **70:** 2492, 1973.
62. Jones, D. D., Haug, A., Jost, M., and Graber, D. R. *Arch. Biochem. Biophys.* **135:** 296, 1969.
63. Henry, S. A. H., and Keith, A. D. K. *Chem. Phys. Lipids* **7:** 245, 1971.
64. Barnett, R. E. B., and Grisham, C. M. G. *Biochim. Biophys. Res. Comm.* **48:** 1362, 1972.
65. Eletr, S., and Keith, A. D. *Proc. Natl. Acad. Sci.* (U.S.) **69:** 1353, 1972.
66. Metcalfe, J. C. In *The Dynamic Structure of Membranes*, D. F. H. Wallach and H. Fischer, eds. Heidelberg: Springer-Verlag, 1971, p. 202.
67. Metcalfe, J. C., and Burgen, A. S. U. *Nature* **220:** 587, 1968.
68. Colley, C. M. C., Metcalfe, S. M., Turner, B., and Burgen, A. S. U. *Biochim. Biophys. Acta* **233:** 720, 1971.
69. Wallach, D. F. H. *Proc. Natl. Acad. Sci.* (U.S.) **61:** 868, 1968.
70. Changeux, J. P., Thiery, J., Tung, Y., and Kittel, C. *Proc. Natl. Acad. Sci.* (U.S.) **57:** 335, 1967.
71. Metcalfe, J. C. In *A Symposium on Calcium and Cellular Function*, A. W. Cuthbert, ed. London: Macmillan, 1970, p. 219.
72. Metcalfe, J. C., Metcalfe, S., and Engelman, D. M. *Biochim. Biophys. Acta* **241:** 412, 1971.
73. Razin, S. *Biochim. Biophys. Acta* **265:** 241, 1972.
74. Wallach, D. F. H. *The Plasma Membrane.* Heidelberg Science Library **18.** New York: Springer-Verlag, 1972.
75. Wallach, D. F. H. In *The Dynamic Structure of Cell Membranes*, D. F. H. Wallach and H. Fischer, eds. Heidelberg: Springer-Verlag, 1971, p. 181.
76. Hsia, J. C., Chen, W. L., Long, R. A., Wong, L. T., and Kalow, W. *Biochem. Biophys. Res. Commun.* **48:** 1273, 1972.
77. Hsia, J. C., Chen, W. L., Long, R. A., Wong, L. T., and Kalow, W. *Proc. Natl. Acad. Sci.* (U.S.) **69:** 3412, 1972.

CHAPTER 9

PERSPECTIVES

Introduction

At this writing, we know little more about the molecular organization of membranes than we did about nucleic acids thirty years ago. Indeed, we can glean much from the history of DNA "molecular biology" for future explorations of membrane structure and function; e.g., when first purified, nucleic acids emerged as a tetranucleotide: cytosine-guanine-adenosine-thymidine (C—G—A—T), with a molecular weight of about 1500 daltons. Clearly, such a small molecule could provide no chemical framework for genetics; large molecules, such as proteins or polysaccharides, appeared as fitter candidates. But when gentle extraction procedures were applied, the tetranucleotides were found to be fragments arising from enormously long molecules, deoxyribonucleic acid (DNA). This led to the discovery that genetic information is encoded in the giant DNA strands through appropriate sequencing of four codons: A, T, C, and G.

We believe that present methods of preparing and purifying membranes and their components obscure their true organization and biological role just as the real nature of DNA was long hidden by the methods used to isolate it. We currently isolate membranes from cells, usually not really pure, rarely in a unique physiologic state, almost always in a foreign environment. We treat the cell surface as a preconceived entity, shear it beyond quantitative recovery and often beyond recognition and purify it, i.e., separate it from other organelles and soluble compounds, by various physical and biochemical procedures, some of which could well perturb subtle but biologically critical associations. While the resultant preparations may retain some of the micromorphologic appearances of native membranes and preserve some membrane properties, few explore the possibility or probability that major functional alterations have been wrought, particularly when "loosely bound" functional macromolecules and other regulatory agents become eluted.

A proliferation of review and theoretical literature addresses diverse aspects of membrane structure, particularly the validity of the "paucimolecular" model of Danielli and Davson (1) relative to other alternatives. Indeed, we have seen how this model forms the working hypothesis of much elegant membrane experimentation. We do not want to concern ourselves much with models, but do wish to discuss the interactions of diverse membrane components. For this, we must deal in some detail with the "paucimolecular" hypothesis, which, despite its ambiguities, has infiltrated the minds of many investigators as a fact rather than a possibility.

We believe that the only reasonable approach to the clarification of membrane-related problems derives from an acceptance of our ignorance. Indeed, we suspect major variations from cell to cell, between different membranes of a single cell and diverse domains of a single membrane, possibly the only common features being certain architectural homologies.

The "Paucimolecular" Model

The traditional Danielli model (1) is the first real "molecular-biological" hypothesis. It envisages cellular membranes as phosphatide bilayers, whose surfaces bind membrane proteins through polar interactions with membrane lipids. This concept derived from the following observations.

(a) The permeabilities and electrical properties of certain cells' surfaces are compatible with the notion that the surface membranes are largely lipid.

(b) Amphiphilic lipids naturally associate as monomolecular or bimolecular films.

(c) Gorter and Grendel (2) reported that the lipids extracted from erythrocytes could be spread into a monolayer with twice the area of the intact cells.

(d) Danielli and Harvey (3) showed that the tension at the surfaces of certain cells is lower than that at oil-water interfaces, which, in view of the fact that adsorption of protein to oil droplets markedly lowers their interfacial tension, was taken to indicate that membrane surfaces are protein coated.

The latest form of the Danielli hypothesis, the "unit membrane" concept (4), draws additional support from several other areas of investigation, many of which we have discussed above. However, the following have profoundly influenced the design of modern experiments.

(a) The birefringence of erythrocytes (5–7) and myelin (8) can be interpreted to indicate that these membranes contain elongated molecules, whose axes are preferentially arranged perpendicular to the membrane surfaces. Since this "form birefringence" is perturbed by organic solvents, and since it is also prominent in phosphatide bilayers, one can consider it compatible with the arrangement of hydrocarbon chains proposed in the Danielli model.

(b) Small-angle X-ray diffraction patterns of fresh peripheral nerve myelin show this to consist of concentric layers, 180 A apart in mammalian nerve; other lamellar membrane systems yield concordant information. Such X-ray patterns compared with those of mixed membrane lipids, phosphatides, and phosphatide-cholesterol mixtures suggest that the myelin identity periods represent two bimolecular lipid leaflets, each with its surfaces coated by protein—as in the Danielli model. More recent data reveal halving of the identity period, and this supports the concept of repeating lipid bilayers, each about 51 A wide.

(c) Thinking about cellular membranes has been heavily influenced by their electron microscopic appearance which, in stained preparations, is that of a trilaminar structure with two outer electron-dense regions. All cellular membranes and also artificial phosphatide bilayers exhibit this "unit membrane" characteristic (4). Stained sections of myelin show the "unit membranes" arranged in concentric lamellae and spaced at intervals fitting the low-angle X-ray periods rather well, if one allows for changes occurring during the preparation of sections for electron microscopy. Moreover, freeze-cleave electron microscopy suggests that membranes possess a natural cleavage plane in the apolar region within the center of a lipid bilayer (9). The micromorphology of myelin is, thus, generally compatible with the Danielli hypothesis. Finally, the classical observations of Geren and others (10) show convincingly that myelin originates from the plasma membranes of Schwann cells or oligodendrocytes. These studies are central to the concept that myelin can serve as a model for cellular membranes in general and is essential to the "unit-membrane" hypothesis.

The "unit membrane" hypothesis continues to be seductive in its simplicity, *but we consider none of the evidence cited in its support as unambiguous* or permitting detailed conclusions as to the molecular organization of membranes. We summarize this contention as follows.

(a) One cannot interpret the movements and transport of diverse solutes through cellular membranes in terms of simple lipid diffusion barriers.

(b) Newer, precise impedance measurements are not consistent with plasma membranes as continuous lipid bilayers. Moreover, the impedances of artificial lipid bilayers exceed those of natural membranes.

(c) The original experiments of Gorter and Grendel underestimated both the surface area and lipid content of erythrocytes (11). Moreover, experiments of this type cannot be interpreted unambiguously, since it is not known how tightly lipids are packed in a membrane. Thus, Barr et al. (12) show that the monolayer area of erythrocyte lipids can vary from twice the erythrocyte surface area, at low surface pressure, to equal area at collapse pressure. Engelman (13) attempts to avoid this problem and, therefore, calculates the apolar volumes of erythrocyte membranes, using generally accepted values for phospholipid and neutral lipid content, the known volumes of CH_3, CH_2, and CH groups in acyl side chains and assuming an average fatty acid to contain 17.5 carbons and 1.26 double bonds. He calculates the apolar volume of cholesterol from its density and adds this to the value computed for the phospholipids, to give the total apolar volume of the membrane. Assuming a uniform lipid distribution, the area available per phospholipid molecule would be (the volume occupied by the two fatty acid chains relative to the total apolar volume) $\times$ (the cell surface area). If all the lipids occupy an area exactly equal to that assumed for the cell surface, 35.5 ± 2 A^2 would be available per phospholipid and 23.0 ± 3 A^2 per cholesterol, giving a sum for the two of 58.5 A^2, or 117 A^2 in a mixed bilayer. This is greater than expected from the data of Rand and Luzatti (14), suggesting that about 20% of the surface area allowed for lipids might be occupied by nonlipid elements; more recent calculations (15) bring this closer to 50%.

However, all area measurements on cell surfaces are suspect, since cell surfaces exhibit many large protrusions and therefore cannot be considered *smooth* at the molecular level. Thus, if a 10,000 A radius, smooth sphere, has a surface area of $4\pi \times 10^8$ A^2, a similar-sized sphere, with a granularity due to 10 A radius hemispheres (i.e., below electron microscopic resolution), packed over $\frac{3}{4}$ of the surface, will have a surface area of about $7\pi \times 10^8$ A^2.

(d) The surface tension arguments of Danielli and Harvey (3) lack relevance, since phosphatides, which are the principal lipids of most membranes, exhibit very low interfacial tensions.

(e) To date, membrane birefringence measurements still only indicate that these structures contain molecules whose long axes oriented preferentially perpendicular to the membrane surface. This qualitative information allows no conclusions as to the *degree* of order within the membrane, or as to what is causing the birefringence. The disordering effect of organic solvents does not permit any unique interpretation, since these agents can denature proteins. In myelin, the birefringence probably does reflect primarily the orientation of the lipid since there it comprises 80% of the dry mass. However, in other membranes, with 60–70% protein, the observed intrinsic birefringence could arise wholly or in part from membrane protein, e.g., helical peptide segments, with axes preferentially arranged perpendicular to the membrane surface; indeed, such arrays have been demonstrated in model systems (16).

(f) X-ray diffraction studies of membranes deal primarily with lamellar systems, myelin in particular. The results with myelin are compatible with the Danielli-Robertson model, but other interpretations are possible, since the low-angle diffractions only give an estimate of the intensity distribution perpendicular to the membrane planes, but do not define the molecular origins of such distributions. Moreover, the low-angle diffraction patterns of chloroplast lamellae do not appear to fit the "unit-membrane" concept. Finally, recent X-ray studies on the plasma membrane of erythrocytes and other cells, packed centrifugally into lamellar arrays, indicate that while some proportion of these membranes may be a lipid bilayer, an appreciable proportion of the membrane core must be protein in nature. On the basis of the membrane shrinkage after phospholipase C, Finean calculates at least 30% of the erythrocyte membrane core to be interrupted by protein (17), and Engleman places this value at close to 50% for the membranes of *A. laidlawii* (15).

(g) The generalized "unit membrane" relies heavily upon the interpretation of the electron microscopic images of cellular membranes, and the only solid argument for extrapolating myelin X-ray data to other membranes derives from the morphologic continuity between myelin and satellite-cell plasma membranes. But the composition of myelin distinguishes it from other cellular membranes. Protein, which comprises 60–70% of the dry mass of most membranes, accounts for only 20% in myelin. The lipid composition of myelin is also atypical. Thus, cholesterol accounts for 40% and cerebroside for 15% of its lipid, contrasted with about 10% and trace amounts, respectively, in most other membranes. Unfortunately, we know very little about the composition of satellite-cell plasma membranes, but if it is identical to that of myelin, these membranes must be very different from other cellular membranes.

However, since all cellular membranes have the same trilaminar appearance in stained sections, one must ask what this represents in molecular terms. This matter has been appraised critically by Korn (11) who shows that (a) various cellular membranes differ so much in width that they cannot be taken to represent a single type of molecular arrangement; (b) the trilaminar image of the inner mitochondrial membrane and myelin remains essentially unchanged after extraction of nearly all lipid; (c) OsO_4 and permanganate probably form both electron-dense adducts with unsaturated fatty acid and precipitates at water-membrane interfaces, thus allowing no firm conclusions as to the native molecular architecture of membranes; (d) freeze-cleave and freeze-etch electron microscopy show globular structures of macromolecular dimensions within the apolar cores of membranes other than myelin.

We therefore question the validity of the evidence on which the Danielli-Robertson hypothesis is based. Moreover, there is considerable information that is not compatible with this model.

(a) The Danielli-Robertson model requires *ionic* bonding between membrane proteins and lipids. However, although certain bacterial membranes dissociate into lipoprotein subunits upon reduction of environmental ionic strength (18), in general membrane lipids can rarely be separated from membrane proteins by manipulation, which influences only ionic interactions.

(b) An abundance of evidence indicates that apolar mechanisms dominate the interactions between membrane proteins and lipids.

(c) The Danielli-Robertson model implies that a substantial portion of membrane protein is in the β-conformation. But infrared spectroscopy of diverse membranes fails to reveal such a general pattern.

(d) As discussed, measurements of the optical activity of various plasma membranes suggest that a considerable, if disputed, proportion of membrane peptide is in α-helical conformation. But the localization of protein proposed in the "unit-membrane" hypothesis is not compatible with a large proportion of helix in the membrane proteins.

(e) The nuclear magnetic resonance spectra of diverse membranes show that the polar residues of membrane phospholipids and sphingolipids lie in an aqueous environment, but also point to close association between the hydrocarbon chains of membrane lipids and the membrane protein, which remain intact unless membrane proteins become denatured. These data do not fit the "unit-membrane" hypothesis.

(f) Plasma membranes are typified by a host of specific, genetically controlled functions, but assembly of the Davson-Danielli membrane derives from the aggregation of membrane *phosphatides* by principally entropic forces and provides no basis for the biological order in plasma membranes.

We feel that current knowledge does not permit formulation of any specific membrane model, and it is unlikely that there is a unique membrane structure. As suggested by many of the physical studies described above, many membranes probably contain regions with the "unit-membrane" structure, but we suspect that most membranes contain organized assemblies of lipoproteins, endowed with architectural specializations making them suitable for their membrane location. We shall now address this matter in some detail. In this, we proceed with the view that the key to understanding membrane function lies in the genetically coded specificity of their *proteins*.

The Multiplicity of Membrane Proteins

We know that most membranes contain many proteins. The question of "how many" can hardly be generally specified, depending so critically on detection sensitivity. This matter has been examined by one of us in an immunologic context (19), which showed that a smooth, spherical cell with a 5 μm radius will contain about 20,000 different proteins. Allowing each protein a generous surface area of 50×50 A = 2500 A would indicate a total surface area of 5×10^7 A^2, 1% of the 5×10^9 A^2 surface area of the hypothetical cell. Such a calculation raises the admittedly drastic possibility that the surface of each cell bears not only plasma membrane components *proper*, but could even carry at least one copy of all protein components of the cell interior. In this speculation, the plasma membrane bears a map of its cell individuality and reflects the cells' phenotype, possibly proportionally.

Certainly, a simple listing of all known plasma membrane functions attributable to proteins would yield hundreds of different entities. However, the contribution of each to the total membrane mass may be small. Thus, the number of ouabain-sensitive ATPase sites on human erythrocytes ranges in the vicinity of only 300/cell (20), a minute proportion of the total protein present. However, the same enzyme would be much more dominant, e.g., in the membranes of the electric organ of *Electroplax*.

At the other extreme, we know of numerous instances where a biomembrane contains one protein that dominates. Thus, the membranes of arbor viruses possess only one protein (21), which is a glycoprotein in some strains (22). Most bacterial membranes reveal an extremely complex protein pattern on SDS-PAGE, but 90% of the protein in the membranes of *Bacillus* PP, a strain of *B. megatherium* KM, consists of an entity with apparent mol wt ~32,000, which seems to constitute only a minor component of the parent strain (23).

Animal cell membranes also generally reveal a complex protein composition, but there is at least one prominent exception, the membranes of retinal rod outer segments, where rhodopsin accounts for nearly 90% of the protein mass (24). Most animal membranes occupy an intermediate position. Thus, in human erythrocyte membranes, six major SDS-PAGE bands account for ~ 70% of the protein staining—and certainly the bulk of the protein mass (25).

Several authors have presented various unifying hypotheses, suggesting that all membranes comprise assemblies of certain basic "structural" polypeptides and lipids into which diverse functional units become integrated. This reasoning began with the "structural protein" concept of Green and associates (26), but this notion has not held up under further study (27). Similarly, the "tektin" hypothesis of Mazia and Ruby (28) cannot be considered as experimentally founded. Finally, the "mini-protein" (approx. mol wt 6000), proposed as a general membrane structural unit by Laico et al. (29), has been refuted by the original authors (24).

On the contrary, the proteins of most membranes, even after removal of possible nonmembranous components, exhibit an extraordinary diversity according to SDS-PAGE (which unfortunately remains the only general analytical technique for membrane proteins) with apparent molecular weights ranging from 15,000 to 300,000 daltons. However, within this diversity lie certain consistent patterns. Thus, SDS-PAGE analyses of erythrocyte ghosts from different species, if not identical, certainly exhibit striking similarities (30). In contrast, the lymphocyte plasma membranes from the same species reveal SDS-PAGE patterns entirely different from those of erythrocyte membranes, but closely similar to each other (31).

In view of the lack of support for the "structural protein" concept, as well as the multiplicity and diversity of membrane proteins, we must assume that the proteins of a given membrane possess certain *structural homologies*, which permit their interaction with the lipids of that membrane, as well as its other protein constituents. This concept has been proposed previously by one of us (32) as well as by others (33), but we wish to discuss its substance here.

We consider "structural homologies" of membrane proteins to derive from their amino acid sequence and apply the term to explain two phenomena:

(a) general, nonspecific membrane affinity, perhaps arising from favorable interactions with membrane lipids;

(b) specific affinity of a given protein for a given membrane or membrane domain. Here we take cognizance of the fact that certain membrane proteins may be incorporated into, or excluded from, one type of membrane but not another. For example, we do not find *host* membrane proteins in *arbor virus membranes*, although the virus buds from the host cell via the plasma membrane. Indeed, this *protein specificity* typifies all virus membranes and contrasts strikingly to the relative *lack of lipid selectivity* in such membranes. On the other hand, the situation resembles what is often found within one cell, or even one membrane organelle. Thus, Schnaitman (34) reports 12 SDS-PAGE bands for the outer membranes of rat liver mitochondria, 23 for inner membranes, and none unambiguously common.

Many other examples can be cited, but all point to a high degree of protein selectivity in biomembranes. One cannot explain this fact in metabolic terms and is thus left with two alternatives.

First, specific biomembranes form through assembly of protein units matching optimally to each other and allowing for optimal lipid-protein interaction. The contact areas participating in these processes may comprise rather small portions of a whole molecule but can be considered identical for different components of a multimacromolecular complex—analogous to the invariant portions of immunoglobulins. The multipartite assembly process, in which diverse molecular entities participate, can be thought as analogous to the assembly of tobacco mosaic virus in the presence of its RNA.

In proposing this view, we suggest that the native association of membrane components can be best described thermodynamically, occurring through primarily apolar associations, the principal driving forces being entropic. Reasoning in analogy with the "renaturation" of proteins, we postulate that a mixture of monomeric membrane components will, under the appropriate conditions, spontaneously assemble to form a native membrane, governed by the minimization of free energy. The experimental and theoretical basis of this reasoning comes originally from Kauzman (35, 36). However, as stressed by Klotz (37), all biopolymers behave fundamentally the same way in aqueous media and in response to specific perturbants. He reasons that since the specific interacting groupings vary, for example in polypeptides and nucleic acids, the chemical or structural features of the macromolecules do not principally determine these interactions. Rather, the action of perturbing molecules must be through a modification of solvent structure, and thereby, interaction of solvent and biopolymer. It is known that ionic and/or hydrogen bonds cannot stabilize the specific conformation that nucleic acids and proteins assume in *aqueous* media, but hydrophobic associations can provide this energy (35, 38, 39). Here the "driving force" comes from the ordering of water in the vicinity of apolar groups causing a thermodynamically unfavorable loss of entropy. The apolar groups tend to rearrange and cluster together, excluding water and decreasing the free energy of the system as a whole. Such hydrophobic associations also tend to stabilize peptide hydrogen bonds within the apolar areas, due to the exclusion of water, which is a competitor for hydrogen bond formation.

The importance of apolar associations in the structure and function of macromolecules is now so well established that it requires no further comment here. However, we do wish to explore present information as to the possible mechanisms by which these and other interactions foster the formation and stabilization of biomembranes. To this effect, we shall further emphasize the role of membrane proteins and their interactions with each other and membrane lipids. In this we stress again that much evidence suggests that membrane lipids exist at least partially in a bilayer; however, the clue to the specific roles of biomembranes lies in their proteins and lipid-protein interactions.

Second, an alternative view holds that the assembly of natural membranes proceeds with preexisting membranes acting as *templates*. This notion derives from the general observation that new membrane formation occurs on or in relation to existing membrane systems. The template concept carries additional interest since, if interpreted strictly, it suggests the possibility of an alternative genetic code.

However, we cannot subscribe to a well-defined template hypothesis because first, certain membrane-enclosed viruses assemble at sites remote from preexisting cellular membranes. Second, enveloped viruses that do mature at preexisting membranes invariably exhibit a *protein* composition distinct from that of the host cell membranes (40). Perhaps "template" is the wrong term, and one should really speak of "support" or "framework," but then the previously noted thermodynamic considerations again dominate.

A *final* complication, which we have mentioned repeatedly, relates to the probability that a macromolecule, which assumes a "membrane" configuration in its membrane site, most likely will not retain this state when in solution, and in the process may lose its "membrane identity."

What Qualifies a Protein as a Membrane "Core Protein"

Overall Polarity and/or Hydrophobicity

There can be little doubt that hydrophobic associations play a major role in the interactions of many proteins with their membranes.

Various authors, including most recently Capaldi and Vanderkooi (41), have suggested that the clue lies in the low *polarity* of "intrinsic" membrane proteins. They have defined protein polarity as the sum of the residue mole percentages of polar amino acids. Asp, Asn, Glu, Lys, His, Arg, Ser, and Thr were considered polar and other amino acids apolar. By these criteria, 85% of 205 soluble proteins had "polarities" of $47 \pm 6\%$, whereas 47% of 19 membrane proteins gave polarities of less than 40% (minimum 30%). Suggestive though this calculation may be, it does not allow very definite conclusions, partly because "polarity" should at least be defined as the ratio polar volume:nonpolar volume (42), which is very different from the criterion employed by Capaldi and Vanderkooi (41). Besides, such calculations define amino side chains as "polar" or "apolar" in an all-or-nothing fashion; thus is lysine more, less, or equally as polar as aspartic acid; or is phenylalanine more, less, or equally as apolar as leucine? Also, as recognized by Fischer (42), polarity classifications such as his depend strongly on the geometry of the protein. Finally, the existence of amide and isopeptide linkages is ignored.

A more reasonable assessment of a protein's average polar or apolar quality derives from Tanford's studies on the solubilities of free amino acids in solvents of different polarity (43) and from Bigelow's computations (44). In this approach, each amino acid residue is assigned an experimentally determined "hydrophobicity" value, $H\Phi$, varying between 0.45 and 3.00 kcal/residue for Thr and Trp, respectively. From these data and the amino acid composition of a protein, its average hydrophobicity can be computed. Bigelow (44) shows that average hydrophobicities of soluble proteins range from 0.440 to 2.020 kcal/residue with 50% of protein $H\Phi$ values between 1.0 and 1.2 kcal/residue.

Several aspects of Tanford's "hydrophobicity" approach and the "polarity" classification appear worthy of additional comment.

First, the hydrophobicity scale comprises a set of experimentally determined values, rather than an arbitrary all-or-none classification.

Table 9.1 Amino Acid Composition of Certain Membrane Preparations

Amino Acid	Human Erythrocyte Ghosts[a]	Endoplasmic Reticulum[b,f] Ascites Carcinoma	Plasma Membrane[b,f] Ascites Carcinoma	Liver Bile Fronts[c]	Liver Smooth Microsomal[d] Membranes	Myelin
Lys	5.2	6.5	6.3	7.2	6.3	5.8
His	2.4	2.1	2.6	2.6	—	2.3
Arg	4.5	5.2	4.7	5.2	3.6	4.0
NH_3	6.9	10.8	14.7	12.4	15.3	—
Asp	8.5	8.7	8.8	9.3	10.0	6.1
Glu	12.2	10.6	10.1	12.0	10.6	7.1
Thr	5.9	5.4	5.5	5.3	5.4	6.4
Ser	6.3	6.2	6.6	6.0	4.4	10.4
Pro	4.3	5.4	5.2	4.9	6.6	1.1
1/2 Cys	1.1	trace	trace	0.9	0.06	3.8
Met	2.0	2.5	2.7	2.3	—	1.0
Gly	6.7	7.7	8.5	7.8	7.3	10.9
Ala	8.2	7.6	7.8	8.0	7.2	9.6
Val	7.1	6.7	6.6	6.6	6.5	5.9
Ile	3.3	5.1	6.1	5.1	4.7	4.5
Leu	11.3	10.0	10.1	9.6	9.9	8.8
Tyr	2.4	3.4	3.1	2.7	—	3.1
Phe	4.2	4.8	4.8	4.5	2.8	4.0
Trp	2.5	1.5	1.5	—	0.2	5.1

[a] From Rosenberg and Guidotti (51).
[b] From Wallach and Zahler (32).
[c] From Takeuchi and Terayama (83).
[d] From Manganiello and Phillips (84).
[e] From Van den Heuvel (85).
[f] Values of NH_3, Ser, Thr, Tyr, and Trp corrected for decomposition of these amino acids during hydrolysis.

Note: Values presented are the number of residues per 100 residues.

Table 9.2 Average Protein Hydrophobicities, $H\Phi$, of Some Membrane Isolates[a]

Fraction	Average Hydrophobicity (kcal/residue)
Erythrocyte ghosts[b]	1.068
Plasma membrane (Ehrlich[c] ascites carcinoma)	1.041
Smooth endoplasmic reticulum (Ehrlich ascites carcinoma)[c]	1.023
Smooth endoplasmic reticulum (liver)[d]	0.940
Myelin[e]	1.000

[a] Computed according to Tanford (42).
[b] Computed from Rosenberg and Guidotti (51).
[c] Computed from Wallach and Zahler (32).
[d] Computed from Manganiello and Phillips (84).
[e] Computed from Van den Heuvel (85).

Note: 50% of soluble proteins have $H\Phi$ values ranging between 1.0 and 1.2 kcal/residue.

Secondly, Tanford (43) considers Arg and Thr as weakly nonpolar, while Capaldi and Vanderkooi (41) classify them as polar.

Third, lysine, with its aliphatic chain, exhibits a high $H\Phi$ value (1.5 kcal/residue), but is considered polar by Capaldi and Vanderkooi (41). This is particularly important since Lys constitutes a prevalent residue.

One of us has previously applied Tanford's approach to diverse membranes (45), and we have now extended these calculations to the membranes listed in Table 9.1 (except for liver bile fronts, where no adequate Trp data exist). The results are listed in Table 9.2. We conclude that when one applies the most stringent criteria available, *one cannot distinguish membrane proteins as a class by virtue of hydrophobicity or polarity*.

In connection with this topic, we want to stress that, while known three-dimensional structures of proteins indicate that charged and polar residues project to the surface of the macromolecule, *the widely believed complementary assertion—that all apolar amino acid residues lie buried in the interior—is definitely incorrect*. In the case of lysozyme, for example, 17% of the apolar groups projects into water, and another 50% is at least partially exposed (46). In subtilisin also, 50% of the apolar groups is accessible to water (47). Presumably, the concentration of polar residues on the surfaces of water-soluble proteins suffices, under most conditions, to prevent any aggregation that might be fostered by the many apolar groups that are also surface located.

The data in Table 9.1 show several features of note.

(a) Where determined, significant —NH_3 is found, suggesting that a considerable fraction of Asp and Glu may be amidated and that the

Table 9.3 Amino Acid Composition of Some Membrane Proteins That Are Easily Eluted into Aqueous Solvents and Others That Resist Extraction

	Elutable			Nonelutable		
Amino Acid	Human RBC Spectrin[a]	S. Faecalis ATPase[b]	Bovine F_1 ATPase[c]	Bovine Retinal Rhodopsin[d]	Bovine Myelin Proteolipid[e]	Murein-Lipoprotein[f]
Lys	6.7	6.1	6.2	4.3	4.5	8.8
His	2.6	1.7	1.7	1.7	2.3	—
Arg	5.8	4.5	5.8	2.6	2.6	7.0
Asp	10.9	10.0	7.9	6.4	4.0	24.5
Glu	20.5	13.0	11.7	8.9	5.9	8.8
Thr	3.6	6.7	5.8	7.2	8.4	3.5
Ser	4.1	6.3	6.2	5.1	5.2	16.5
Pro	2.4	3.9	4.2	5.5	2.8	—
1/2 Cys	1.1	0.3	0.4	2.1	2.9	—
Met	1.7	2.3	2.1	3.4	1.3	3.5
Gly	4.9	8.7	9.2	6.8	10.4	—
Ala	9.2	8.4	10.4	8.5	11.9	15.8
Val	4.7	6.8	7.5	8.5	7.4	7.0
Ile	4.0	6.2	6.2	5.5	5.1	1.8
Leu	12.4	9.3	8.7	8.5	11.5	7.0
Tyr	2.0	3.3	2.9	4.7	4.4	1.8
Phe	3.0	3.1	2.9	8.1	8.2	—
Trp	—	—	0	2.1	1.8	—
NH_3	—	—	—	—	—	8.0

[a] From Marchesi et al. (86).
[b] From Schnebli et al. (87).
[c] From Racker (75).
[d] From Hiller and Lawrence (59).
[e] From Wolfgram and Kotorii (88).
g From Braun and Bosch (54).

Note: Results presented as residues per 100 residues. The hydrophobicities (excluding Trp values) of spectrin, streptococcus ATPase, bovine F_1 ATPase, rhodopsin, bovine proteolipid, and murein lipoprotein are computed as 0.96, 1.10, 1.09, 1.23, 1.21, and 0.73 kcal/mole, using the approach of Tanford (43) and Bigelow (44). Clearly, the chosen "elutable" proteins exhibit lower hydrophobicities than the nonelutable species, although all fall within the 1.00–1.20 kcal/residue range found for most *soluble* proteins. The only notable exception is murein-lipoprotein with its very low hydrophobicity. Clearly, hydrophobicity yields only a blurred image, whose significance depends on sequence analyses.

proteins may not be as acidic as it otherwise appears. The implications of this observation have been discussed by Wallach and Zahler (32).

(b) Except in myelin, $\frac{1}{2}$ Cys is low, consistent with the paucity of —S—S— bonds in membrane proteins, suggested by SDS-PAGE.

(c) There are no dominant amino acids or groups of amino acids.

(d) Unusual amino acids appear lacking.

(e) Proline levels do not contraindicate high helicity.

Except for the possibly significant high amidation, and its conceivable role in permitting protein–phosphatide ion pairing, the overall amino acid analyses of membranes have not proved informative. Since all of the membranes analyzed contain many proteins, this is hardly surprising. However, one cannot logically assume that individual membrane proteins will exhibit drastic singularities of amino acid composition, without simultaneously invoking other proteins that deviate equally dramatically in an opposite direction. However, subtle differences might be anticipated and have been detected.

Thus, if one groups membrane proteins into those that can be eluted by *ionic manipulations* and those requiring *detergents* or *organic* solvents for extraction (Table 9.3), one finds that in some cases the elutable proteins exhibit distinctly *lower hydrophobicity* than the more tightly bound species. For example, ionically elutable erythrocyte-spectrin, *S. faecalis* ATPase, and bovine F_1 ATPase exhibit HΦ values of 0.96, 1.10, and 1.09 kcal/residue, respectively, while the "nonelutable" proteins rhodopsin and bovine myelin proteolipid yield hydrophobicities of 1.23 and 1.21 kcal/residue, respectively (Table 9.3). Notably, murein lipoprotein exhibits an unexpectedly *low* hydrophobicity 0.73 kcal/mole. We shall comment further on the last, but here stress that except for this protein, *both* the "elutable" and "nonelutable" proteins show hydrophobicities in the 0.9–1.2 kcal/residue range typical for most *water-soluble* proteins.

We thus feel that the amino acid analyses of individual membrane components will very likely teach little more than the analyses of whole membranes. On the other hand, we feel secure that essential information will derive from peptide mapping and/or amino acid analyses of individual membrane proteins. Indeed. the first approach has been and continues to be under vigorous study (48).

Amino Acid Distribution and Sequence

Penetration of apolar side chains into a lipid bilayer

One of the models proposed to account for apolar associations of proteins with membranes suggests that the proteins lie at the membrane surfaces and extend apolar residues. However, a little work with space-filling models shows that even the largest apolar amino acid side chain,

Trp, would only extend into the glycerol region of a phosphatide bilayer and could not reach the fatty acid region. Haydon and Taylor (49) discuss this matter in detail and also show that such a lipid-protein interaction is energetically improbable and sterically impossible. One must, therefore, consider mechanisms by which *part or all of a membrane protein, rather than its apolar side chains, can penetrate into or through the apolar membrane core.*

Concentration of apolar residues at one end of a polypeptide chain

A relatively simple model for apolar protein membrane associations, and involving penetration of the protein into the membrane, is suggested by the structure of *mellitin,* the membrane-active polypeptide from bee venom (Fig. 9.1). Here the first 20 residues from the amino terminal comprise principally hydrophobic entities, while the C-terminus contains a sequence of six polar and/or charged residues.

1 5 10
Gly Ile Gly Ala Val Leu Lys Val Leu Thr Thr Gly Leu Pro--

15 16 20
--Ala Leu Ile Ser Trp Ile Lys Arg Lys Arg Gln Gln

Fig. 9.1 Amino acid sequence of mellitin. Major mellitin component; mol wt 2848. Calculated hydrophobicity 1.25. The boxed-in portion indicates the apolar segment. From Dayhoff (50).

Such a protein could intercalate its apolar residues among the relatively liquid fatty acid chains of a phosphatide bilayer in a random array. Alternatively, the first 14 residues (up to the proline) could form an α-helical, "greasy plug" (Fig. 9.2) still allowing the polar segment to protrude to the membrane surface. Other conformational possibilities clearly exist, and nonpolar interactions between individual apolar segments appear eminently feasible. Indeed, mellitin aggregates hydrophobically in water to form a tetramer (50).

One of us (48) has suggested that this model may explain the properties of the major proteins in human erythrocyte membranes (51).

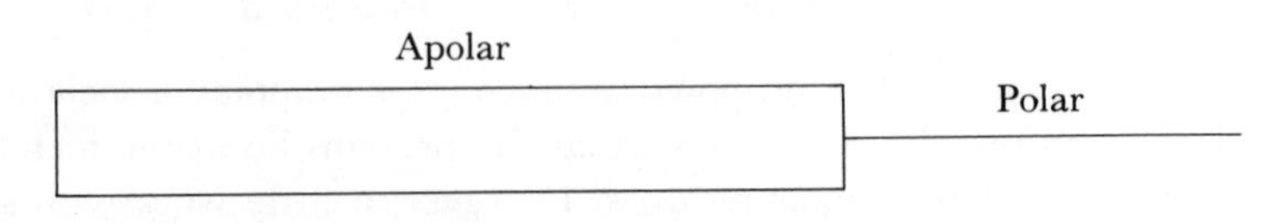

Fig. 9.2 Schematic of mellitin.

This point is illustrated in Table 9.4, giving the amino acid composition of the intact glycoprotein, the soluble peptide segment released by mild proteolysis of intact cells and the insoluble peptide fraction remaining after proteolysis. Very clearly, the protease accessible portion (bearing the entire sugar moiety) contains a large proportion of low-hydrophobicity amino acids, giving it an average $H\Phi$ of only 0.6 kcal/residue, compared with 0.93 kcal/residue for the whole protein. The very high average hydrophobicity of the "residue," 1.33 kcal/residue, is concordant with the concept that this hydrophobic portion "anchors" the protein to the membrane.

Table 9.4 Amino Acid Composition of the Major Human Erythrocyte Membrane Glycoprotein and Its Fragments

	Sialoglycoprotein[a]		
Amino[b] Acid	Intact	Soluble Peptide	Insoluble Peptide
Lys	3.5	4.3	2.0
His	3.8	4.9	3.8
Arg	4.1	3.3	3.9
Asp	6.0	7.6	2.4
Glu	10.0	4.7	7.0
Thr	13.8	23.8	6.2
Ser	13.6	23.8	6.0
Pro	6.5	4.0	4.5
1/2 Cys	0	0	0
Met	0	0	2.1
Gly	6.8	3.5	10.7
Ala	6.8	6.1	7.9
Val	7.7	4.9	8.1
Ile	4.5	2.8	14.3
Leu	4.5	1.6	11.6
Tyr	3.6	1.5	2.4
Phe	3.5	0	5.1
Trp	—	—	—

[a] From Winzler (48).
[b] Values are in residues per 100 residues. This protein contains 64% carbohydrate including most of the cell membranes' sialic acid in conjunction with galactose, mannose, fucose, *N*-acetyl-galactosamine, and *N*-acetyl-glucosamine. These comprise oligoheterosaccharides bound to the frequent threonine and serine residues usually via the *N*-acetylgalactosamine.

Note: The computed hydrophobicities are 0.93 kcal/residue for the intact protein, 0.60 kcal/residue for the soluble fragment, and 1.33 kcal/residue for the insoluble peptide segment.

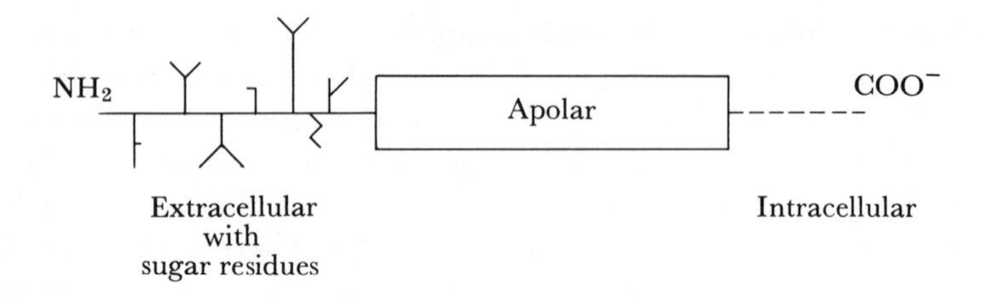

Fig. 9.3 Schematic of major erythrocyte glycoprotein.

More detailed analyses, by Segrest et al. (52), however, show the matter to be more complex. These authors, and others, argue that the amino terminal segment, bearing the carbohydrate, lies extracellularly, while the apolar region occupies the hydrophobic membrane core, and a *polar carboxyl terminal* protrudes into the cytoplasmic domain (Fig. 9.3).

The authors have purified their protein, using the lithium iodosalicylate procedure (Chap. 2) and suggest the tentative amino acid sequence for the apolar segment shown in Fig. 9.4. As expected from prior data, the sequence abounds in apolar residues. No potentially ionizable residues occur for a stretch of 23 residues (11–35). However, two *Glu* residues (9, 11) and two *Arg* residues (35, 36) give the apolar stretch (a) two polar ends and (b) an asymmetric charge distribution.

No prolines occur in the 23-residue apolar segment, which could thus be in α-helical array. The authors point out that such a helical rod would

1 5 10 15
Val-Gln-*Leu*-Ala-His-His-*Phe*-Ser-Glu-*Ile*-Glu-*Ile*-Thr-*Leu*-Ile-
Pro *Pro* *Ala*

16 20 25 30
-Gly-*Phe*-Gly-*Val*-*Met*-*Ala*-Gly-*Val*-*Ile*-Gly-*Thr*-*Ile*-*Leu*-*Leu*-*Ile*-
Val

31 35 40 45
-Ser-*Tyr*-Gly-*Ile*-Arg-Arg-*Leu*-*Ile*-Lys-Lys-Ser-*Pro*-Ser-Asp-*Val*-

46 50
-Lys-*Pro*-Leu-*Pro*-Ser-*Pro*-

Fig. 9.4 Tentative amino acid sequence for the apolar portion of the major glycoprotein in human erythrocyte membranes. [From Segrest et al (52).] Hydrophobic residues are italicized. Presumably because of less than ideal purity, the major residue at position 4, 6, 15, and 16 cannot be defined absolutely.

extend for about 35 A, the approximate width of the hydrocarbon moiety of a phosphatide bilayer.*

Specialized Amino Acid Sequences

We have described how concentration of the apolar residues at one end of a polypeptide chain will make such a "hydrophobic terminus" into a membrane anchor. However, another structural principle can serve the same purpose, and requires our attention because (a) it provides the versatility required in complex membranes and (b) it constitutes an architectural device known to occur in many nonmembrane proteins.

Certain sequences of polar and apolar amino acids permit a polypeptide chain with an unremarkable amino acid composition to fold into a secondary structure (e.g., α-helix) with a *polar face* and an opposite *apolar face*. Tertiary structuring and/or quaternary interactions will then *foster association of polar-with-polar and apolar-with-apolar faces*. This process naturally depends strongly upon the solvent, and, in *water*, minimal free energy and maximal entropy obtain when the *polar* faces *orient into the solvent* and the apolar ones into the structural core, i.e., exactly in anology with the two halves of a phosphatide bilayer. Many well-documented examples of this situation exist, the most fascinating being *hemoglobin*. The peptide chains of all vertebrate myoglobins and hemoglobins fold into virtually identical tertiary structures, one of whose most prominent features is the almost total exclusion of polar residues from the interior of the molecules. Only 9 out of more than 140 residues are the same in all the globin chains analyzed so far. However, in all of these, 30 residues are always nonpolar and make up the hydrophobic core of these proteins. The apolar sites occur at intervals of 3.6 residues along α-helical segments, giving each helix a hydrophobic face (53).

This structural device, illustrated for the H-helix of human hemoglobin in Fig. 9.5, occurs in numerous, partly helical proteins, whose structure has been elucidated by X-ray crystallography; it constitutes an important principle of protein structure. The amino acid analyses of Braun and Bosch (54) further demonstrate it for the murein-lipoprotein of the outer *E. coli* membrane.

Continuing our earlier thermodynamic reasoning, we suggest that substitution of a low dielectric solvent—e.g., the hydrocarbon chains of phosphatides—will induce association of the polar surfaces at the in-

* Further data on the amino acid sequence and possible structure of the major glycoprotein in human erythrocyte ghosts is given by Segrest et al. (J. P. Segrest, I. Kahane, R. L. Jackson, and V. T. Marchesi, *Arch. Biochem. Biophys.* **155:** 167, 1973). The data support the concept that the hydrophilic N- and C-terminals extend externally and intracellularly respectively, being linked by an apolar peptide segment anchored in the lipophilic membrane core.

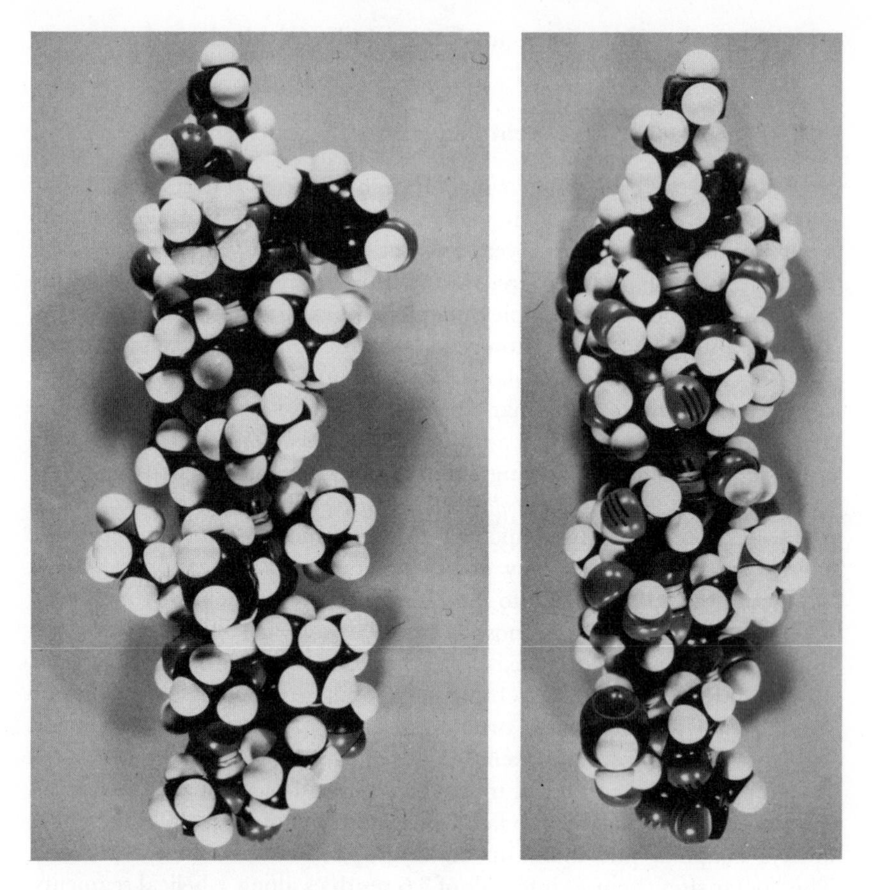

Fig. 9.5 Space-filling model of the H-helix of human hemoglobin β-chains. *Left:* apolar face; *right:* polar face.

terior of the folded polypeptide, giving it a hydrophobic perimeter. This model is fully analogous to the reversal of molecular orientation of amphipathic lipids, when transferred from an aqueous dispersion to an organic solvent (55). In the former, the molecules tend to align in bilamellar arrays with the *polar* groups of each bilayer oriented away from each other into the solvent. In the latter, the *hydrocarbon* chains of each bilayer point away from each other into the solvent and the polar groups cluster in the bilayer interior.

Thus, appropriate combinations of amino acid sequence, 2°, 3°, and perhaps quaternary structures, can endow protein with the hydrophobic surfaces required for apolar membrane-protein associations, *without* an unusually high average hydrophobicity and/or concentration of apolar residues in a given segment of a polypeptide chain. Interestingly, the best-studied models, e.g., the hemoglobins and lysozyme, exhibit the polar-apolar separation in α-helical segments; this also appears in murein-lipoprotein. However, β- and/or irregular conformations should yield a similar effect with the appropriate amino acid sequences.

Clearly, a protein with a fully apolar perimeter, while suited for residence in a membrane's apolar core, cannot be considered ideally structured to deal with events involving the aqueous intra- and extra-cellular compartments.

As proposed before by one of us (32, 45, 56), such traffic would require (a) more or less cylindrical proteins, with (b) *hydrophobic perimeters,* (c) *polar moieties* projecting into one or both aqueous spaces, and (d) possibly, polar channels running through the protein assembly—analogous to the "pore" penetrating the hemoglobin tetramer.

Murein-lipoprotein

The cell walls of *E. coli* comprise a rigid layer, stabilized by murein (peptidoglycan, mucopeptide) consisting of polysaccharide chains cross-linked by short peptide segments. About 10^5 lipoprotein molecules are linked to this net in any one cell. The full amino acid sequence of this protein, as well as the murein and lipid attachment sites, has now been published (54, 57). As shown in Fig. 9.6, the *N*-terminus consists of three almost identical sequences, and starting with the third residue, apolar residues occur constantly at every 3rd–4th position in a consistently alternating set. In an α-helical conformation, this would provide murein-lipoprotein with a polar and an apolar face, as in hemoglobins, *despite a very low hydrophobicity* ($H\Phi = 0.73$ kcal/residue)! The authors find that despite the low $H\Phi$, the protein aggregates hydrophobically once murein

```
Lipid
 |1                5                  10                  15
(Ser) Ser Asn Ala Lys Ile Asp Glu Leu Ser Ser Asp Val Gln Thr--
                   20                  25
  --Leu Asn Ala Lys Val Asp Glu Leu Ser Asn Asp Val Asn Ala--
                               30
                            --Met Arg Ser Asp Val Gln Ala Ala Lys
```

Fig. 9.6 *N*-terminal portions of murein-lipoprotein. [From Braun and Bosch (54).] The sequence is presented to emphasize the structural design. The first 29 residues comprise two almost identical halves with only 5 very conservative substitutions and a deletion possibly of a Lys or Arg. Apolar residues occur at every 3rd or 4th position which, in an α-helix, would give the structure a polar and an apolar face, as in the H-helices of hemoglobins (53) (Fig. 9.5).

is cleaved off. They attribute this to the bound lipid, but also consider the proposition first raised by one of us (35, 45, 56) that the protein is folded with the polar residues interiorized and the apolar faces free to associate with other apolar cell wall components.

Clearly, much more can be learned from this model. In addition, however, we urgently need amino acid *sequence* analyses on diverse purified membrane proteins, in order to determine what, if any, structural principles apply to all membrane proteins.

Protein-Carbohydrate Associations

Most cellular membranes are associated with diverse carbohydrates to a greater or lesser extent. This may be in the form of a mucopolysaccharide "glycocalyx" coating many cells' plasma membranes (58) and/or as membrane-associated glycoprotein and glycolipid. The last topic has been recently reviewed by one of us (51).

The carbohydrate moiety of membrane-associated glycoproteins can often be released in the form of soluble glycopeptides, allowing considerable clarification of the oligosaccharide structure. This appears rather similar to that of soluble glycoproteins in the following respects.

(a) The oligosaccharides are bound either *N*-glycosidically to Asn or *O*-glycosidically to a Ser or Thr hydroxyl.

(b) More than one heterosaccharide may occur on a given membrane or membrane protein.

(c) The sugar moieties are small heterosaccharides, usually with specific monosaccharide sequences.

(d) The major constituent monosaccharides are fucose, galactose, glucose, mannose, *N*-acetylgalactosamine, *N*-acetylglucosamine, and *N*-acetylneuraminic acid.

Two major membrane glycoproteins of very different character have been extensively studied. These are (a) the *M*N-glycoprotein of human erythrocytes (48) and (b) bovine rhodopsin (59).

The former contains about 64% carbohydrate by weight, bears most of the membrane's sugar and all of its *N*-acetylneuraminic acid. In its chemically extracted form it has a molecular weight of 31,400, but SDS-PAGE data suggest it may migrate as a dimer in that system. One cannot yet be certain that this substance is a single homogeneous molecule, particularly since it exhibits multiple biological functions, such as binding of kidney-bean phytohemagglutinin, *M*N-antibody, and influenza virus. However, these functions probably reflect the diverse specificities of different heterosaccharides attached to a single polypeptide chain. Many of these heterosaccharides are tetrasaccharides consisting of two molecules of *N*-acetylneuraminic acid, a galactose, and an *N*-acetylgalactosamine linked *O*-glycosidically to Ser or Thr. Some oligosaccharides appear incomplete.

At least one of the heterosaccharides, *N*-glycosidically linked to the peptide chain, exhibits greater complexity. Kornfeld and Kornfeld (60) propose the following structure:

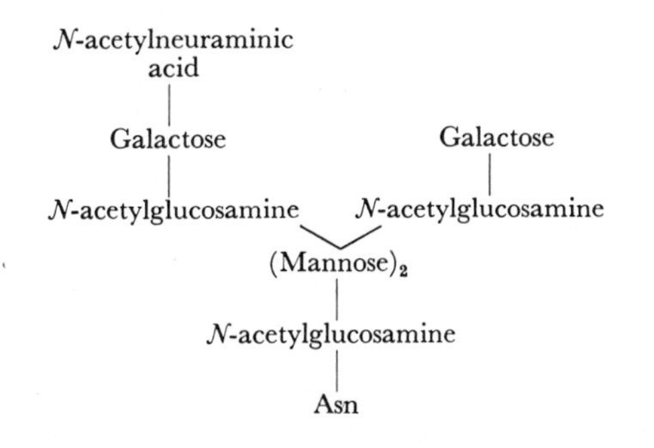

This heterosaccharide strongly binds kidney-bean phytohemagglutinin.

Other carbohydrate side chains of this protein have also been characterized and attest to the complexity of this molecule or group of molecules. This is in great contrast to our other example, bovine rhodopsin. Here only 4% of the mass derives from carbohydrate. Each 28,000 dalton polypeptide chain contains one heterosaccharide joined *N*-glycosidically via *N*-acetylglucosamine and Asn. The other sugars comprise two molecules of *N*-acetylglucosamine and three of mannose.

Specific Protein-Lipid Associations

In the earlier chapters and previous sections of this chapter, we have treated lipid-protein interactions in a rather general way, although we did point out in Chapters 1 and 2 that some very specific associations have been uncovered. Nearly all lipid-protein associations are noncovalent. However, in the murein-lipoprotein of *E. coli* (Fig. 9.6), lipid is linked covalently to the *N*-terminal serine (54).

As noted earlier, erythrocytes of different species vary substantially in lipid composition, but this species specificity resides in the phosphatides, which are "tightly" associated with membrane proteins according to rather empirical extractability criteria (61). This fact is illustrated for phosphatidylcholine (PC) and sphingomyelin (S) in Table 9.5. Kramer et al. (62) have explored this matter further. They separated ghost proteins from lipids by gel permeation on lipophilic sephadex, using 2-chlorethanol, added traces of radioactive PC and S to the respective lipid fractions, and let the proteins and lipids recombine through dialytic replacement of the solvent by aqueous buffer. Their analyses of the

Table 9.5 Phosphatidylcholine and Sphingomyelin Composition of Human and Sheep Erythrocyte Ghosts[a]

	Total Phosphatides			"Loosely Bound"			"Tightly Bound"		
Species	%PC[b]	%S[c]	PC/S	%PC	%S	PC/S	%PC	%S	PC/S
Human	39	37	1.05	24	22	1.1	40.5	28	1.4
Sheep	1	63	0.02	5	24	0.2	4.5	57	0.08

[a] From Roelofson (61).
[b] Phosphatidylcholine.
[c] Sphingomyelin.

"recombinates," purified by differential gradient centrifugation, clearly show retention of the *sphingomyelin preference* of the sheep membrane proteins.

As often reviewed (63), many membrane-bound enzymes lose their activity upon lipid extraction, but can be reactivated by restoring the lipid. In this, the conditions of the lipid is rather critical; the more finely dispersed the better.

This matter was initially studied very extensively in mitochondria (64). Thus, most of the mitochondrial phosphatide can be extracted by cold aqueous acetone (Chap. 2), with associated loss of respiratory activity. The latter could be restored by addition of aqueously dispersed unfractionated mitochondrial phosphatides, purified lecithin, or purified diphosphatidylglycerol. The last was most effective at low lipid binding, but at saturation this specificity is lost.

In contrast to this relative lack of specificity in enzyme–lipid protein interaction, very high or absolute levels of selectivity have been documented in many instances. Thus, rat liver 5′-*nucleotidase* exhibits a very high sphingomyelin binding (65). Also mitochondrial D(−)*β-hydroxybutyric acid dehydrogenase* shows an *absolute* requirement for phosphatidylcholine (66). A fascinating aspect of this case is the fact that the lipid-induced activation clearly depends at least in part on the apolar moiety, since only unsaturated lecithins provide full activation.

Several enzymes associated with bacterial membranes show a total lipid dependence of activity, as well as very high degrees of selectivity. Thus, the constitutive, sugar-specific phosphotransferases of *E. coli* exhibit an absolute requirement for phosphatidylglycerol (67). Similarly, *S. aureus* isoprenoid alcohol phosphokinase requires either phosphatidylglycerol or diphosphatidylglycerol for activity (68).

Rothfield and associates (69) utilize an elegant means for the reconstitution of active bacterial glycosyl transferase systems at air-water interfaces. They initially form a monolayer of phosphatidylethanolamine at a buffer-air interphase. Bacterial lipopolysaccharide added subsequently to the subphase leads to the formation of a composite film, signaled by an increase in surface pressure. By determining molecular surface areas and the molar ratios of phosphatidylethanolamine and lipopolysaccharide, Rothfield and associates (69) conclude that the two components of the composite layer lie side to side, with the fatty acid chains oriented into the air and perpendicular to the film surface.

When the glycosyl transferase is introduced into the subphase underlying the binary film, a ternary layer forms as indicated by a further rise of surface pressure, presumably due to partial penetration of the binary film by the enzyme. This is accompanied by reconstitution of enzymatic function, as demonstrated by the transfer of ^{3}H-galactose to the lipopolysaccharide from UDP-^{3}H-galactose in the subphase.

Comments on the Disposition of Proteins in Intact Membranes

The many powerful physical techniques now available to the membrane biologist have not as yet revealed much about the disposition of proteins within membranes (70). However, this and related topics constitute burning issues and raise fascinating questions: Do membranes constitute protein-lipid mosaics? If so, how large are these? Are they stable? Which is the mosaic frame, lipid or protein? What orders of symmetry prevail perpendicular and/or parallel to the membrane plane? Many other questions can be asked, but assuredly the answers will not be identical for all membranes.

As noted before, many physical experiments indicate that some of the phospholipid and other membrane lipid of most membranes approximates a lipid bilayer. But proteins constitute the major component of membrane mass and, despite engaging models, we know little of protein-lipid interactions. Certainly, few now generalize the early notions of protein monolayers electrostatically linked to the head groups of phosphatides, even though cytochrome c is so linked to the inner mitochondrial membrane. Indeed, most would agree that membrane "core" proteins are bound to their membranes by hydrophobic associations. But where do they lie and how do they associate?

Recent evidence obtained by freeze-cleaving electron microscopy (71) indicates that some membrane areas of considerable extent exhibit high degrees of order. Further, a major concept of membrane biology, that of Changeux and associates (72, 73), postulates cooperative interactions over wide membrane domains, mediated by membrane proteins.

Much evidence suggests that biomembranes are not symmetrical normal to the plane of the membrane, and many of the symmetry considerations are rigorously discussed by Changeux et al. (72, 73) and should be considered together with Mitchell's "chemiosmotic" hypothesis of oxidative phosphorylation (74). According to this, the functional components of the inner mitochondrial membrane are asymmetrically arrayed across the permeability barrier, sustaining a proton gradient, which serves as energy source for ATP synthesis. Considerable dispute continues regarding the validity of Mitchell's and Changeux' hypotheses, but these notions have stimulated new experimental approaches, which do demonstrate transmembrane asymmetry (75, 76).

Several authors suggest that membrane proteins possess considerable mobility tangential to the plane of the membrane. Thus, Blasie and Worthington (77) show that the rhodopsin of bovine retinal rod outer segments exhibits considerable lateral mobility. Moreover, Frye and Eddidin (78) found that the species-specific antigens of Sendai-virus induced heterokaryous monitored by the binding of specific fluorescent antibody intermingled over the surface rather rapidly, and the more rapidly, the higher the temperature. Similar studies (79–81) show that the membrane immunoglobulins and histocompatibility antigens of lymphocytes appear to possess lateral mobility. Fascinating though they appear, we consider molecular interpretations, e.g., the "fluid mosaic" membrane model (82), premature on the basis of such evidence. First, fluorescence microscopy is a low resolution (~ 500 nm) indicator; *secondly*, the surfaces of most cells elaborate into mobile microvilli; *third*, what may appear as mobility could represent reversible *association-dissociation equilibria*. *Finally*, if some proteins appear to move relatively freely in the apparent membrane plane, others, e.g., the protein of enveloped viruses, definitely remain highly localized and do not commingle with host proteins either in the host or the virus membrane.

Numerous recent probe experiments, cited in Chapters 7 and 8, indicate that penetrating membrane proteins are surrounded by shells of immobilized membrane lipid, as predicted in (32, 45). If one assumes that membrane lipids constitute homogeneous solvents for membrane proteins, such shells could be considered analogous to solvation layers. However, in most membranes lipids are heterogeneous and there the shells may consist of specific, tightly bound lipids. Since the proportion of immobilized lipid appears to lie near 20% and since estimates for the proportion of protein in membrane cores now approach 50%, one must consider the possibility that as much as 60% of "membrane core" is solid. Even if one assumes random distribution of solid in an otherwise fully fluid phase, the system as a whole could not exhibit the lateral diffusion properties of a pure, fluid lipid system as assumed in (78–82).

References

1. Danielli, J. F., and Davson, H. *J. Cell. Comp. Physiol.* **5**: 495, 1935.
2. Gorter, E., and Grendel, F. *J. Exp. Med.* **41**: 439, 1925.
3. Danielli, J. F., and Harvey, E. N. *J. Cell. Comp. Physiol.* **5**: 483, 1935.
4. Robertson, J. D. *Biochem. Soc. Symp.* **16**: 3, 1959.
5. Schmidt, W. J. *Z. Zellforsch.* **23**: 261, 1936.
6. Schmidt, W. J. *Die Doppelbrechung von Karyoplasma, Zytoplasma und Metaplasma.* Berlin: Gerüder Bornsträger, 1937.
7. Schmidt, W. O. *Cold Spring Harbor Symposia.* Long Island, N.Y.: Long Island Biolog. Assoc., 1936, vol. 4, p. 7.
8. Schmitt, F. O., Bear, R. S., and Ponder, E. *J. Cell. Comp. Physiol.* **9**: 89, 1936.
9. Branton, D. *Phil. Trans. Roy. Soc.* B, **261**: 133, 1971.
10. Geren, B. B. *Exp. Cell Res.* **7**: 558, 1954.
11. Korn, E. D. *Fed. Proc.* **28**: 6, 1969.
12. Barr, R. S., Deamer, W., and Cornwell, D. G. *Science* **153**: 1010, 1966.
13. Engelman, D. M. *Nature* **223**: 1279, 1969.
14. Rand, R. P., and Luzatti, V. *Biophys. J.* **8**: 125, 1968.
15. Engelman, D. M. *Chem. Physics of Lipids* **8**: 298, 1972.
16. Malcolm, B. R. *Biopolymers* **9**: 911, 1970.
17. Finean, J. B. *Chem. Physics of Lipids* **8**: 279, 1972.
18. Brown, D. D. *J. Mol. Biol.* **12**: 491, 1965.
19. Wallach, D. F. H. *J. Theor. Biol.* **39**: 321, 1973.
20. Dunham, P. B., and Hoffman, J. F. *Proc. Natl. Acad. Sci.* (U.S.) **66**: 939, 1970.
21. Straus, J. H., Jr., Burge, B. W., and Darnell, J. E., Jr. *J. Mol. Biol.* **47**: 437, 1970.
22. Burge, B. W., and Straus, J. H., Jr. *J. Mol. Biol.* **47**: 449, 1970.
23. Patterson, P. H., and Lennarz, W. J. *Biochem. Biophys. Res. Comm.* **40**: 408, 1970.
24. Dryer, W. J., Papermaster, D. S., and Kühn, H. *Ann. N.Y. Acad. Sci.* **195**: 61, 1972.
25. Fairbanks, G., Steck, T. L., and Wallach, D. F. H. *Biochemistry* **10**: 2606, 1971.
26. Richardson, S. H., Hultin, H. O., and Green, D. E. *Proc. Natl. Acad. Sci.* (U.S.) **50**: 821, 1963.
27. Green, D. E., Haard, N. F., Lenaz, G., and Silman, H. I. *Proc. Natl. Acad. Sci.* (U.S.) **60**: 277, 1968.
28. Mazia, D., and Ruby, A. *Proc. Natl. Acad. Sci.* (U.S.) **61**: 1005, 1968.
29. Laico, M. T., Ruoslahti, E. I., Papermaster, D. S., and Dryer, W. J. *Proc. Natl. Acad. Sci.* (U.S.) **67**: 120, 1970.
30. Knüfermann, H., Schmidt-Ullrich, R., Ferber, E., Fischer, H., and Wallach, D. F. H., in *Erythrocytes, Thrombocytes, Leukocytes,* E. Gerlach, K. Moser, E. Deutsch, and W. Wilmanns, eds. Stuttgart: G. Thieme, 1973, p. 12.
31. Schmidt-Ullrich, R., Ferber, E., Knüfermann, H., Fischer, H., and Wallach, D. F. H. *Biochim. Biophys. Acta.*, in press.
32. Wallach, D. F. H., and Zahler, P. H. *Proc. Natl. Acad. Sci.* (U.S.) **56**: 1552, 1966.
33. Changeux, J. P. In *Regulatory Functions of Biological Membranes,* J. Järnefelt, ed. Amsterdam: Elsevier, 1968, p. 115.

34. Schnaitman, C. A. S. *Proc. Natl. Acad. Sci.* (U.S.) **63:** 412, 1969.
35. Kauzman, W. K. *Advan. Protein Chem.* **14:** 1, 1959.
36. Eisenberg, D., and Dauzman, W. K. *The Structure and Properties of Water.* New York: Oxford University Press, 1969.
37. Klotz, I. M. K. *Fed. Proc.* **24:** 525, 1965.
38. Scheraga, H. A. *J. Phys. Chem.* **65:** 1071, 1965.
39. Scheraga, H. A., Nemethy, G., and Steinberg, J. Z. *J. Biol. Chem.* **237:** 2560, 1962.
40. Klenck, H. D. In *Biomembranes,* D. Chapman and D. F. H. Wallach, eds. London: Academic Press, 1973, p. 145.
41. Capaldi, R. A. C., and Vanderkooi, G. V. *Proc. Natl. Acad. Sci.* (U.S.) **69:** 930, 1972.
42. Fischer, H. *Proc. Natl. Acad. Sci.* (U.S.) **51:** 1285, 1964.
43. Tanford, C. *J. Am. Chem. Soc.* **84:** 4240, 1962.
44. Bigelow, C. B. *J. Theoret. Biol.* **16:** 187, 1967.
45. Wallach, D. F. H., and Gordon, A. *Fed. Proc.* **27:** 1263, 1968.
46. Brown, W. J., North, A. T., Phillips, D. C., Brew, K., Vanaman, T. C., and Hill, R. L. *J. Mol. Biol.* **42:** 65, 1969.
47. Wright, C. S., Alden, R. A., and Kraut, J. *Nature* **221:** 235, 1969.
48. Winzler, R. J. In *Red Cell Membrane Structure and Function,* G. A. Jamieson and T. J. Greenwalt, eds. Philadelphia: Lippincott, 1969, p. 157.
49. Haydon, D. A. H., and Taylor, J. T. *J. Theoret. Biol.* **4:** 281, 1963.
50. *Atlas of Protein Sequence and Structure,* M. D. Dayhoff, ed. Washington, D.C.: National Biomedical Research Foundation, 1972, p. 223.
51. Rosenberg, S. A., and Guidotti, G. *J. Biol. Chem.* **243:** 1985, 1968.
52. Segrest, J. P., Jackson, R. L., Andrews, E. P., and Marchesi, V. T. *Biochem. Biophys. Res. Comm.* **44:** 390, 1971.
53. Perutz, M. F., Kendrew, J. D., and Watson, H. C. *J. Mol. Biol.* **13:** 669, 1965.
54. Braun, V., and Bosch, V. *Eur. J. Biochem.* **28:** 51, 1972.
55. Becher, P. *Emulsions: Theory and Practice.* New York: Reinhold, 1965, ch. 2.
56. Wallach, D. F. H. *Heidelberg Science Library* **18:** 1972.
57. Braun, V., and Bosch, V. *Proc. Natl. Acad. Sci.* (U.S.) **970:** 1972.
58. Parsons, D. F., and Subjeck, J. R. *Biochim. Biophys. Acta* **265:** 85, 1972.
59. Heller, J., and Lawrence, M. A. *Biochemistry* **9:** 864, 1970.
60. Kornfeld, R. K., and Kornfeld, S. K. *J. Biol. Chem.* **245:** 2536, 1970.
61. Roelofson, B., De Gier, J., and Van Deenen, L. L. M. *J. Cell. Comp. Physiol.* **63:** 233, 1964.
62. Kramer, R. K., Schlatterer, C., and Zahler, P. H. *Biochim. Biophys. Acta* **282:** 146, 1972.
63. Triggle, D. J. In *Recent Progress in Surface Science,* J. F. Danielli, A. C. Riddiford, and M. D. Rosenberg, eds. New York: Academic Press, 1970, vol. 3, p. 273.
64. Fleischer, S., Brierly, G., Klouwen, H., and Slautterback, D. B. *J. Biol. Chem.* **237:** 3264, 1962.
65. Widnell, C., and Unkelless, J. C. *Proc. Natl. Acad. Sci.* (U.S.) **61:** 1050, 1968.
66. Jurtschuk, P., Jr., Sekuzu, I., and Green, D. E. *J. Biol. Chem.* **238:** 3595, 1963.
67. Kundig, W., and Roseman, S. *J. Biol. Chem.* **246:** 1407, 1971.
68. Higashi, Y., and Strohminger, J. L. *J. Biol. Chem.* **245:** 3691, 1970.

69. Rothfield, L. I. In *The Dynamic Structure of Cell Membranes*, D. F. H. Wallach and H. Fischer, eds. Heidelberg: Springer-Verlag, p. 165.
70. Wallach, D. F. H. *Biochim. Biophys. Acta* **265:** 61, 1972.
71. McNutt, J. S., Hershberg, B. H., and Weinstein, R. S. *J. Cell Biol.* **51:** 805, 1970.
72. Changeux, J. P., Thiery, J., Tung, Y., and Kittel, C. *Proc. Natl. Acad. Sci.* (U.S.) **57:** 335, 1967.
73. Changeux, J. P., Blumenthal, R., Kasai, M., and Podleski, T. In *Molecular Properties of Drug Receptors*, R. Porter and M. O'Connor, eds. London: J. A. Churchill, 1970, p. 197.
74. Mitchell, P. *Nature* **191:** 144, 1961.
75. Racker, E. Essays in *Biochem.* **6:** 1, 1970.
76. Steck, T. L., Weinstein, R. S., Straus, J. H., and Wallach, D. F. H. *Science* **168:** 255, 1970.
77. Blasie, J. K. B., and Worthington, C. R. W. *J. Mol. Biol.* **39:** 417, 1969.
78. Frye, L. D. F., and Eddidin, M. J. E. *J. Cell Sci.* **7:** 319, 1970.
79. Taylor, R. B., Duffus, P. H., Raff, M. C., and de Petris, S. *Nature* **233:** 225, 1971.
80. Loor, F., Forri, L., and Pernis, B. *Europ. J. Immunol.* **2:** 203, 1972.
81. Kourilsky, F. M., Silvestre, D., Levy, J. P., Dausset, J., Nicolai, M. G., and Senik, A. *J. Immunol* **106:** 454, 1971.
82. Singer, S. J., and Nicolson, G. *Science* **175:** 720, 1972.
83. Takeuchi, M., and Terayama, H. *Exptl. Cell Res.* **40:** 32, 1965.
84. Manganiello, V. C., and Phillips, A. H. *J. Biol. Chem.* **240:** 3951, 1965.
85. Van den Heuvel, F. *J. Am. Chem. Soc.* **42:** 481, 1965.
86. Marchesi, S. L., Steers, E., and Marchesi, V. T. *Biochemistry* **9:** 50, 1970.
87. Schnebli, H. P., Vatter, A. E., and Abrams, A. *J. Biol. Chem.* **245:** 1122, 1970.
88. Wolfgram, F., and Kotorii, K. J. *J. Neurochem.* **15:** 1281, 1968.

INDEX